Lecture Notes in Control and Information Sciences

Volume 471

About this Series

This series aims to report new developments in the fields of control and information sciences—quickly, informally and at a high level. The type of material considered for publication includes:

1. Preliminary drafts of monographs and advanced textbooks
2. Lectures on a new field, or presenting a new angle on a classical field
3. Research reports
4. Reports of meetings, provided they are
   (a) of exceptional interest and
   (b) devoted to a specific topic. The timeliness of subject material is very important.

More information about this series at http://www.springer.com/series/642

Filippo Cacace · Lorenzo Farina
Roberto Setola · Alfredo Germani
Editors

Positive Systems

Theory and Applications (POSTA 2016)
Rome, Italy, September 14–16, 2016

Springer

Editors
Filippo Cacace
Università Campus Bio-Medico
Rome
Italy

Lorenzo Farina
Sapienza Università di Roma
Rome
Italy

Roberto Setola
Università Campus Bio-Medico
Rome
Italy

Alfredo Germani
Università dell'Aquila
L'Aquila
Italy

ISSN 0170-8643 ISSN 1610-7411 (electronic)
Lecture Notes in Control and Information Sciences
ISBN 978-3-319-54210-2 ISBN 978-3-319-54211-9 (eBook)
DOI 10.1007/978-3-319-54211-9

Library of Congress Control Number: 2017932635

Printed on acid-free paper

This Springer imprint is published by Springer Nature
The registered company is Springer International Publishing AG
The registered company address is: Gewerbestrasse 11, 6330 Cham, Switzerland

Preface

Positive systems are dynamical systems whose state variables are positive (or at least nonnegative) in value at all times. Such exceedingly simple definition is nevertheless full of far-reaching consequences both on theory and applications of dynamical systems. Moreover, positivity of variables is readily available information, stemming directly from the intrinsic nature of the phenomenon of interest. It is therefore not surprising that many researchers, coming from very different areas of dynamical systems and control, joined together in Rome, for the second time from 2003, to give rise to the fifth edition of the POSTA conference (Positive Systems Theory and Applications) in September 14–16, 2016. As a matter of fact, the most intriguing feature of the positive systems "community" is that it is *not* a community! If we look at it clearly, we see an intrinsically *open* group of people where members are not recruited on a permanent but on an occasional basis. In fact, a researcher may face a problem whose solution require to exploit "positivity" to be solved effectively (and often elegantly). This is an absolutely fascinating aspect of positive systems theory: a field of research fueled by contribution from any other field and, as such, and ever-expanding one. If we look at the titles of the talks presented, we will find a large heterogeneity of topics, ranging from epidemiology to anesthesia, from switched or fractional systems to communication systems and also a variety of deep theoretical problems such as the construction robust observers or stability analysis in the presence of time delays. As such, the POSTA conference offered a wonderful opportunity to establish research networks among scholars having similar interests related to positivity.

We all look forward to the next edition of the conference, to be held in Hangzhou (China) in 2018. Over the last two decades Chinese researchers have become very important contributor to positive systems research. We are excited by such historical initiative which will boost research in the field on both quality and quantity.

We wish to thank the International Program Committee for the outstanding work in reviewing the papers thus providing a substantial contribution to the improvement of the quality of the Symposium. Furthermore, we wish to thank the support from Università Campus Bio-Medico di Roma who hosted the conference and especially all the participants to POSTA 2016 for making this meeting a success.

The final remark is dedicated to Profs. Maria Elena Valcher and Jean-Luc Gouzè for their availability, support to the initiative and for enriching the Symposium with their inspired lectures.

Rome, Italy
December 2016

Filippo Cacace
Lorenzo Farina
Roberto Setola
Alfredo Germani

Contents

Part I
Positive Systems Biology

Chapter 1
Persistence, Periodicity and Privacy for Positive Systems in Epidemiology and Elsewhere

Oliver Mason, Aisling McGlinchey and Fabian Wirth

Abstract We first recall and describe some recently published results giving sufficient conditions for persistence and the existence of periodic solutions for switched SIS epidemiological models. We extend the result on the existence of persistent switching signals in two ways. We establish uniform strong persistence where previous work only guaranteed weak persistence; we replace the hypothesis that there exists an unstable matrix in the convex hull of the linearized systems with the weaker assumption that the JLE is positive. In the final section of the chapter, the issue of data privacy for positive systems is addressed.

Keywords Switched systems · SIS models · Persistence · Joint Lyapunov exponent · Differential privacy

1.1 Introduction and Outline

Mathematical models based on differential equations have long played an important role in epidemiology and population biology, dating back to the early, seminal work of Kermack, McKendrick and others. The point has been well made before that mathematical models are of particular importance in epidemiology as they allow researchers to investigate the feasibility and effectiveness of containment strategies through simulation and theoretical analysis; experimental investigation is neither

O. Mason (✉)
Department of Mathematics & Statistics/Hamilton Institute, Maynooth University and Lero, The Irish Software Research Centre, Kildare, Ireland
e-mail: oliver.mason@nuim.ie

A. McGlinchey
Department of Mathematics & Statistics, Maynooth University and Lero, The Irish Software Research Centre, Kildare, Ireland
e-mail: aisling.mcglinchey.2009@mumail.ie

F. Wirth
Faculty of Computer Science and Mathematics, University of Passau, Innstrasse 33, 94032 Passau, Germany
e-mail: fabian.lastname@uni-passau.de

F. Cacace et al. (eds.), *Positive Systems*, Lecture Notes in Control and Information Sciences 471, DOI 10.1007/978-3-319-54211-9_1

feasible nor ethical in this setting. Since the early work in mathematical epidemiology, certain questions have occupied a central role in the development of the subject. It is arguable that the two most central issues concern the existence and stability of disease free equilibria and determining conditions for the disease becoming endemic in the population [1]. These questions are addressed using techniques from dynamical systems theory and, as the models studied have become more realistic and sophisticated, new approaches have been brought to bear on the problems. In particular, it is necessary to develop methods of analysis that can be applied to models incorporating uncertainty and stochastic effects, heterogeneous contact patterns, as well as time-variation in parameters and delay. Much of the work described in this chapter is motivated by this overall programme.

In [2], a simple SIS model for disease propagation in a population with multiple groups was described. The population is first stratified into groups and then each group is further divided into two epidemiological classes: susceptibles and infectives. New infectives can be generated by contacts between different groups and the infection rates as well as curing and birth/death rates can vary between classes. The authors of [2] showed that the spectral abscissa of the matrix of the linearized system can be used as a threshold parameter for the onset of endemic behaviour under a combinatorial irreducibility assumption on the matrix. Recently, this work was extended in the paper [3] in which a switched SIS model was studied.

The model considered in [3] incorporated both time variation and uncertainty and showed that the Joint Lyapunov Exponent (JLE) of the linearized inclusion can be used to determine the stability of the disease free equilibrium DFE. Moreover, it was shown that provided the convex hull of the linearized system matrices contains an unstable matrix, there exists a switching signal with respect to which the disease persists in every group. This work left two questions open: (i) can the condition on the convex hull of the system matrices be replaced with the (weaker) assumption that the JLE is positive? (ii) is the persistence uniform? A major contribution of this chapter is to provide answers to these questions.

The signals in applications such as epidemiology often contain sensitive personal information and it is important to develop analysis techniques that respect the privacy of individuals. A number of recent papers within the field of control have begun to address the interplay between control and privacy. In the final section of the chapter, we will focus on the work described in [4] in which the design of differentially private observers was considered: a motivating example in that paper was a simple epidemiological model. Many systems in which privacy arises as a concern are positive systems, so it seems entirely natural to ask whether or not a differentially private positive observer can be constructed. Our aim is to describe some novel questions for the positive systems community arising from the interplay between privacy and positivity.

1.2 Notation, Definitions and Preliminary Results

Throughout this chapter, we denote by $\mathbb{R}^n$ the vector space of all n-tuples of real numbers and by $\mathbb{R}^{n\times n}$ the space of all $n \times n$ matrices with real entries. For two vectors x, y in $\mathbb{R}^n$, the notation $x \geq y$ means that $x_i \geq y_i$ for $1 \leq i \leq n$; $x > y$ means that $x \geq y$, $x \neq y$; finally $x \gg y$ means that $x_i > y_i$ for $1 \leq i \leq n$. Similar notation is used for matrices. We denote by $\mathbb{R}^n_+$ the nonnegative orthant of $\mathbb{R}^n$:

$$\mathbb{R}^n_+ := \{x \in \mathbb{R}^n \mid x \geq 0\}.$$

For a matrix $A \in \mathbb{R}^{n\times n}$, $\sigma(A)$ denotes the spectrum of A and we denote by $\mu(A)$ the spectral abscissa of A, $\mu(A) := \max\{\text{Re}(\lambda) \mid \lambda \in \sigma(A)\}$. For a set S in $\mathbb{R}^{n\times n}$, conv(S) denotes the convex hull of S.

For an autonomous nonlinear system whose right hand side is Lipschitz,

$$\dot{x} = f(x), \quad x(0) = x^0, \tag{1.1}$$

we denote by $x(t, x^0)$ the unique solution with $x(0, x^0) = x^0$. In the case where f is C^1 on a neighbourhood of $\mathbb{R}^n_+$ (as will be the case throughout here), it is well known that the system (1.1) is order preserving if the Jacobian of f is Metzler in $\mathbb{R}^n_+$. For background on monotone or order-preserving systems, we refer the reader to [5].

1.2.1 An Autonomous Multi-group SIS Model

We briefly recall the core SIS model of [2] which motivates our work. We consider a population that is divided into n groups; each group is then sub-divided into susceptibles and infectives and we denote the number of susceptibles in group i by $S_i(t)$ and the number of infectives in group i by $I_i(t)$. The rate at which susceptibles in group i are infected by infectives from group j is β_{ij}; the curing rate for infectives in group i is γ_i and the birth and death rates for group i are both given by μ_i. Following standard mass-action kinetics, the core model takes the form:

$$\dot{S}_i(t) = \mu_i N_i - \mu_i S_i(t) - \sum_{j=1}^{n} \beta_{ij} \frac{S_i(t) I_j(t)}{N_i} + \gamma_i I_i(t)$$

$$\dot{I}_i(t) = \sum_{j=1}^{n} \beta_{ij} \frac{S_i(t) I_j(t)}{N_i} - (\gamma_i + \mu_i) I_i(t).$$

The population of each group, N_i, is constant and if we focus on the dynamics of the fraction $x_i(t) = \frac{I_i(t)}{N_i}$ of infectives in each group the system simplifies to the compact form

$$\dot{x}(t) = (-D + B)x(t) + \operatorname{diag}(x(t))Bx(t). \tag{1.2}$$

Here the matrix D is diagonal with entries $\alpha_i = \gamma_i + \mu_i$ along the main diagonal and B has entries $b_{ij} = \frac{\beta_{ij} N_j}{N_i}$. It is assumed throughout that $\alpha_i > 0$ for all i. It is easy to see that the origin is always an equilibrium for (1.2) corresponding to the disease free equilibrium (DFE). Moreover, the compact set

$$\Sigma_n := \{x \in \mathbb{R}^n_+ \mid x \leq \mathbf{1}\}$$

is invariant under (1.2) and for every initial condition $x^0 \in \Sigma_n$, there exists a unique solution $x(t, x^0)$ of (1.2) defined for all $t \geq 0$ with $x(0, x^0) = x^0$.

The two key results from [2] concerning stability of the DFE and endemic behaviour for (1.2) are recalled below.

Theorem 1.1 *Let B be an irreducible matrix. Then the DFE of (1.2) is globally asymptotically stable if and only if $\mu(-D + B) \leq 0$.*

The next result characterises possible endemic behaviour of (1.2).

Theorem 1.2 *Let B be irreducible. There exists an endemic equilibrium $\bar{x}$ in int$(\mathbb{R}^n_+)$ if and only if $\mu(-D + B) > 0$. Furthermore, in this case $\bar{x}$ is asymptotically stable and has region of attraction containing $\Sigma_n \backslash \{0\}$.*

1.2.2 Persistence

Our later results will be concerned with persistence for a switched version of the model (1.2). Persistence for a semiflow on a state space X is usually defined with respect to a function $\eta : X \to \mathbb{R}_+$. We next recall the definitions of weak and strong persistence and the uniform versions of both [6].

Definition 1.1 A semiflow $\phi : X \times \mathbb{R}_+ \times X$ is weakly persistent if

$$\limsup_{t \to \infty} \eta(\phi(t, x)) > 0 \quad \forall x \in X \text{ with } \eta(x) > 0.$$

The semiflow $\phi : X \times \mathbb{R}_+ \times X$ is uniformly weakly persistent if there is some $\varepsilon > 0$ such that:

$$\limsup_{t \to \infty} \eta(\phi(t, x)) > \varepsilon \quad \forall x \in X \text{ with } \eta(x) > 0.$$

The corresponding definitions of strong and uniform strong persistence replace the lim sup with lim inf.

1.2.3 Extension to Switched/Differential Inclusion Model

The major focus of [3] was to extend the study of the model described above to allow for switching and uncertainty in the system parameters. Before recalling the relevant results, we introduce appropriate concepts of weak and strong persistence for switched systems.

Consider a switched system

$$\dot{x}(t) = f_{\sigma(t)}(x),\ x(0) = x^0 \tag{1.3}$$

defined on a state space $X \subseteq \mathbb{R}^n_+$ where $\{f_1, \ldots, f_m\}$ is a given set of functions, assumed to be sufficiently smooth so that unique solutions of (1.3) exist on $[0, \infty)$ for every fixed σ and x^0 is a measurable switching signal. For our purposes, X will denote the box Σ_n defined earlier.

We only give the precise formulation for uniform strong persistence here due to space limitations. The corresponding definitions for strong and non-uniform persistence are easy to see.

If there is some $\varepsilon > 0$ and a switching signal σ such that $\eta(x^0) > 0$ implies $\liminf_{t\to\infty} \eta(x(t, x^0, \sigma)) > \varepsilon$, we refer to σ as a uniformly strongly persistent switching signal.

We now briefly recall the most relevant results of [3] to our current presentation.

We start with a finite set of diagonal matrices $\{D_1, \ldots, D_m\}$ with positive diagonal entries and a set $B_1, \ldots, B_m$ of nonnegative matrices. Each pair D_i, B_i corresponds to one SIS system of the form (1.2). The switched model is then given by

$$\dot{x}(t) = (-D_{\sigma(t)} + B_{\sigma(t)})x(t) - \mathrm{diag}(x(t))B_{\sigma(t)}x(t). \tag{1.4}$$

We denote by $\mathscr{M}$ the set of matrices

$$\mathscr{M} := \{-D_i + B_i \mid 1 \leq i \leq m\}.$$

The key idea in [3] was to replace the spectral abscissa of the linearized matrix $-D + B$ with the corresponding joint Lyapunov exponent of the linearized switched system/inclusion. We now briefly recall the definition of this concept.

Let $A_i = -D_i + B_i$ for $i = 1, \ldots, m$.

For each switching signal σ and $t \geq 0$, the evolution operator $\Phi_\sigma(t)$ is given by the solution of the matrix differential equation:

$$\dot{\Phi}_\sigma(t) = A_{\sigma(t)}\Phi_\sigma(t),\ \Phi(0) = I.$$

We let $\mathscr{H}_t$ denote the set of all time evolution operators for time t and then define the operator semigroup

$$\mathscr{H} := \bigcup_{t\geq 0} \mathscr{H}_t,$$

setting $\mathscr{H}_0 = \{I\}$. The growth rate of the switched system at time t is defined by

$$\rho_t(\mathscr{M}) := \sup_{\Phi_\sigma \in \mathscr{H}_t} \frac{1}{t} \log \|\Phi_\sigma(t)\|.$$

Finally, the joint Lyapunov exponent (JLE) of the linearized system is:

$$\rho(\mathscr{M}) = \lim_{t \to \infty} \rho_t(\mathscr{M}).$$

Essentially, the JLE defined here represents a natural generalisation of the spectral abscissa of a single matrix to the context of switched linear systems.

In order to properly set context for our results here, we need to recall two of the main facts established in [3] for the switched epidemic model. The first of these establishes a sufficient condition for the DFE to be globally asymptotically stable with respect to arbitrary switching signals.

Theorem 1.3 *Consider the switched system (1.4) and the associated set $\mathscr{M}$ of matrices. Assume that conv($\mathscr{M}$) contains an irreducible matrix. The DFE of (1.4) is uniformly globally asymptotically stable if and only if $\rho(\mathscr{M}) \leq 0$.*

While the previous theorem establishes a condition for the DFE of the switched model to be globally asymptotically stable, the next result from [3] provides a condition for the existence of a persistent switching signal for (1.4).

Proposition 1.1 *Consider the switched SIS model (1.4) and assume that every B_i is irreducible. Assume that there exists some $R \in$ conv($\mathscr{M}$) with $\mu(R) > 0$. Then there exists a switching signal σ such that for all $x^0 > 0$, $1 \leq i \leq n$*

$$\liminf_{t \to \infty} \; x_i(t, x^0, \sigma) > 0.$$

We may summarise what the previous two results establish in the following way:

- if conv($\mathscr{M}$) contains an irreducible matrix and the JLE $\rho(\mathscr{M}) \leq 0$ the DFE is GAS and the disease dies out.
- if all the matrices B_i are irreducible and $\mu(M) > 0$ for some $M \in$ conv($\mathscr{M}$), there exists a switching signal which is strongly persistent with respect to every function $\eta(x) = |x_i|$, $1 \leq i \leq n$.

Several questions arise naturally here. A first question is whether the switching signal above can be chosen so as to ensure *uniform strong persistence*. It is well known that while the existence of an unstable matrix in conv($\mathscr{M}$) ensures that $\rho(\mathscr{M}) > 0$, there is in general a gap between the two conditions [7]. This raises the question of whether persistence can be established under the weaker assumption that $\rho(\mathscr{M}) > 0$. In the next section of the chapter we shall present a number of results addressing these issues.

1.3 Uniform Persistence and the JLE

In this section, we present some novel results and observations that address some of the issues mentioned at the close of the previous section. We first consider the case where the system (1.4) is 2-dimensional, corresponding to a population with two groups.

1.3.1 The 2-Group Case

To begin, we recall the following fact from [8].

Proposition 1.2 *Consider a switched linear system*

$$\dot{x}(t) = A_{\sigma(t)}x(t), \tag{1.5}$$

where $\sigma : [0, \infty) \to \mathcal{M} \subseteq \mathbb{R}^{2\times 2}$ *for a finite set* $\mathcal{M}$ *of Metzler matrices. Then (1.5) is globally uniformly asymptotically stable if and only if conv(*$\mathcal{M}$*) consists of Hurwitz matrices.*

Consider now the system (1.4) and suppose that all of the matrices B_i are irreducible. If $\rho(\mathcal{M}) > 0$, then this will still be true if we replace each matrix $-D_i + B_i$ by $-D_i + B_i - \varepsilon I$ for $\varepsilon > 0$ sufficiently small, by continuity of the JLE. It now follows from Proposition 1.2 that there exists some matrix M in conv$\{-D_i + B_i - \varepsilon I \mid 1 \leq i \leq m\}$ with $\mu(M) \geq 0$. It is easy to see that $\hat{M} = M + \varepsilon I$ is in conv$(\mathcal{M})$ and $\mu(\hat{M}) > 0$. Putting these simple observations together, we get the following result.

Proposition 1.3 *Consider the switched system (1.4) and suppose that* $n = 2$ *and that each matrix* B_i *is irreducible. Then:*

(i) if $\rho(\mathcal{M}) \leq 0$*, the DFE is globally asymptically stable;*
(ii) if $\rho(\mathcal{M}) > 0$*, there exists switching signal* σ *that is strongly persistent with respect to* $\eta_i(x) = |x_i|$ *for* $1 \leq i \leq 2$*.*

1.3.2 Uniform Strong Persistence

In the next result, we show that under the same hypotheses as in Proposition 1.1 we can conclude the existence of a *uniformly strongly persistent switching signal.*

Proposition 1.4 *Consider the switched SIS model (1.4) and assume that each* B_i *is irreducible. Assume that there exists some* $R \in$ *conv(*$\mathcal{M}$*) with* $\mu(R) > 0$*. Then there exists some* $\varepsilon > 0$ *and a switching signal* σ *such that for all* $x^0 > 0$, $1 \leq i \leq n$

$$\liminf_{t\to\infty} x_i(t, x^0, \sigma) > \varepsilon.$$

Proof In the proof of Proposition 1.1 in [3] (where it appears as Proposition 6.1), it is shown that there exists a periodic switching signal σ with period $T = \frac{1}{N_0}$ for some $N_0 \in \mathbb{N}$ with the following properties.

(i) There exists some $v \gg 0$ and $\delta > 0$ such that $x(1, v, \sigma) \gg v$ and $x_i(t, v, \sigma) \geq \delta$ for all $t \geq 0$.
(ii) For any λ with $0 < \lambda < 1$ and $t \geq 0$, $x(t, \lambda v, \sigma) \geq \lambda x(t, v, \sigma)$.
(iii) As each constituent vector field is irreducible, standard results from [5] show that $x(t, x^0, \sigma) \gg 0$ for all $t > 0$ and $x^0 > 0$. In particular for all $x^0 > 0$, $x(1, x^0, \sigma) \gg 0$.

It is a simple rephrasing of (i) to say that there is some $\alpha > 1$ such that $x(1, v, \sigma) \geq \alpha v$. Let $x^0 \gg 0$ be given. We claim that there is some time T such that $x(T, x^0, \sigma) \geq v$.

As in the proof of Proposition 6.1 in [3], we can find some $0 < \lambda < 1$ such that $x^0 \geq \lambda v$. Then, using (ii), $x(1, \lambda v, \sigma) \geq \lambda x(1, v, \sigma) \geq \alpha\lambda v$. If $\alpha\lambda \geq 1$, then $x(1, \lambda v, \sigma) \geq v$ and we are done. Otherwise, $\alpha\lambda < 1$ and again using (ii), combined with our choice of σ and the monotonicity of the constituent systems, we have

$$\begin{aligned} x(2, x^0, \sigma) &= x(1, x(1, x^0, \sigma), \sigma) \\ &\geq x(1, \alpha\lambda v, \sigma) \\ &\geq \alpha\lambda x(1, v, \sigma) \\ &\geq \alpha^2\lambda v. \end{aligned}$$

Iterating and using the periodicity of σ together with the order-preserving property of each constituent vector field, we find that eventually there is some T such that $\alpha^T\lambda \geq 1$ and hence $x(T, x^0, \sigma) \geq v$. It now follows from the monotonicity of the constituent systems that $x(T + t, x^0, \sigma) \geq x(t, v, \sigma)$ for $t \geq 0$ and hence that $\liminf_{t\to\infty} x_i(t, x^0, \sigma) \geq \delta$ for $1 \leq i \leq n$.

It only remains to consider the case of $x^0 > 0$ but $x^0 \not\gg 0$. It follows from (iii) and the above argument that in this case also, $\liminf_{t\to\infty} x_i(t, x^0, \sigma) \geq \delta$ for $1 \leq i \leq n$. This completes the proof.

1.3.3 Uniform Weak Persistence and the JLE

The results of the previous subsections show that there will exist persistent switching signals when the convex hull of the linearized system matrices contains an unstable matrix. However, there is a gap in general between the two conditions:

(A) $\exists M \in \text{conv}(\mathcal{M})$ with $\mu(M) > 0$;
(B) $\rho(\mathcal{M}) > 0$.

We now ask what can be said about persistence when we make the weaker assumption (B).

Theorem 1.4 *Consider the switched SIS model (1.4). Assume that $\rho(\mathcal{M}) > 0$ and that each B_i is irreducible. Then there exists a switching signal σ that is uniformly weakly persistent with respect to $\eta(x) = \max_i |x_i|$.*

Remark Combining this result with Theorem 1.3, we see that for switched SIS models with irreducible B_i:

- $\rho(\mathcal{M}) \le 0$ implies DFE is globally asymptotically stable;
- $\rho(\mathcal{M}) > 0$ implies there exists a uniformly weakly persistent switching signal.

Outline of Proof:

We argue by contradiction. So, suppose that no uniformly weakly persistent switching signal exists. This would mean that for all $\varepsilon > 0$, and all switching signals σ, there would exist a solution $x(t, x^0, \sigma)$ with $\eta(x^0) > 0$ and

$$\limsup_{t\to\infty} \eta(x(t, x^0, \sigma)) < \varepsilon.$$

Choose $\varepsilon > 0$ so that the JLE of the matrices

$$\hat{\mathcal{M}} := \{-D_1 + (1-\varepsilon)B_1, -D_2 + (1-\varepsilon)B_2, \ldots, -D_n + (1-\varepsilon)B_n\}$$

is still positive. This can be done as the JLE is continuous with respect to the Hausdorff metric on compact sets of Metzler matrices by [9].

Next, we write Φ_σ for the evolution operators corresponding to $\hat{\mathcal{M}}$. As $\rho(\hat{\mathcal{M}}) > 0$, there is some $T > 0$ and some σ such that $\|\Phi_\sigma(T)\| = e^{\alpha T}$ where $\alpha > 0$. Consider the periodic switching signal σ_1 constructed from this σ by setting $\sigma_1(t) = \sigma(t)$ for $0 \le t < T$ and $\sigma(t+T) = \sigma(t)$ for all $t \ge 0$.

By assumption there is some solution of the SIS model for this switching signal and some $T_1 > 0$ such that $\eta(x(t, x^0, \sigma)) < \varepsilon$ for all $t \ge T_1$. Choose a positive integral multiple kT of T such that $kT > T_1$. Then for all $t \ge kT$,

$$\dot{x}(t) \ge (-D_{\sigma(t)} + (1-\varepsilon)B_{\sigma(t)})x(t). \tag{1.6}$$

As the matrices B_i are irreducible, it follows from [5] that $x(kT) \gg 0$. Moreover, as the evolution operator is nonnegative, we can choose some vector $v > 0$ such that $\|\Phi_\sigma(kT)v\| = e^{k\alpha T}\|v\|$ with $v \le x(kT)$. It now follows that for $p = 1, 2, \ldots$,

$$\eta(x(pkT, x^0, \sigma)) \ge e^{pk\alpha T}\|v\|$$

which clearly contradicts $\eta(x(t, x^0, \sigma)) < \varepsilon$. We conclude that there is a uniformly weakly persistent switching signal as claimed.

1.4 Privacy and Positive Systems

Monitoring population variables in order to determine whether or not a disease outbreak is likely to become an epidemic is a key aspect of epidemiological modelling in real world situations. In a recent paper [4], an interesting application of observer design motivated by *syndromic surveillance* methods for public health was considered. Specifically, a simple SIR model with output was considered, whose output consists of variables being used to monitor the level of disease in the population. This could be the number of tweets or blog posts about the disease for instance and the core idea is to design observers that can track the actual epidemiological variables based on the measured output.

It is important to address the privacy concerns of individuals who are contributing the data being measured in such a system. While many frameworks for privacy protection have been proposed in the data science and computing communities in the recent past, those based on information theoretic foundations and differential privacy [10, 11] appear the most suitable for dynamic situations and control applications. With this in mind, Le Ny introduced the problem of constructing a Luenberger observer that is differentially private in [4]. In the remainder of this section, our purpose is to describe the core idea behind the design of such observers and to highlight some novel and interesting questions for the field of positive systems that arise here.

Focussing on the essential details, the core question considered in [4] can be described as follows. We have a discrete time system with measured outputs of the form:

$$\begin{aligned} x_{k+1} &= f_k(x_k) \\ y_k &= g_k(x_k), \end{aligned} \tag{1.7}$$

and we wish to construct a simple Luenberger observer $\mathscr{L}$ of the form:

$$z_{(k+1)} = f_k(z_k) + L_k(y_k - g(z_k))$$

to asymptotically track the state x_k of (1.7). This is of course not a new problem. The novelty arises when some of the signals contain sensitive information in application areas such as epidemiology, population dynamics and social networks. In such a scenario, the problem is to construct observers that also guarantee that the mapping from a sensitive signal to the eventual (released) output of the observer satisfies an appropriate differential privacy constraint.

The original formulation of differential privacy for databases considered records belonging to a set D and modeled databases as vectors $\mathbf{d}$ in D^n. Two such vectors satisfy the adjacency relation $\mathbf{d} \sim \mathbf{d}'$ if they differ in exactly one component (the hamming distance between them is exactly 1). A query is a mapping Q from D^n to some output space E. Differential privacy aims to protect the privacy of individuals by supplying randomised answers to a query so that the distribution of answers differs little when any one user changes their entry. Formally, for a query Q, an ε differentially private mechanism is a set of random variables $X_{Q,\mathbf{d}} \in D^n$ taking values in E such that

$$\mathbb{P}(X_{Q,\mathbf{d}'} \in A) \leq e^{\varepsilon}\mathbb{P}(X_{Q,\mathbf{d}} \in A)$$

for any $\mathbf{d} \sim \mathbf{d}'$ and any measurable subset A of E. In a system theoretic setting, we replace the database space with a set of sensitive signals, and a query corresponds to the mapping between signal spaces defined by a system. When dealing with dynamic scenarios, hamming distance is often not an appropriate notion of adjacency.

In [4], the following definition of adjacency was adopted. $K > 0$ and $0 < \alpha < 1$ are given real constants; two sequences of measured values y, y' are adjacent, $y \sim y'$, if there is some k_0 such that

$$\begin{cases} y_k = y'_k & \forall k \leq k_0 \\ \|y_k - y'_k\| \leq K\alpha^{k-k_0} & \forall k > k_0. \end{cases}$$

Each entry y_k, y'_k lies in $\mathbb{R}^p$ and $\|\cdot\|$ can be any norm on $\mathbb{R}^p$. For simplicity, we will consider the l_1 norm. The output signal is considered sensitive (it may concern online activity of individuals for instance) and the aim is to release a differentially private perturbation of the observer state, $z_{(k)}$, of the form $\hat{z}_{(k)} = z_{(k)} + \delta_{(k)}$ where $\delta_{(k)}$ is an appropriate noise signal, chosen so that the mechanism mapping y to $\hat{z}$ is differentially private. Based on earlier work in [12], it is shown that this can be achieved by taking $\delta_{(k)}$ to be appropriate Laplacian or Gaussian random variables/vectors, whose variance depends on the *sensitivity* of the system mapping y to z. If $y \sim y'$ and we denote the corresponding states of the observer by z, z', then the l_1 sensitivity of the system is given by

$$\sup_{y \sim y'} \|z - z'\|_1, \tag{1.8}$$

where, in a slight abuse of notation, $\|z - z'\|_1 = \sum_{k=0}^{\infty} \|z_k - z'_k\|_1$.

The work of [4] and similar papers raises many very interesting questions for systems theory in general, and positive systems in particular. First, many of the motivating applications arise in area such as social networks and epidemiology, both of which naturally fall within the realm of positive systems. The question of how to design observers that preserve the positivity of the signals in the system and the impact that this might have on the accuracy of the outputs has not yet been addressed. Of course, positive systems possess many special properties that give a particular flavour to many fundamental questions, including that of observer design [13]. While realistic models will require an analysis for nonlinear models, the remainder of our discussion will focus on the linear case in the interest of highlighting some significant questions without muddying the waters with technical detail.

So consider a linear system with output of the form:

$$\begin{aligned} x_{k+1} &= Ax_k \\ y_k &= Cx_k, \end{aligned} \tag{1.9}$$

where both $A \in \mathbb{R}_+^{n\times n}$ and $C \in \mathbb{R}_+^{p\times n}$ are nonnegative. A Luenberger observer would take the form

$$z_{k+1} = Az_k + L(y_k - Cz_k) = (A - LC)z_k + Ly_k.$$

Even in the simple linear case, certain questions/challenges naturally suggest themselves.

- Characterise the minimal possible l_1 sensitivity of the system $\mathscr{L}$ where the observer system is required to be itself positive.
- Characterise the minimal l_1 sensitivity (for positive systems) without imposing the positivity constraint on the observer.
- Can we design a positive differentially private observer; here we are requiring that the noise added to z is truncated so as to ensure that the noisy signal remains positive.
- In reducing sensitivity, we can achieve ε differential privacy with less noise. Can we characterise explicitly the impact this has, on the speed of convergence of the observer?

The above questions represent early steps in a programme to develop a foundation for differentially private observer design for positive systems. Extensions to time-varying and nonlinear systems will certainly be necessary. However, we feel that this is a topic of sufficient practical importance and theoretical interest to merit being brought to the attention of the positive systems community.

Acknowledgements This work was supported, in part, by Science Foundation Ireland grant 13/RC/2094 and co-funded under the European Regional Development Fund through the Southern & Eastern Regional Operational Programme to Lero—the Irish Software Research Centre (http://www.lero.ie)

References

1. van den Driessche, P., Watmough, J.: Reproduction numbers and sub-threshold endemic equilibria for compartmental models of disease transmission. Math. Biosci. **180**, 29–48 (2002)
2. Fall, A., Iggidr, A., Sallet, G., Tewa, J.: Epidemiological models and Lyapunov functions. Math. Model. Nat. Phenom. **2**, 62–68 (2007)
3. Ait-Rami, M., Bokharaie, V.S., Mason, O., Wirth, F.: Stability criteria for SIS epidemiological models under switching policies. Discret. Contin. Dyn. Syst. Ser. B **19**(9), 2865–2887 (2014)
4. J. Le Ny, Privacy-Preserving Nonlinear Observer Design Using Contraction Analysis. In: Proceedings IEEE 54th Annual Conference on Decision and Control (CDC) (2015)

5. Smith, H.: Monotone Dynamical Systems. American Mathematical Society (1995)
6. Smith, H., Thieme, H.: Dynamical Systems and Population Persistence. American Mathematical Society (2011)
7. Fainshil, L., Margaliot, M., Chigansky, P.: On the stability of positive linear switched systems under arbitrary switching laws. IEEE Trans. Autom. Control **54**(4), 897–899 (2009)
8. Gurvits, L., Shorten, R., Mason, O.: On the stability of switched positive linear systems. IEEE Trans. Autom. Control **52**, 1099–1103 (2007)
9. Mason, O., Wirth, F.: Extremal norms for positive linear inclusions. Linear Algebra Appl. **444**, 100–113 (2014)
10. Dwork, C.: Differential Privacy. In: Proceedings of the International Colloquium on Automata, Languages and Programming, pp. 1–12. Springer (2006)
11. Holohan, N., Leith, D., Mason, O.: Differential privacy in metric spaces: numerical categorical and functional data under the one roof. Inform. Sci. **305**, 256–268 (2015)
12. Le Ny, J., Pappas, G.J.: Differentially private filtering. IEEE Trans. Autom. Control **59**(2), 341–354 (2014)
13. Hardin, H., van Schuppen, J.H.: Observers for linear positive systems. Linear Algebra Appl. **425**, 571–607 (2007)

Chapter 2
Control of Anesthesia Based on Singularly Perturbed Model

Sophie Tarbouriech, Isabelle Queinnec, Germain Garcia and Michel Mazerolles

Abstract This chapter deals with the control of anesthesia taking into account the positivity together with the upper limitation constraints of the variables and the target interval tolerated for the depth of anesthesia during a surgery. Due to the presence of multiple time scale dynamics in the anesthesia model, the system is re-expressed through a singularly perturbed system allowing to decouple the fast dynamics from the slow ones. Differently from general approaches for singularly perturbed systems, the control objective is then to control and accelerate the fast subsystem without interest in modifying the slow dynamics. Thus, a structured state feedback control is proposed through quasi-LMI (linear matrix inequalities) conditions. The characterization of domains of stability and invariance for the system is provided. Associated convex optimization issues are then discussed. Finally, the theoretical conditions are evaluated on a simulated patient case.

Keywords Control of anesthesia · BIS · Positive constraints · Singularly perturbed system · State feedback · LMI

2.1 Introduction

The principle of general anesthesia and drug delivery control during surgery corresponds to the suspension of consciousness (hypnosis), pain (analgesia) and movement (immobility). Indeed, to address these three main actions, a combination of drugs is

S. Tarbouriech (✉) · I. Queinnec · G. Garcia
LAAS-CNRS, Université de Toulouse, CNRS Toulouse, France
e-mail: tarbour@laas.fr; sophie.tarbouriech@laas.fr

I. Queinnec
e-mail: isabelle.queinnec@laas.fr

G. Garcia
e-mail: germain.garcia@laas.fr

M. Mazerolles
Département D'anesthésie-réanimation, CHU Toulouse, 31059 Toulouse, Cedex, France
e-mail: mazerolles.m@chu-toulouse.fr

F. Cacace et al. (eds.), *Positive Systems*, Lecture Notes in Control
and Information Sciences 471, DOI 10.1007/978-3-319-54211-9_2

used. In this chapter we focus on the hypnosis problem only, with Propofol used as hypnotic drug. Even if this is an old problem (notion of closed-loop control appeared in the fifties), it remains largely open. Actually, medical practices remain yet in open loop and several researchers from the control community have been concerned with such applications and suggested advanced control techniques to move from open-loop control by the anesthesiologist to closed-loop control [1]. Hence, the control of the anesthetic state of a patient consists in adjusting the perfusion of hypnotics based on clinical indicators such as heart rate, blood pressure and BIS (Bispectral index). The control of anesthetic drugs injection for maintaining an adequate anesthetic state during surgery has been studied through several approaches. Among them, one can, for example, cite the use of PID controllers [2], adaptive control [3], model predictive control [4], LPV modeling and control [5], bifurcation analysis [6] and set-theoretic tools [7].

As for many biological systems, the design of an adequate control law should take into account some physical aspects such as patient variability, positivity constraints, output measurement availability, the presence of multiple time scales in the dynamics. Indeed, the dynamics of the drug evolution in the patient's body is usually described by a pharmacokinetic positive model with multiple time scales. In this chapter, we use to represent this difference the framework of singularly perturbed systems [8]. Hence, the compartmental system describing the anesthesia model is re-expressed through a singularly perturbed system allowing to decouple the fast dynamics (blood, effect site) from the slow ones (muscles, fat). Differently from general approaches for singularly perturbed systems, the control objective is then to control and accelerate the fast subsystem without interest in modifying the slow dynamics. Furthermore, the control design has to take into account the positivity together with the upper limitation constraints of the variables during a surgery. Thus, based on the results in [9, 10], a structured state feedback control is proposed through theoretical matrix inequalities, which constitutes the main contribution of the chapter. The characterization of domains of stability and invariance for the system is provided by using some relaxation schemes in order to obtain linear matrix inequalities (LMI) conditions. Associated convex optimization issues are then discussed.

The chapter is organized as follows. Section 2.2.1 presents the compartment-based model, for which the presence of multiple time scale dynamics is pointed out. Then, the system is re-expressed through a singularly perturbed system allowing to decouple the fast dynamics from the slow ones. The general problem formulation is summarized in Sect. 2.2.2 and the theoretical conditions allowing to design the structured state feedback controller are provided in Sect. 2.3.1. Associated algorithms are then proposed in Sect. 2.3.2 in order to exhibit numerical solutions. Section 2.4 presents the patient case considered in order to illustrate the effectiveness, the drawback and the trade-off of the proposed solution. Finally, some concluding remarks in Sect. 2.5 end the chapter.

Notation. For a matrix P in $\mathbb{R}^{n \times n}$, the notation $P > 0$ ($P \geq 0$) means that P is symmetric positive (semi) definite. For a vector $x \in \mathbb{R}^n$, the notation $x \geq 0$ means that all the components of the vector are nonnegative. The superscript 'T' stands for matrix transposition, and the notation He(P) stands for $P + P^T$. The symbols I and 0 represent the identity and the zero matrices of appropriate dimensions.

2.2 Patient Model and Problem Formulation

2.2.1 Patient Model

It is well accepted that the model used to describe the evolution of drugs in a patient's body is a Pharmacokinetic/Pharmacodynamic (PK/PD) model, which is based on a three-compartment model [11]. Such a PK/PD model describes the distribution of the drugs between three compartments (blood, muscles and fat). Furthermore, the effect of drugs on the patient is expressed throughout the effect site, which represents the action of drugs on the brain and is related to the concentration in the central compartment through a first order dynamics [5, 12].

Hence, the compartmental model representing the circulation of the drug in the body can be written as follows[1]:

$$\dot{x}_{an} = A_0 x_{an} + B_0 u_{an} \tag{2.1}$$

with

$$A_0 = \begin{bmatrix} -(a_{10}+a_{12}+a_{13}) & a_{21} & a_{31} & 0 \\ a_{12} & -a_{21} & 0 & 0 \\ a_{13} & 0 & -a_{31} & 0 \\ a_{e0}/V_1 & 0 & 0 & -a_{e0} \end{bmatrix};\ B_0 = \begin{bmatrix} 1 \\ 0 \\ 0 \\ 0 \end{bmatrix} \tag{2.2}$$

In the vector $x_{an} = [x_1\ x_2\ x_3\ x_4]'$, x_1, x_2, x_3 are the masses in grams of the drug in the different compartments (blood, fat, muscle), x_4 is the effect site concentration and u_{an} is the infusion rate in g/min of the anesthetic. The parameters $a_{ij} \geq 0,\ \forall i \neq j,\ i, j = 1, 2, 3,$ are the transfer rates of the drug between compartments. The parameter a_{10} represents the rate of elimination from the central compartment. These parameters are functions of the patient characteristics (weight, age, height, ...). Several empirical models give the relation between those parameters and patient's characteristics [13]. One can cite, for example, the models of [14] or [15] related to Propofol (hypnotic drug) and Remifentanil (analgesic drug), respectively, to define a typical patient and to build uncertain models to represent the inter-patient variability.

Moreover, the depth of anesthesia indicator widely used by clinicians is the *BIS* (the bispectral index), which is a signal derived from the EEG analysis. *BIS* quantifies the level of consciousness of a patient from 0 (no cerebral activity) to around 100 (fully awake patient). The relationship between the concentration at the effect site x_4 and the *BIS* can be described empirically by a decreasing sigmoid function [16]:

---

[1]The time dependence is omitted for simplicity of the notation.

$$BIS(x_4) = BIS_0(1 - \frac{x_4^\gamma}{x_4^\gamma + EC_{50}^\gamma}), \tag{2.3}$$

BIS_0 is the BIS value of an awake patient typically set to 100, EC_{50} corresponds to drug concentration associated with 50% of the maximum effect and γ is a parameter modeling the degree of non-linearity. Typical values for these parameters are $EC_{50} = 3.4\,\mu g/ml$ and $\gamma = 3$. Let us stress that the chosen three-compartment model (2.1) is one of the possible compartment models. Its simplicity and its good representativity have motived our choice even if there exist other models of different complexity for the Propofol—*BIS* relationship [1].

Finally, it is important to note that the state and the input of system (2.1) have to be positive, that is to respect the following constraints:

$$\begin{array}{l} x_{an} \geq 0 \\ u_{an} \geq 0 \end{array} \tag{2.4}$$

It is then important to observe that the system (2.1) and (2.4) enters in the class of positive systems. Furthermore, note that matrix A_0 is a Metzler matrix [17].

2.2.2 Problem Formulation

One important fact regarding model (2.1) resides in the difference of dynamics: indeed, the dynamics of metabolism and circulation of Propofol in the central compartment and the site effect is ten times faster than in muscles, and even a hundred times faster than in fat. A classical way to address this kind of problem is to describe the system as a singularly perturbed system [8]. Hence, based on a singularly perturbed description [10], the blood and the effect site parts are gathered in the fast subsystem and the muscle and the fat parts in the slow subsystem. Then, system (2.1) can be rewritten as follows:

$$\begin{bmatrix} \dot{\bar{x}} \\ \varepsilon\dot{\bar{z}} \end{bmatrix} = \begin{bmatrix} A_{11} & A_{12} \\ A_{21} & A_{22} \end{bmatrix} \begin{bmatrix} \bar{x} \\ \bar{z} \end{bmatrix} + \begin{bmatrix} B_1 \\ B_2 \end{bmatrix} u_{an} \tag{2.5}$$

with

$$\begin{array}{l} A_{11} = \begin{bmatrix} -a_{21} & 0 \\ 0 & -a_{31} \end{bmatrix}; A_{12} = \begin{bmatrix} a_{12} & 0 \\ a_{13} & 0 \end{bmatrix}, B_1 = \begin{bmatrix} 0 \\ 0 \end{bmatrix} \\ A_{21} = \begin{bmatrix} \varepsilon a_{21} & \varepsilon a_{31} \\ 0 & 0 \end{bmatrix} = \varepsilon A_{21}^0; A_{22} = \begin{bmatrix} -\varepsilon(a_{10} + a_{12} + a_{13}) & 0 \\ \varepsilon a_{e0}/V_1 & -\varepsilon a_{e0} \end{bmatrix} = \varepsilon A_{22}^0 \\ B_2 = \begin{bmatrix} \varepsilon \\ 0 \end{bmatrix} = \varepsilon B_2^0 \end{array} \tag{2.6}$$

where $\varepsilon > 0$, $\bar{x}$ corresponds to the slow state and $\bar{z}$ corresponds to the fast state. ε takes small values and corresponds to the perturbation parameter. Furthermore, the BIS is rewritten in this case as:

$$BIS(\bar{z}_2) = BIS_0(1 - \frac{\bar{z}_2^{\gamma}}{\bar{z}_2^{\gamma} + EC_{50}^{\gamma}}), \tag{2.7}$$

where $\bar{z}_i$, $i = 1, 2$ are the components of vector $\bar{z}$.

The following assumption holds.

Assumption 1 Matrix A_{22}^0 is non singular. Matrix A_{22} is non singular for any $\varepsilon > 0$.

In most of studies addressing the control design for singularly perturbed systems, the goal is to control the slow dynamics as the crucial problem [8]. In the case of the depth of anesthesia, the most important objective is the control of the fast dynamics because the regulation of the BIS is a direct function of the concentration at the effect site and thus of the fast dynamics on which the administered drug directly acts.

Moreover, during a surgery, the BIS must be brought then maintained close to 50, or at least in an interval between 40 and 60. Due to relation (2.7) describing the relation between the BIS and the effect site concentration, it follows that for the BIS equal to 50% of BIS_0 the effect site concentration must be equal to EC_{50}. Then, the computation of the associated equilibrium point $\bar{x}_e$, $\bar{z}_e$ satisfying $\dot{\bar{x}}_e = 0$ and $\dot{\bar{z}}_e = 0$ gives:

$$\begin{aligned} \bar{z}_{e1} &= V_1 \bar{z}_{e2} \\ \bar{z}_{e2} &= EC_{50}(\tfrac{BIS_0}{BIS_e} - 1)^{1/\gamma} \\ \bar{x}_{e1} &= \tfrac{a_{12}}{a_{21}} \bar{z}_{e1} \\ \bar{x}_{e2} &= \tfrac{a_{13}}{a_{31}} \bar{z}_{e1} \\ \bar{u}_e &= a_{10} \bar{z}_{e1} \end{aligned} \tag{2.8}$$

where BIS_e denotes the desired value of the BIS at the equilibrium and $\bar{x}_i$, $i = 1, 2$ are the components of vector $\bar{x}$.

Hence, we can define the error model around the equilibrium with $x = \bar{x} - \bar{x}_e$, $z = \bar{z} - \bar{z}_e$ and $u = u_{an} - \bar{u}_e$:

$$\begin{bmatrix} \dot{x} \\ \varepsilon \dot{z} \end{bmatrix} = \begin{bmatrix} A_{11} & A_{12} \\ A_{21} & A_{22} \end{bmatrix} \begin{bmatrix} x \\ z \end{bmatrix} + \begin{bmatrix} B_1 \\ B_2 \end{bmatrix} u \tag{2.9}$$

with matrices defined in (2.6).

The problem we intend to solve can be summarized as follows.

Problem 2.1 Find a structured control gain K:

$$K = \begin{bmatrix} 0 & K_f \end{bmatrix}, K_f \in \mathbb{R}^{2\times 2} \tag{2.10}$$

such that:

1. The system (2.6)–(2.9) controlled through the control law $u = K_f z$ is asymptotically stable;
2. The positivity of x_{an} and u_{an} is ensured, or equivalently due to the change of variables around the equilibrium point $\begin{bmatrix} x \\ z \end{bmatrix} \geq - \begin{bmatrix} \bar{x}_e \\ \bar{z}_e \end{bmatrix}$ and $u \geq -\bar{u}_e$.

Note that to address Problem 2.1, the state of the fast subsystem is assumed to be available.

2.3 Main Conditions

In order to solve Problem 2.1, we consider in the sequel a procedure in two main steps: (1) we design the structured control gain to ensure the closed-loop asymptotic stability; and (2) we provide an analysis of the solution to ensure the constraints satisfaction.

2.3.1 *Theoretical Conditions*

Let us introduce the following notation:

$$\begin{array}{l} A_s = A_{11} - A_{12}A_{22}^{-1}A_{21} = A_{11} - A_{12}(A_{22}^0)^{-1}A_{21}^0 \\ B_s = B_1 - A_{12}A_{22}^{-1}B_2 = B_1 - A_{12}(A_{22}^0)^{-1}B_2^0 \end{array} \tag{2.11}$$

From (2.9), the slow subsystem can be derived by considering $\varepsilon = 0$ and expressing z as a function of x and u, which are denoted by x_s and u_s, that is from Assumption 1:

$$z_s = -A_{22}^{-1}(A_{21}x_s + B_2 u_s) = -(A_{22}^0)^{-1}(A_{21}^0 x_s + B_2^0 u_s) \tag{2.12}$$

In (2.12), z_s can be interpreted as the slow part of z. By replacing z_s in the original system, the slow dynamics reads:

$$\dot{x}_s = A_s x_s + B_s u_s \tag{2.13}$$

with A_s and B_s defined in (2.11). Similarly to define the fast dynamics, the vector x is considered as constant (that is $x = x_s$ and $\dot{z}_s = 0$) and we denote by $z_f = z - z_s$ and $u_f = u - u_s$ the fast part of the state and the control, respectively. Then, the fast dynamics reads:

$$\dot{z}_f = A_{22}^0 x_f + B_2^0 u_f \tag{2.14}$$

If the slow control u_s and the fast one u_f are determined, the complete control law is given by $u = u_s + u_f$.

Then by using a Lyapunov-based approach and adapting the results of [9] and [10], we can state the following conditions to solve item 1 of Problem 2.1.

Theorem 2.1 *If there exist two symmetric positive definite matrices* $W_s \in \mathbb{R}^{2\times 2}$, $W_f \in \mathbb{R}^{2\times 2}$ *and a matrix* $S_f \in \mathbb{R}^{1\times 2}$ *satisfying the following inequalities:*

$$He(A_{22}^0 W_f + B_2^0 S_f) < 0 \tag{2.15}$$

$$He(A_s W_s - B_s(I + S_f W_f^{-1}(A_{22}^0)^{-1} B_2^0)^{-1} S_f W_f^{-1}(A_{22}^0)^{-1} A_{21}^0 W_s) < 0 \tag{2.16}$$

then the control gain as defined in (2.10) with $K_f = S_f W_f^{-1}$ *solves item 1 of Problem 2.1.*

Proof This result is based on the use of Theorem 4 in [9] adapted to our case. Thus, we want to find a symmetric positive definite matrix $W_0 \in \mathbb{R}^{4\times 4}$ and a matrix $S_0 \in \mathbb{R}^{1\times 4}$ such that

$$A_\varepsilon W_0 + W_0 A_\varepsilon^T + B_\varepsilon S_0 + S_0^T B_\varepsilon^T < 0 \tag{2.17}$$

where from (2.6)

$$A_\varepsilon = \begin{bmatrix} A_{11} & A_{12} \\ \frac{A_{21}}{\varepsilon} & \frac{A_{22}}{\varepsilon} \end{bmatrix} = \begin{bmatrix} A_{11} & A_{12} \\ A_{21}^0 & A_{22}^0 \end{bmatrix}; B_\varepsilon = \begin{bmatrix} B_1 \\ \frac{B_2}{\varepsilon} \end{bmatrix} = \begin{bmatrix} B_1 \\ B_2^0 \end{bmatrix} \tag{2.18}$$

By developing each terms of the matrix at the right-hand side of relation (2.17) and by using arguments as in [9], matrices W_0 and S_0 can be described as follows:

$$\begin{aligned} W_0 &= \begin{bmatrix} W_s & -(A_{21}^0 W_s + B_2^0 S_s)^T (A_{22}^0)^{-T} \\ \star & W_f + (A_{22}^0)^{-1}(A_{21}^0 W_s + B_2^0 S_s) W_s^{-1} (A_{21}^0 W_s + B_2^0 S_s)^T (A_{22}^0)^{-T} \end{bmatrix} \\ S_0 &= \begin{bmatrix} S_s & S_f - S_s W_s^{-1} (A_{21}^0 W_s + B_2^0 S_s)^T (A_{22}^0)^{-T} \end{bmatrix} \end{aligned} \tag{2.19}$$

where W_f, S_f are solutions to relation (2.15) and W_s, S_s solutions to

$$\mathrm{He}(A_s W_s + B_s S_s) < 0 \tag{2.20}$$

That corresponds to characterize a gain $K = S_0 W_0^{-1}$ such that $A_\varepsilon + B_\varepsilon K$ is Hurwitz. From (2.19), one gets the following expression of K:

$$\begin{aligned} K &= \begin{bmatrix} S_s & S_f \end{bmatrix} \begin{bmatrix} W_s^{-1} & 0 \\ W_f^{-1}(A_{22}^0)^{-1}(A_{21}^0 W_s + B_2^0 S_s) W_s^{-1} & W_f^{-1} \end{bmatrix} \\ &= \begin{bmatrix} S_s W_s^{-1} + S_f W_f^{-1}(A_{22}^0)^{-1}(A_{21}^0 W_s + B_2^0 S_s) W_s^{-1} & S_f W_f^{-1} \end{bmatrix} \end{aligned} \tag{2.21}$$

In order to obtain a gain K structured as in (2.10), one has to satisfy:

$$S_s W_s^{-1} + S_f W_f^{-1}(A_{22}^0)^{-1}(A_{21}^0 W_s + B_2^0 S_s) W_s^{-1} = 0$$

or equivalently

$$S_s + S_f W_f^{-1} (A_{22}^0)^{-1} (A_{21}^0 W_s + B_2^0 S_s) = 0$$

which corresponds to

$$(I + S_f W_f^{-1} (A_{22}^0)^{-1} B_2^0) S_s + S_f W_f^{-1} (A_{22}^0)^{-1} A_{21}^0 W_s = 0 \tag{2.22}$$

By denoting $K_f = S_f W_f^{-1}$, one can remark that relation (2.15) is equivalent to verify

$$\text{He}((A_{22}^0 + B_2^0 K_f) W_f) < 0$$

that is matrix $(A_{22}^0 + B_2^0 K_f)$ is Hurwitz and therefore non singular. Then, one can observe that matrix $(I + S_f W_f^{-1} (A_{22}^0)^{-1} B_2^0)$ is also non singular by using the inverse matrix definition of $(A_{22}^0 + B_2^0 K_f)^{-1}$. Hence, relation (2.22) reads

$$S_s = -(I + S_f W_f^{-1} (A_{22}^0)^{-1} B_2^0)^{-1} S_f W_f^{-1} (A_{22}^0)^{-1} A_{21}^0 W_s \tag{2.23}$$

From (2.23), if relation (2.16) holds then relation (2.20) is verified. □

As mentioned before, we need at this stage to ensure the satisfaction of item 2 of Problem 2.1. Actually, considering the controller issued from Theorem 2.1, we have to provide a stability analysis of the original system (2.1)–(2.2) by considering that the input can saturate as follows: $u_{an} = sat(Kx_{an})$). Rather than addressing the problem in a linear framework (saturation not allowed), it is preferable to consider the problem in the saturated allowed framework. Depending on the controller designed the global asymptotic stability (GAS) or the local asymptotic stability (LAS) of the closed-loop system is achevied [18]. This is detailed in the following section.

2.3.2 Computational Issues

The main drawback of Theorem 2.1 resides in the fact that relation (2.16) is nonlinear in the decision variables due to the presence of products between some variables, relation (2.15) being linear. Hence, the lack of linearity of this condition makes it not computationally tractable to obtain a solution to Problem 2.1 [19]. However, some relaxation steps can be proposed. Note that the first inequality (2.15) is linear in the decision variables W_f, S_f. The second inequality (2.16) is nonlinear in the decision variables W_s, W_f, S_f but becomes linear in W_s if W_f and S_f are fixed. Hence, one can consider the following first algorithm regarding the controller design procedure.

Algorithm 1

1. Select a desired decay rate for the fast subsystem with parameter $\mu_f > 0$.
2. Compute $K_f = S_f W_f^{-1}$ stabilizing and improving the decay rate of the fast subsystem by solving

$$\mathrm{He}(A_{22}^{0} W_f + B_2^0 S_f + \mu_f W_f) < 0 \tag{2.24}$$

3. Feasibility problem. Find W_s solution to

$$\mathrm{He}(A_s W_s - B_s(I + K_f (A_{22}^0)^{-1} B_2^0)^{-1} K_f (A_{22}^0)^{-1} A_{21}^0 W_s) < 0 \tag{2.25}$$

4. If (2.25) is feasible, then $K = \begin{bmatrix} 0 & K_f \end{bmatrix}$ stabilizes the closed-loop system by acting on fast dynamics.
   If not, then decrease the decay rate parameter μ_f and go back to step 2.

Remark 2.1 System (2.1) being open-loop stable, there always exists a solution to the feasibility linear problem (2.24)–(2.25) with $K_f = 0$. Then, there always exists a μ_f small enough such that, for a controller issued from step 2, the LMI condition in step 3 is feasible.

From Algorithm 1, we have in hand the stabilizing controller, and we can now manage the constraints. A first direction could be to adapt the conditions provided in [10] to our current problem. Due to the difficulties encountered to deal with the nonlinearities appearing in the conditions, we decided here to propose an alternative route by providing analysis conditions based on tools issued from [18, 20], by using the toolbox SATAW-Tool.[2]

Algorithm 2

1. Given the value of K_f (and therefore of K) resulting from Algorithm 1.
2. Global asymptotically stability (GAS) case. Find a symmetric positive definite matrix $W \in \mathbb{R}^{4\times4}$ and a diagonal positive definite matrix $S \in \mathbb{R}^{1\times1}$ solution to the feasibility problem:

$$\begin{bmatrix} W(A_\varepsilon + B_\varepsilon K)^T + (A_\varepsilon + B_\varepsilon K)W & B_\varepsilon S - WK^T \\ SB_\varepsilon^T - KW & -2S \end{bmatrix} < 0 \tag{2.26}$$

3. Local asymptotic stability (LAS) case. If the global case is unfeasible, given $u_0 = \bar{u}_e$, find a symmetric positive definite matrix $W \in \mathbb{R}^{4\times4}$, a diagonal positive definite matrix $S \in \mathbb{R}^{1\times1}$, a matrix $Z \in \mathbb{R}^{1\times4}$ and a positive scalar γ solution to the optimization problem:

$$\begin{gathered} \min \quad -trace(W) + \gamma \\ \text{s.t.} \\ \begin{bmatrix} W(A_\varepsilon + B_\varepsilon K)^T + (A_\varepsilon + B_\varepsilon K)W & B_\varepsilon S - WK^T - Z^T \\ SB_\varepsilon^T - KW - Z & -2S \end{bmatrix} < 0 \\ \begin{bmatrix} W & Z^T \\ Z & \gamma u_0^2 \end{bmatrix} \geq 0 \end{gathered} \tag{2.27}$$

[2]http://homepages.laas.fr/queinnec/satawtool.html.

The objective of the optimization criterion considered in step 3 of Algorithm 2 is to maximize the region

$$\mathscr{E}(W,\gamma) = \left\{ \begin{bmatrix} x \\ z \end{bmatrix} \in \mathbb{R}^4; \begin{bmatrix} x \\ z \end{bmatrix}^T W^{-1} \begin{bmatrix} x \\ z \end{bmatrix} \leq \gamma^{-1} \right\} \tag{2.28}$$

which is a region of invariance and asymptotic stability for the closed-loop system.

Remark 2.2 The global condition does not depend on any bound u_0 and formally allows that non-symmetric bounds may be applied in practice. It also means that any initial condition may be applied, and, typically, formally guarantees that the controller may be applied from the patient awake state. On the other hand, the local condition is directly related to the bound $u_0 = \bar{u}_e$, which means that, formally, $0 \leq u_{an} \leq 2\bar{u}_e$. Moreover, only initial states belonging to the set $\mathscr{E}(W,\gamma)$ should be considered.

Remark 2.3 One could also be interested in guaranteeing that, once the *BIS* enters the interval [40, 60], it remains inside this interval. Such a constraint could be added in the problems (2.26) and (2.27) through the additional condition:

$$EWE' \leq \gamma z_{2M}^2 \tag{2.29}$$

with $E = \begin{bmatrix} 0 \; 0 \; 0 \; 1 \end{bmatrix}$ and $z_{2M} = \max(z_{2min}, z_{2max})$ corresponding to the bounds on the effect site concentration $-z_{2min} \leq z_2 \leq z_{2max}$ issued from the change of variables around the equilibrium point. However, this would result in drastically reducing the size of the region of invariance and asymptotic stability for the closed-loop system and would prevent to consider the patient awake state as initial state.

2.4 Simulations

To illustrate the approach let us consider a patient with the following characteristics: woman, 49 years old, 68 kg and 172 cm. It corresponds to the system matrices defined with:

$$\left[\begin{array}{c|c} A_{11} & A_{12} \\ \hline A_{21}^o & A_{22}^o \end{array}\right] = \left[\begin{array}{cc|cc} -0.068 & 0 & 0.138 & 0 \\ 0 & -0.004 & 0.077 & 0 \\ \hline 0.068 & 0.003 & -0.389 & 0 \\ 0 & 0 & 0.042 & -0.456 \end{array}\right], \quad \left[\begin{array}{c} B_1 \\ \hline B_2^o \end{array}\right] = \left[\begin{array}{c} 0 \\ 0 \\ \hline 1 \\ 0 \end{array}\right]$$

For a target *BIS* of 50, the equilibrium point and associated input are given by

$$\bar{x}_e = \begin{bmatrix} 69.5776 \; 809.2000 \end{bmatrix}, \quad \bar{z}_e = \begin{bmatrix} 36.7608 \; 3.4000 \end{bmatrix}, \quad \bar{u}_e = 6.7519$$

and the open-loop spectrum of system (2.1) is equal to

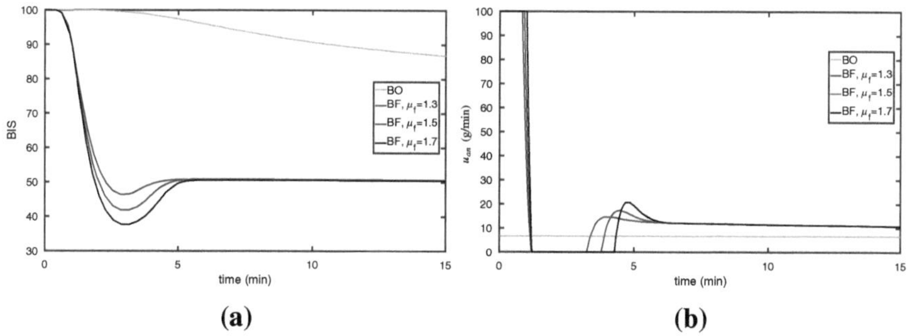

Fig. 2.1 Time simulation of the open-loop and saturated closed-loop systems with controllers K2, K3 and K4

$$\lambda_{bo} = \{-0.002, \ -0.043, \ -0.415, \ -0.456\}$$

First, Algorithm 1 allows to provide solution to the first objective of Problem 2.1. Considering various pole-placement constraints, the problem is feasible and one obtains:

$$\begin{array}{lll}
\text{K1}: & \mu_f = 1.2 & K_f = \begin{bmatrix} -3.3124 & -73.3247 \end{bmatrix} \\
& & \lambda_{bf} = \{-0.003, \ -0.067, \ -2.079 \pm 0.672i\} \\
\text{K2}: & \mu_f = 1.3 & K_f = \begin{bmatrix} -3.9172 & -102.4763 \end{bmatrix} \\
& & \lambda_{bf} = \{-0.003, \ -0.067, \ -2.382 \pm 0.780i\} \\
\text{K3}: & \mu_f = 1.5 & K_f = \begin{bmatrix} -3.6796 & -118.7038 \end{bmatrix} \\
& & \lambda_{bf} = \{-0.003, \ -0.067, \ -2.263 \pm 1.317i\} \\
\text{K4}: & \mu_f = 1.7 & K_f = \begin{bmatrix} -4.3694 & -167.6553 \end{bmatrix} \\
& & \lambda_{bf} = \{-0.003, \ -0.067, \ -2.608 \pm 1.560i\}
\end{array}$$

where λ_{bf} denotes the closed-loop spectrum. Then Algorithm 2 is used to check if the closed-loop saturated system is globally asymptotically stable or, if not, if the system may be initialized from the patient awake state. It results that the saturated closed-loop system is GAS with controller K1, but only LAS with controllers K2, K3 and K4. With the controller K2, $-x_{ane}$ belongs to the associated set $\mathscr{E}(W, \gamma)$ and the controller may be safely applied with the patient initially awake. On the other hand, with the controllers K3 and K4, $-\bar{x}_e$ belongs to the associated set $\mathscr{E}(W, \gamma)$ but only a percentage of $-\bar{z}_e$ belongs to the set (40% with K3, and 7% with K4).

Numerical simulations are plotted in Fig. 2.1. In Fig. 2.1a, one can see the evolution of *BIS* from the patient awake state to the reference 50, in open-loop (green) and in closed-loop with controllers K2 (blue), K3 (red) and K4 (black). The overshoot with controller K4 in not desirable as the *BIS* goes temporarily below the bound 40. Figure 2.1b presents the associated input.

2.5 Conclusion

Taking benefit from the singularly perturbed systems framework, the fast and slow dynamics present in the compartmental system have been separated. With the aim at accelerating the fast dynamics, the design of a structured state feedback controller has been first proposed. Second, some relaxation schemes associated to convex optimization problems allowed to guarantee the satisfaction of the constraints.

This work lets some questions open. In particular, one would be interested with more complete conditions not only to deal with the fast dynamics but also to guarantee that the constraints are satisfied and to initialize the system to the patient awake conditions, in order to mathematically validate the medical strategy from induction to maintenance. This will be the subject of future works.

References

1. Lemos, J.M., Caiado, D.V., Costa, B.A., Paz, L.A., Mendonca, T.F., Esteves, S., Seabra, M.: Robust control of maintenance-phase anesthesia. IEEE Control Syst. Mag. **34**(6), 24–38 (2014)
2. Soltesz, K., Hahn, J.-O., Dumont, G.A., Ansernimo, J.M.: Individualized PID control of depth of anesthesia based on patient model identification during the induction phase of anesthesia. In: IEEE Conference on Decision and Control and European Control Conference (CDC-ECC) Orlando USA, pp. 855–860, December 2011
3. Haddad, W.M., Hayakawa, T., Bailey, J.: Adaptive control for non-negative and compartmental dynamical systems with application on general anesthesia. Int. J. Adapt. Control Signal Process. **17**(3), 209–235 (2003)
4. Gentilini, A., Schaniel, C., Morari, M., Bieniok, C., Wymann, R., Schnider, T.: A new paradigm for the closed-loop intraoperative administration of analgesics in humans. IEEE Trans. Biomed. Eng. **49**(4), 289–299 (2002)
5. Beck, C.L.: Modeling and control of pharmacodynamics. Eur. J. Control **24**, 33–49 (2015)
6. Zhusubaliyev, Z.T., Medvedev, A., Silva, M.M.: Nonlinear dynamics in closed-loop anesthesia: pharmacokinetic/pharmacodynamic model under PID-feedback. In: American Control Conference, pp. 5496–5501, Portland, USA (2014)
7. Fiacchini, M., Queinnec, I., Tarbouriech, S., Mazerolles, M.: Invariant based control of induction and maintenance phases for anesthesia. In: 6th IFAC Conference on Foundations of Systems Biology in Engineering (FOSBE), MAgdeburg, Germany, October 2016
8. Kokotovic, P.V., Khalil, H.K., O'Reilly, J.: Singular Perturbation Methods in Control; Analysis and Design. Academic, New York (1986)
9. Garcia, G., Tarbouriech, S.: Control of singularly perturbed systems by bounded control. In: American Control Conference, Denver, USA, pp. 4482–4487 (2003)
10. Lizarraga, I., Tarbouriech, S., Garcia, G.: Control of singularly perturbed systems under actuator saturation. In: 16th World IFAC Congress, pp. 243–248, Prague, Czech Republic (2005)
11. Derendorf, H., Meibohm, B.: Modeling of pharmacokinetic/pharmacodynamic (pk/pd) relationships: concepts and perspectives. Pharm. Res. **16**(2), 176–185 (1999)
12. Haddad, W.M., Chellaboina, V., Hui, Q.: Nonnegative and Compartmental Dynamical Systems. Princeton (2010)
13. Coppens, M.J., Eleveld, D.J., Proost, J.H., Marks, L.A., Van Bocxlaer, J.F., Vereecke, H., Absalom, A.R., Struys, M.M.: An evaluation of using population pharmacokinetic models to estimate pharmacodynamic parameters for propofol and bispectral index in children. Anesthesiology **115**(1), 83–93 (2011)

14. Minto, C.F., Schneider, T.W., Shafer, S.L.: Pharmacokinetics and pharmacodynamics of remifentanil. Model application. Anesthesiology **86**(1), 24–33 (1997)
15. Schnider, T.W., Minto, C.F., Gambus, P.L., Anderson, C., Goodale, D.B., Young, E.: The influence of method of administration and covariates on the pharmacokinetics of propofol in adult volunteers. J. Am. Soc. Anesthesiol. **88**(5), 1170–1182 (1998)
16. Bailey, J.M., Haddad, W.M.: Drug dosing control in clinical pharmacology. IEEE Control Syst. Mag. **25**(2), 35–51 (2005)
17. Berman, A., Neumann, M., Stern, R.J.: Nonnegative Matrices in Dynamic Systems. Wiley-Interscience, John Wiley and Sons, New York, USA (1989)
18. Tarbouriech, S., Garcia, G., Gomes da Silva Jr., J.M., Queinnec, I.: Stability and Stalization of Linear Systems with Saturating Actuators. Springer (2011)
19. Boyd, S., Ghaoui, L.E., Feron, E., Balakrishnan, V.: Linear Matrix Inequalities in System and Control Theory. Society for Industrial and Applied Mathematics, June (1997)
20. Zabi, S., Queinnec, I., Tarbouriech, S., Garcia, G., Mazerolles, M.: New approach for the control of anesthesia based on dynamics decoupling. In: 9th IFAC Symposium on Biological and Medical Systems (BMS 2015), Berlin, Germany (2015)

Chapter 3
Interval Observers for SIR Epidemic Models Subject to Uncertain Seasonality

Pierre-Alexandre Bliman and Bettina D'Avila Barros

Abstract Epidemic models describe the establishment and spread of infectious diseases. Among them, the SIR model is one of the simplest, involving exchanges between three compartments in the population, that represent respectively the number of susceptible, infective and recovered individuals. The issue of state estimation is considered here for such a model, subject to seasonal variations and uncertainties in the transmission rate. Assuming continuous measurement of the number of new infectives per unit time, a class of interval observers with estimate-dependent gain is constructed and analyzed, providing lower and upper bounds for each state variable at each moment in time. The dynamical systems that describe the evolution of the errors are monotonous. Asymptotic stability is ensured by appropriate choice of the gain components as a function of the state estimate, through the use of a common linear Lyapunov function. Numerical experiments are presented to illustrate the method.

Keywords Interval observer · Uncertain systems · Monotone systems · Linear Lyapunov functions · SIR model · Mathematical epidemiology

3.1 Introduction, Presentation of the SIR Model

The SIR model with vital dynamics, see e.g. [4, 11], is one of the most elementary compartmental models of epidemics. It describes the evolution of the relative proportions of three classes of a population of constant size, namely the susceptibles S, capable of contracting the disease and becoming infective; the infectives I, capable of transmitting the disease to susceptibles; and the recovered R, permanently immune after healing. This model is as follows:

P.-A. Bliman (✉) · B. D'Avila Barros
Escola de Matemática Aplicada, Fundação Getulio Vargas, Rio de Janeiro - RJ, Brazil
e-mail: pierre-alexandre.bliman@inria.fr

P.-A. Bliman
Lab. J.-L. Lions UMR CNRS 7598, Sorbonne Universités, Inria, UPMC Univ Paris 06, Paris, France
e-mail: barrosbettina@gmail.com

F. Cacace et al. (eds.), *Positive Systems*, Lecture Notes in Control and Information Sciences 471, DOI 10.1007/978-3-319-54211-9_3

$$\dot{S} = \mu - \mu S - \beta SI \tag{3.1a}$$
$$\dot{I} = \beta SI - (\mu + \gamma)I \tag{3.1b}$$

The natural birth and mortality rate is μ (the disease is supposed not to induce supplementary death rate), γ is the recovery rate, while β represents the transmission rate per infective. All these parameters are positive. We consider here *proportions* of the population, and more precisely that $S + I + R \equiv 1$. Notice that the dynamics of R (given by $\dot{R} = \gamma I - \mu R$, that guarantees that $\dot{S} + \dot{I} + \dot{R} \equiv 0$) may be omitted, as the total population size is constant.

When the parameters are constant, the evolution of the solutions of system (3.1) depends closely upon the ratio $\mathscr{R}_0 := \frac{\beta}{\mu+\gamma}$ [4, 11]. The disease-free equilibrium ($S = 1, I = R = 0$) always exists. When $\mathscr{R}_0 < 1$, it is the only equilibrium and it is globally asymptotically stable. It becomes unstable when $\mathscr{R}_0 > 1$, and an asymptotically stable endemic equilibrium then appears.

On the contrary, when the parameters are time-varying, complicated dynamics may occur [12]. We are interested here in estimating the value of the three populations, a first step paving the way for epidemic outbreak forecasting. We use techniques of interval observers, including output injection, in the spirit e.g. of [7, 8, 14]. The dynamics of the obtained error equation is seen as a linear uncertain time-varying positive system, whose asymptotic stability is ensured through the search of a common *linear* Lyapunov function and adequate choice of the gain as function of the state estimate.

The hypotheses on the model and some qualitative results are presented in Sect. 3.2. The considered class of observers is given in Sect. 3.3, with some a priori estimates and technical results. The main result is provided in Sect. 3.4, where the asymptotic error corresponding to certain gain choice is quantified. Last, illustrative numerical experiments are shown in Sect. 3.5.

3.2 Hypotheses on the Model and Preliminaries

We consider in the sequel that the transmission rate is subject to uncertain seasonal variations. It is known that relatively modest variations of this type have the capacity to induce large amplitude fluctuations in the observed disease incidence. This seems due to harmonic resonance, the seasonal forcing exciting frequencies close to the natural near-equilibrium oscillatory frequency [6].

One assumes that the transmission rate β is bounded by two functions $\beta_\pm$, available in real-time (all functions are supposed locally integrable):

$$\beta_-(t) \leq \beta(t) \leq \beta_+(t) \quad \text{for a.e. } t \geq 0 \tag{3.2}$$

(typically with $0 < \liminf_{t\to+\infty} \beta_-(t) \leq \limsup_{t\to+\infty} \beta_+(t) < +\infty$).

Our goal is to estimate lower and upper bounds for the three subpopulations. The unique available measurement is supposed to be the incidence rate $y = \beta SI$, i.e. the number of new infectives per time unit (accessible through epidemiological surveillance). With representation (3.1), y is not a state component, contrary e.g. to [3, 5]. One sees easily that with this output, the system is detectable, but *not* observable at the disease-free equilibrium (where $I = 0$).

The following result provides qualitative estimates of its solutions.

Lemma 3.1 *Assume $S(0) \geq 0$, $I(0) \geq 0$ and $S(0) + I(0) \leq 1$. Then the same properties hold for any $t \geq 0$. The same is true with strict inequalities.*

Proof Integrating (3.1a), (3.1b) yields

$$I(t) = I(0)e^{\int_0^t (\beta(\tau)S(\tau)-(\mu+\gamma)).d\tau}$$

$$S(t) = S(0)e^{-\int_0^t (\mu+\beta(\tau)I(\tau)).d\tau} + \mu \int_0^t e^{-\int_\tau^t (\mu+\beta I)}.d\tau$$

which show that $S(t), I(t) \geq 0$ for any $t \geq 0$; while integrating the differential inequality $\dot{S} + \dot{I} \leq \mu(1 - S - I)$ shows that $1 - (S(t) + I(t)) \geq 0$ for any $t \geq 0$. The same formulas provide the demonstration in the strict inequality case. □

3.3 A Class of Nonlinear Observer Models

As preparation for the upcoming study, we explore now the following class of observers for system (3.1):

$$\dot{\hat{S}} = \mu - \mu\hat{S} - y + k_S(t)(y - \beta_S\hat{S}\hat{I}) \tag{3.3a}$$

$$\dot{\hat{I}} = y - (\mu + \gamma)h\hat{I} + k_I(t)(y - \beta_I\hat{S}\hat{I}) \tag{3.3b}$$

where the time-varying gain components $k_S(t), k_I(t)$ are yet to be defined.

Lemma 3.2 *Suppose that for some $\varepsilon > 0$,*

$$k_S(t) \geq 1 \textit{ whenever } \hat{S}(t) \leq \varepsilon, \qquad k_I(t) \geq -1 \textit{ whenever } \hat{I}(t) \leq \varepsilon. \tag{3.4}$$

Assume $\hat{S}(0) \geq 0$, resp. $\hat{I}(0) \geq 0$. Then, for any $t \geq 0$, $\hat{S}(t) \geq 0$, resp. $\hat{I}(t) \geq 0$.

Proof Verify directly that, under assumption (3.4), $\dot{\hat{S}} \geq 0$, resp. $\dot{\hat{I}} \geq 0$, in the neighborhood of a point where $\hat{S} = 0$, resp. $\hat{I} = 0$. This proves the result. □

Last, the following technical result will be needed.

Lemma 3.3 *Define* $e_S := S - \hat{S}$, $e_I := I - \hat{I}$. *Then,*

$$\begin{pmatrix} \dot{e}_S \\ -\dot{e}_I \end{pmatrix} = \begin{pmatrix} -(\mu + k_S\beta_S h\hat{I}) & k_S\beta_S S \\ k_I\beta_I h\hat{I} & -(\mu+\gamma+k_I\beta_I S) \end{pmatrix} \begin{pmatrix} e_S \\ -e_I \end{pmatrix} + SI \begin{pmatrix} k_S(\beta_S - \beta) \\ k_I(\beta - \beta_I) \end{pmatrix} \tag{3.5}$$

Proof One has $\dot{e}_S = -\mu e_S + k_S(\beta_S\hat{S}\hat{I} - y)$ and $\dot{e}_I = -(\mu+\gamma)e_I + k_I(\beta_I\hat{S}\hat{I} - y)$. On the other hand, $\beta_S\hat{S}\hat{I} - y = (\beta_S - \beta)SI + \beta_S S(\hat{I} - I) + \beta_S\hat{I}(\hat{S} - S) = -SI(\beta - \beta_S) - \beta_S S e_I - \beta_S\hat{I}e_S$, and similarly for $\beta_I\hat{S}\hat{I} - y$. One deduces

$$\begin{pmatrix} \dot{e}_S \\ \dot{e}_I \end{pmatrix} = -\begin{pmatrix} \mu + k_S\beta_S\hat{I} & k_S\beta_S S \\ k_I\beta_I\hat{I} & k_I\beta_I S + \mu + \gamma \end{pmatrix} \begin{pmatrix} e_S \\ e_I \end{pmatrix} - SI \begin{pmatrix} k_S(\beta - \beta_S) \\ k_I(\beta - \beta_I) \end{pmatrix} \tag{3.6}$$

and finally (3.5) when using $-e_I$ instead of e_I. □

Observe that system (3.6) appears monotone for *nonpositive* gains, which is detrimental to its stability. This is not the case with system (3.5), which is used in the sequel to construct interval observers.

3.4 Error Estimates for Interval Observers

Notice first that system (3.1) is not evidently, or transformable into, a monotone system. The instances of (3.3) presented in the next result provide a class of interval observers with guaranteed speed of convergence.

Theorem 3.1 *Consider the two* independent *systems*

$$\dot{S}_+ = \mu - \mu S_+ - y + k_{S+}(t)(y - \beta_-(t)S_+I_-) \tag{3.7a}$$
$$\dot{I}_- = y - (\mu+\gamma)I_- + k_{I-}(t)(y - \beta_+(t)S_+I_-) \tag{3.7b}$$

$$\dot{S}_- = \mu - \mu S_- - y + k_{S-}(t)(y - \beta_+(t)S_-I_+) \tag{3.8a}$$
$$\dot{I}_+ = y - (\mu+\gamma)I_+ + k_{I+}(t)(y - \beta_-(t)S_-I_+) \tag{3.8b}$$

i. Assume that the gains are nonnegative functions of $S_\pm$, $I_\pm$, *that fulfill* (3.4) *for some* $\varepsilon > 0$. *If* $0 \le S_-(t) \le S(t) \le S_+(t)$ *and* $0 \le I_-(t) \le I(t) \le I_+(t)$ *for* $t = 0$, *then the same holds for any* $t \ge 0$.

ii. If in addition the gain components $k_{S\pm}(t)$, $k_{I\mp}(t)$ *are chosen such that*

$$\beta_-(t)k_{S+}(t) - \rho_+\beta_+(t)k_{I-}(t) = \frac{\rho_+\gamma}{\rho_+I_-(t) + S_+(t)} \tag{3.9a}$$
$$\beta_+(t)k_{S-}(t) - \rho_-\beta_-(t)k_{I+}(t) = \frac{\rho_-\gamma}{\rho_-I_+(t) + S_+(t)} \tag{3.9b}$$

for fixed $\rho_\pm > 0$, *then, writing* $V_+(t) := (S_+(t) - S(t)) + \rho_+(I(t) - I_-(t))$, $V_-(t) := (S(t) - S_-(t)) + \rho_-(I_+(t) - I(t))$, *the state functions* $V_\pm$ *are positive definite when the trajectories are initialized according to point i., and*[1]

$$\forall t \geq 0,\ V_\pm(t) \leq \int_0^t e^{-\int_\tau^t \delta_\pm} \max\{k_{S\pm}(\tau), \rho_\pm k_{I\mp}(\tau)\} S(\tau) I(\tau)(\beta_+(\tau) - \beta_-(\tau))\, d\tau$$
$$+ e^{-\int_0^t \delta_\pm(\tau).d\tau} V_\pm(0), \quad \text{with } \delta_\pm(t) := \mu + \gamma \frac{\rho_\pm I_\mp(t)}{\rho_\pm I_\mp(t) + S_+(t)} \tag{3.10}$$

The proposed observers guarantee that the errors converge exponentially, with speeds $\delta_\pm(t)$ that smoothly vary from μ (in absence of infectives: $I_\pm(t) = 0$) to at most $\mu + \gamma$ (in case of outbreak, if $\rho_\pm I_\mp(t) \gg S_+(t)$). Recall that a *positive* linear time-invariant system is asymptotically stable *iff* it admits a *linear* Lyapunov function of the type $V_\pm$ [1, 10, 13]. With this in mind, it may indeed be deduced from the proof (see in particular (3.11)) that the convergence speed of observer of type (3.7)–(3.8) is bound to be *at most equal to* $\mu + \gamma$ in presence of epidemics, and *cannot be larger than* μ in absence of infectives.[2] Recall that μ is the inverse of the mean life duration, while γ is the inverse of the mean recovery time: typically $\mu \ll \gamma$. Therefore, the observer takes advantage of epidemic bursts to reduce faster the estimation error. Notice that these convergence speeds do not depend upon the values of $\beta_\pm$.

The trade-off between stability and precision is clear from formula (3.10): an intrinsic limitation is evident from the fact that the integral therein is at least equal to $\int_0^t e^{-(\mu+\gamma)(t-\tau)} \max\{k_{S\pm}(\tau), \rho_\pm k_{I\mp}(\tau)\} S(\tau) I(\tau)(\beta_+(\tau) - \beta_-(\tau)) d\tau$ which is guaranteed to vanish only when both gains are zero. On the other hand, for zero gains, the error equation for the susceptibles is $\dot{e}_{S\pm} + \mu e_{S\pm} = 0$ which converges slowly to zero.

Last, observe that, with the estimate-dependent choice of the gain defined in (3.9), the error equations may be non monotone. However they fulfill the positivity and stability properties mentioned in the statement.

Proof of Theorem 3.1. We show the results for system (3.7) only, system (3.8) is treated similarly.

• Introduce the error terms $e_{S+} := S_+ - S$ and $e_{I-} := I - I_-$. Applying Lemma 3.3 to system (3.7) with $\hat{S} = S_+, \hat{I} = I_-, k_S = k_{S+}, \beta_S = \beta_-, k_I = k_{I-}, \beta_I = \beta_+$ (and therefore $e_S = -e_{S+}, e_I = e_{I-}$) yields

---

[1] In accordance with the usual convention, in the following formula the signs $\pm, \mp$ should be interpreted either everywhere with the upper symbols, or everywhere with the lower ones.

[2] We constrain the closed-loop system to be monotone, so not any closed-loop spectrum can be realized.

$$
\begin{aligned}
\begin{pmatrix} \dot{e}_{S+} \\ \dot{e}_{I-} \end{pmatrix} &= -\begin{pmatrix} \dot{e}_S \\ -\dot{e}_I \end{pmatrix} \\
&= -\begin{pmatrix} -(\mu + k_{S+}\beta_- I_-) & k_{S+}\beta_- S \\ k_{I-}\beta_+ I_- & -(\mu+\gamma+k_{I-}\beta_+ S) \end{pmatrix} \begin{pmatrix} e_S \\ -e_I \end{pmatrix} - SI \begin{pmatrix} k_{S+}(\beta_- - \beta) \\ k_{I-}(\beta - \beta_+) \end{pmatrix} \\
&= \begin{pmatrix} -(\mu + k_{S+}\beta_- I_-) & k_{S+}\beta_- S \\ k_{I-}\beta_+ I_- & -(\mu+\gamma+k_{I-}\beta_+ S) \end{pmatrix} \begin{pmatrix} e_{S+} \\ e_{I-} \end{pmatrix} + SI \begin{pmatrix} k_{S+}(\beta - \beta_-) \\ k_{I-}(\beta_+ - \beta) \end{pmatrix}.
\end{aligned}
$$

The previous system may thus be written $\dot{X} = f(t, X)$, for $X := \begin{pmatrix} e_{S+} & e_{I-} \end{pmatrix}^{\mathsf{T}}$ and where the dependence with respect to time comes indirectly through the presence of the other time-varying term. The off-diagonal terms of the Jacobian matrix are respectively $k_{S+}(t)\beta_-(t)S_+(t)$ and $k_{I-}(t)\beta_+(t)I_-(t)$, clearly nonnegative for a.e. $t \geq 0$ due to the hypotheses on the gain components (see Lemma 3.2). The corresponding system is therefore monotone [9, 15], and any solution of (3.7) departing with $e_{S+}(0), e_{I-}(0) \geq 0$ verifies $e_{S+}(t), e_{I-}(t) \geq 0$ for any $t \geq 0$. This proves *i.*

• Writing $X := \begin{pmatrix} e_{S+} & e_{I-} \end{pmatrix}^{\mathsf{T}}$, notice that $V_+(X) := u^{\mathsf{T}}X$, for $u := \begin{pmatrix} 1 & \rho_+ \end{pmatrix}$, and V_+ and $\rho_+ > 0$ as in the statement. When X is initialized with nonnegative values, then this property is conserved (see point *i.*), so V_+ is positive definite and may be considered as a candidate Lyapunov function.

Along the trajectories of (3.7), one has, using δ_+ defined in (3.10),

$$
\begin{aligned}
&\dot{V}_+(X) + \delta_+ V_+(X) \\
&= u^{\mathsf{T}}(\dot{X} + \delta_+ X) \\
&= u^{\mathsf{T}} \begin{pmatrix} \delta_+ - (\mu + k_{S+}\beta_- I_-) & k_{S+}\beta_- S \\ k_{I-}\beta_+ I_- & \delta_+ - (\mu+\gamma+k_{I-}\beta_+ S) \end{pmatrix} X + SIu^{\mathsf{T}} \begin{pmatrix} k_{S+}(\beta - \beta_-) \\ k_{I-}(\beta_+ - \beta) \end{pmatrix} \\
&= \begin{pmatrix} \delta_+ - \mu + (\rho_+ k_{I-}\beta_+ - k_{S+}\beta_-)I_- & \rho_+(\delta_+ - \mu - \gamma) - (\rho_+ k_{I-}\beta_+ - k_{S+}\beta_-)S \end{pmatrix} X \\
&\quad + SI(k_{S+}(\beta - \beta_-) + \rho_+ k_{I-}(\beta_+ - \beta)). \qquad (3.11)
\end{aligned}
$$

Choosing the gain as in (3.9a) gives

$$
\delta_+ - \mu + (\rho_+ k_{I-}\beta_+ - k_{S+}\beta_-)I_- = \frac{\gamma I_-}{I_- + \frac{S_+}{\rho_+}} - \frac{\gamma I_-}{I_- + \frac{S_+}{\rho_+}} = 0
$$

and

$$
\rho_+(\delta_+ - \mu - \gamma) - (\rho_+ k_{I-}\beta_+ - k_{S+}\beta_-)S = -\frac{\gamma S_+}{I_- + \frac{S_+}{\rho_+}} + \frac{\gamma S}{I_- + \frac{S_+}{\rho_+}} \leq 0.
$$

Formula (3.11) then yields $\dot{V}_+(X) + \delta_+ V_+(X) \leq SI(k_{S+}(\beta - \beta_-) + \rho_+ k_{I-}(\beta_+ - \beta)) \leq \max\{k_{S+}, \rho_+ k_{I-}\} SI(\beta_+ - \beta_-)$, which gives (3.10) by integration. This proves point *ii.* and achieves the proof of Theorem 3.1. □

3.5 Numerical Experiments

We consider in the sequel the following parameter values. One takes $\mu = 0.02$/year, $\gamma = \frac{1}{20}$/day $= \frac{365}{20}$/year. The transmission rate $\beta(t)$ is taken as $\beta^*(1 + \eta \cos(\omega t))$, with nominal value β^* such that $\mathscr{R}_0 = \frac{\beta^*}{\mu+\gamma} = 17$, $\eta = 0.4$ and $\omega = 2.4$ rad/year, close to the pulsation of the near-equilibrium natural oscillations. Last, S and I are initialized at 0.06 and 0.001, close to the perturbation-free equilibrium ($\eta = 0$), and the observer initial conditions as 0 and 1 (lower and upper values).

The gains were chosen as follows

$$k_{S_+} = \frac{\gamma}{I_- + \frac{S_\pm}{\rho_+}} \frac{1}{\beta_-}, \qquad k_{I_-} = 0$$

$$k_{S_-} = \max\left(\frac{\gamma}{I_+ + \frac{S_\pm}{\rho_-}} \frac{1}{\beta_+}, \frac{1}{1 + \frac{S_-}{\varepsilon}}\right) \text{ with } \varepsilon = 5 \times 10^{-3}, \quad k_{I_+} = \frac{1}{\rho_- \beta_-}\left(\beta_+ k_{S_-} - \frac{\gamma}{I_+ + \frac{S_\pm}{\rho_-}}\right)$$

in accordance with (3.9). The small parameter ε is introduced in order to ensure that S_- remains positive, according to Lemma 3.2.

• First, essays were realized in the absence of uncertainty on the transmission rate, that is taking $\beta = \beta_- = \beta_+$. Figure 3.1 shows, for $\rho_\pm = 100$, the logarithm of the errors of $S_\pm$, i.e., $\log_{10}(e_{S+}) = \log_{10}(S_+ - S)$ and $\log_{10}(e_{S-}) = \log_{10}(S - S_-)$. We see that it decays with a high speed, whose instantaneous values are between μ and $\mu + \gamma$, as proved in Theorem 3.1.

• We now introduce uncertainty in the transmission rate. We use quite imprecise estimates, namely $\beta_\pm(t) = (1 \pm 0.6)\beta(t)$. Figure 3.2 shows results for $\rho_\pm = 10^2, 10^3, 10^4$.

The convergence of $I_\pm$ towards I is fast, with small residual errors. On the other hand, errors remain present in the estimates of S, illustrating the phenomenon mentioned right after Theorem 3.1.

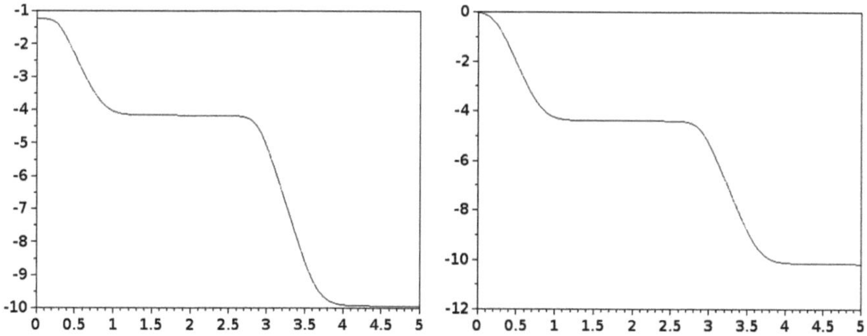

Fig. 3.1 Decimal logarithm of the e_{S-} (*left*) and e_{S+} (*right*) as a function of time (in years)

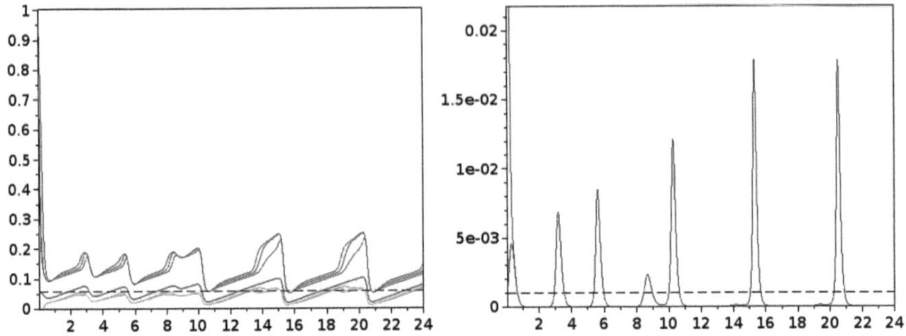

Fig. 3.2 Actual value (*blue*), lower estimates (*green*) and upper estimates (*red*) of S (*left*) and I (*right*) as functions of time (in years). The values at unperturbed equilibrium appear as dashed lines (color figure online)

3.6 Conclusion

A family of SIR model with time-varying transmission rate has been considered. For these models, a class of interval observers has been proposed, assuming that the rate of new infectives is continuously measured and that the transmission rate is uncertain and limited by (time-varying) known lower and upper bounds. It has been shown that these observers ensure fast convergence to the exact values in absence of uncertainty. For uncertain transmission rates, analytical bounds have been provided for the estimation errors.

To improve these results, we plan to consider in the future higher-order observers, and to apply the techniques of bundles of interval observers introduced in [2]. Also, processing real experimental data should allow to assess the interest of the proposed method.

References

1. Berman, A., Plemmons, R.J.: Nonnegative matrices in themathematical sciences. Classics in Applied Mathematics, vol. 9 (1979)
2. Bernard, O., Gouzé, J.-L.: Closed loop observers bundle for uncertain biotechnological models. J. Process Control **14**(7), 765–774 (2004)
3. Bichara, D., Cozic, N., Iggidr, A.: On the estimation of sequestered infected erythrocytes in Plasmodium falciparum malaria patients. Math. Biosci. Eng. (MBE) **11**(4), 741–759 (2014)
4. Capasso, V.: Mathematical Structures of Epidemic Systems, vol. 88. Springer (1993)
5. Diaby, M., Iggidr, A., Sy, M.: Observer design for a schistosomiasis model. Math. Biosci. **269**, 17–29 (2015)
6. Dietz, K: The incidence of infectious diseases under the influence of seasonal fluctuations. In: Mathematical Models in Medicine, pp. 1–15. Springer (1976)
7. Efimov, D., Raïssi, T., Chebotarev, S., Zolghadri, A.: Interval state observer for nonlinear time varying systems. Automatica **49**(1), 200–205 (2013)

8. Gouzé, J.-L., Rapaport, A.: Interval observers for uncertain biological systems. Ecol. Modell. **133**(1), 45–56 (2000)
9. Hirsch, M.W.: Stability and convergence in strongly monotone dynamical systems. J. Reine Angew. Math **383**, 1–53 (1988)
10. Horn, R.A., Johnson, C.R.: Topics in Matrix Analysis. Cambridge University Press, New York (1991)
11. Keeling, M.J., Rohani, P.: Modeling Infectious Diseases in Humans and Animals. Princeton University Press (2008)
12. Kuznetsov, YuA, Piccardi, C.: Bifurcation analysis of periodic SEIR and SIR epidemic models. J. Math. Biol. **32**(2), 109–121 (1994)
13. Mason, O., Shorten, R.: Quadratic and copositive Lyapunov functions and the stability of positive switched linear systems (2007)
14. Meslem, N., Ramdani, N., Candau, Y.: Interval observers for uncertain nonlinear systems. Application to bioreactors. In: 17th IFAC World Congress, Seoul, Korea, pp. 9667–9672 (2008)
15. Smith, H.L.: Monotone dynamical systems: an introduction to the theory of competitive and cooperative systems, vol. 41. Mathematical Surveys and Monographs. American Mathematical Society (1995)

Chapter 4
Analysis of a Reaction-Diffusion Epidemic Model

B. Cantó, C. Coll, S. Romero-Vivó and E. Sánchez

Abstract A model of an epidemic is introduced to describe an indirect transmission of the disease through the density of pathogens in the environment. The scenario of an emerging disease in a contaminated environment is assumed and the possibility that an initial infection can spread in the population living in that environment is analyzed.

Keywords Epidemic model · Stability · Equilibrium points · Basic reproduction number · Discrete-time system

4.1 Introduction

Over the years a large number of models to describe the dynamics of an infection are available in the literature, see e.g. [9] and the references given there. Most of them are models that analyze the temporal behavior of a disease which is not extended in space. However, it is important to examine the consequences of including the spatial effect in the study of the evolution of a disease since the behavior of an epidemic in two dimensions can lead to different results from those in one dimension. The study of how the epidemic evolves according to the position of the individuals is even as important as the analysis of temporal evolution, especially in cases where individuals are confined in enclosed spaces. In particular, in the case of animals that can be confined in cages or pens.

B. Cantó · C. Coll · S. Romero-Vivó (✉) · E. Sánchez
Instituto de Matemática Multidisciplinar, Universitat Politècnica de València, 46022 Valencia, Spain
e-mail: sromero@imm.upv.es

B. Cantó
e-mail: bcanto@mat.upv.es

C. Coll
e-mail: mccoll@mat.upv.es

E. Sánchez
e-mail: esanchezj@mat.upv.es

F. Cacace et al. (eds.), *Positive Systems*, Lecture Notes in Control and Information Sciences 471, DOI 10.1007/978-3-319-54211-9_4

Our aim is to design a mathematical model that analyzes diseases whose transmission occurs through a contaminated environment. Thus, we assume the existence of an emerging disease in a contaminated environment and the possibility that the initial infection can spread in the population living in that environment. Furthermore, it is assumed that the pathogens disperse by the enclosure via a diffusion process. In reality, we do not think that the pathogens are diffusing. We can imagine them as fixed in a grid, with contacts to their nearest neighbors, through which the disease spreads. We assume that the disease spreads due to a certain pathogen found in the environment or in food. The disease agent could persist without a host by absorbing and metabolizing dissolved decomposed organic matter. Sometimes, the infectious agent can infect a host by opportunity on contact or ingestion, and it can multiply within the host. An example of this type of transmission is the zoonotic diseases that have indirect mechanisms of transmission. The route of infection from animals to humans is usually through contaminated food. For example: by ingestion of contaminated water or food (e.g. salmonellosis), by inhalation of contaminated fluids such as feces, urine, milk, etc. (e.g. brucellosis, Hanta virus) or by exposure to contaminated soil or water (e.g. schistosomiasis, leptospirosis). In particular, people usually get salmonellosis by eating contaminated food, such as undercooked chicken or eggs. Some models and results on Salmonella in industrial house hens can be found in [1, 12].

Our model supposes that the population flux at a point x of a spatial domain Ω is function of the density variation in the immediate vecinity of x. For simplicity, we limit ourselves to a one-dimensional habitat $\Omega = (0, L)$. In this case, the population is formed by susceptible individuals and infective individuals, denoted by $S(x, t)$ and $I(x, t)$, respectively, which are functions of a time variable t as well as of spatial variable x, with $t \in \mathbb{Z}$, $t \geq 0$ and the $x \in \mathbb{Z}$, $0 \leq x \leq L$, where L is the width of our environment Ω. In particular, variables $S(x, t)$ and $I(x, t)$ represent population spatial densities rather of susceptible and infected individuals at point $x \in \Omega$ and at time $t > 0$, respectively. The variable $C(x, t)$ represents the environmental contamination spatial density at point $x \in \Omega$ and at time $t > 0$. We also assume the following basic assumptions for these models: each individual has the same probability of catching the disease and the total population $S(x, t) + I(x, t)$ remains constant equal to N.

Furthermore, the parameters $0 < p, \ q, \ s \ < 1$ represent the survival rate of $S(x, t)$, $I(x, t)$ and $C(x, t)$, respectively. And we denote

$$\bar{S}(t) = \int_{\Omega} S(x, t)dx, \ \ \bar{I}(t) = \int_{\Omega} I(x, t)dx, \ \ \bar{C}(t) = \int_{\Omega} C(x, t)dx.$$

The death removal rate is $\mu(x, t)$. The parameter σ denotes the exposition rate and the transmission via contact with the contaminated environment is given by $\sigma C(x, t)S(x, t)$. For different values of this parameter we can get different types of infections.

Moreover, we denote by

$$g(t) = \sigma \int_{\Omega} C(x,t)S(x,t)dx.$$

From definition, note that $g(t) \geq 0$. In the next system, the term $\beta I(x,t)$, $0 < \beta < 1$, represents the density of pathogen produced by infected individuals and α^2 denotes the diffusion coefficient. So, a first mathematical description of the model is given by

$$\begin{aligned} \frac{\partial S}{\partial t} &= (p-1)S - \sigma CS + \mu \\ \frac{\partial I}{\partial t} &= (q-1)I + \sigma CS \\ \frac{\partial C}{\partial t} &= (s-1)C + \beta I + \alpha^2 \frac{\partial^2 C}{\partial x^2}, \end{aligned} \tag{4.1}$$

together with nonflux boundary conditions (no individuals cross the boundary) $\partial_\eta z(x,t) = 0$, in $\partial\Omega \times \mathbb{R}^+$, being $z(x,t) = \text{col}(S(x,t),\ I(x,t),\ C(x,t)) = \text{col}(z_1,\ z_2,\ z_3)$, and giving appropriate initial conditions $z(x,0) = f(x) = \text{col}(f_1(x), f_2(x), f_3(x))$ in Ω, where $f_1(x)$ and $f_2(x)$ are continuous nonnegative functions and $f_3(x)$ is a continuous positive function. Here, Ω is a bounded domain with smooth boundary $\partial\Omega$ and ∂_η denotes the outward normal derivative on $\partial\Omega$. Finally, by the initial model description $\mu = (1-p)S + (1-q)(N-S)$.

At this point we rewrite the previous system as follows

$$\frac{\partial z}{\partial t} - A\frac{\partial^2 z}{\partial x^2} - F(x,t,z) = 0$$

where $z = z(x,t)$, $A = \begin{pmatrix} 0 & 0 & 0 \\ 0 & 0 & 0 \\ 0 & 0 & \alpha^2 \end{pmatrix}$ and $F(x,t,z) = F_1 z + \sigma z_1 z_3 v + (1-q)N u_1$ with

$$F_1 = \begin{pmatrix} -(1-q) & 0 & 0 \\ 0 & -(1-q) & 0 \\ 0 & \beta & -(1-s) \end{pmatrix},\quad v = \begin{pmatrix} -1 \\ 1 \\ 0 \end{pmatrix},\quad u_1 = \begin{pmatrix} 1 \\ 0 \\ 0 \end{pmatrix}.$$

It is clear that our system does not possess a positive definite diffusion matrix. This occurs because there is no diffusion in the z_1 and z_2 directions. All the same, a diffusion matrix is acceptable if it is positive semidefinite. Throughout the literature, there are many works that prove the existence and uniqueness of solutions of similar problems related to system (4.1) subject to suitable boundary and initial conditions, see [3, 4] and the references given there. To establish sufficient conditions for the existence of a nonnegative solution to system (4.1) we use its associated ODE system given by

$$\frac{d\bar{z}}{dt} - F_1\bar{z} = g(t)v + (1-q)NLu_1, \tag{4.2}$$

with the initial conditions $\bar{z}(0) = z_0$. We seek conditions to the existence of nonnegative solutions to (4.2) satisfying $\bar{z}(t) \geq 0,\ t \geq 0$. Since the total population size remains constant we have

$$\int_\Omega S(x,t)dx + \int_\Omega I(x,t)dx = \bar{S}(t) + \bar{I}(t) = LN$$

then $\dfrac{d\bar{S}}{dt} = -\dfrac{d\bar{I}}{dt}$. Using this equality in system (4.2), we have

$$\frac{d\bar{z}}{dt} - F_2\bar{z} = g(t)v, \tag{4.3}$$

with

$$F_2 = \begin{pmatrix} 0 & 1-q & 0 \\ 0 & -(1-q) & 0 \\ 0 & \beta & -(1-s) \end{pmatrix}.$$

Note that, all entries of matrix F_2 are nonnegative except for those on the main diagonal. The real matrices satisfying this property are called Metzler matrices or essentially nonnegative matrices. It is known that, a matrix M is Metzler if and only if e^{Mt} is nonnegative for all $t \geq 0$, and e^{Mt} is strictly positive if and only if M is irreducible, see [2, 14]. Additionally, given a matrix M, we denote by $\rho(M)$ and by $s(M)$ the spectral radius and the spectral abscissa of M, respectively. Recall that, $\rho(M)$ and $s(M)$ are the maximum modulus and the maximun real part of eigenvalues of M, respectively. From Perron-Frobenius theorem for Metzler matrices we have that if M is an irreducible Metzler matrix then $s(M)$ is an algebraically simple eigenvalue of M with the unique strictly positive eigenvector v_s, [6]. Moreover, a matrix M is a Hurwitz matrix if $s(M) < 0$ and the asymptotic stability of a system is followed if its coefficient matrix is Hurwitz. Since matrix F_2 is a Hurwitz matrix, system (4.3) is asymptotically stable. On the other hand, a characterization to ensure that a matrix is Hurwitz is given in [2]. So, $s(M) < 0$ if and only if $-M$ is a non-singular M-matrix, that is, $M = \nu I - A$ where A is a nonnegative matrix with $\rho(M) < \nu$.

To obtain the solution of system (4.3), first, we consider the linear part, so the solution of $\bar{z}' - F_2\bar{z} = 0$ is given by $\bar{z}_h(t) = e^{F_2 t}z_0$ for every admissible z_0, that leads us to the theory of semigroups, see [11]. Now, let us consider $\bar{z}' - F_2\bar{z} = g(t)v$ where $g(t)$ is continuously differentiable then, it follows by a standard perturbation result (see [11]) that there exists a unique mild solution to this equation for every z_0 and for sufficiently short time intervals. This solution $\bar{z}(t)$ is a continuous function solution of the integral equation

$$\bar{z}(t) = T(t)z_0 + \int_0^t T(t-\tau)g(\tau)v\,d\tau$$

where $T(t),\ t \geq 0$ is the semigroup generated by F_2. It is easy to prove that in our case $T(t) = e^{F_2 t}$ and the solution given by

$$\bar{z}(t) = e^{F_2 t} z_0 + \int_0^t e^{F_2(t-\tau)} g(\tau) v d\tau$$

is nonnegative if $z_0 \geq 0$.

4.2 Linear Model

Our main aim is to study the ability of a spatially localized initial infection to propagate into the susceptible population, and the behavior of the solution in a neighborhood of a point of equilibrium, i.e., stability or not, which allow us to deduce whether the disease will disappear, will be endemic or will grow creating a pandemic. So, we shall limit the analysis of the steady states of system (4.1) to classical solutions of the stationary problem

$$A\frac{\partial^2 z}{\partial x^2} + F(x, t, z) = 0$$

for $x \in \Omega$, subject to boundary conditions $\partial_\eta z(x, t) = 0$, in $\partial\Omega \times \mathbb{R}^+$.

Assuming that $(S_f(x), I_f(x), C_f(x))$ is a disease-free equilibrium point (where the variables do not change with time) then, $I_f(x) = 0, \forall x \in \Omega$. Hence, system (4.2) can be reduced to the differential equation $0 = -(1-s)C_f(x) + \alpha^2 C_f''(x)$ whose solution, using the boundary conditions, is $C_f(x) = 0$. On the other hand, the total population size remains constant N, then $S_f(x) = N$. Let us assume that $P_f = (S_f, 0, 0)$ is an equilibrium point. To study the system behavior around the equilibrium point, we linearize the initial system at P_f.

So we would like to find the linear system when (S, I, C) is close to $(S_f, 0, 0)$. To get this $S(x,t)C(x,t) \simeq S_f C_f + S_f C(x,t) + S(x,t) C_f \simeq S_f C(x,t)$. We denote $\hat{X} = X - X_f$, then the new variables are $\hat{z} = (\hat{S}, I, C)$ and a system close to the original nonlinear system is

$$\frac{\partial \hat{z}}{\partial t} - A\frac{\partial^2 \hat{z}}{\partial x^2} = B\hat{z} \tag{4.4}$$

with

$$B = \begin{pmatrix} (q-1) & 0 & -\sigma S_f \\ 0 & (q-1) & \sigma S_f \\ 0 & \beta & (s-1) \end{pmatrix},$$

and nonflux boundary conditions $\partial_\eta \hat{z}(x,t) = 0$, in $\partial\Omega \times \mathbb{R}^+$, and $\hat{z}(x,0) = \hat{f}(x)$. Now, we consider the subsystem involving only I and C. In this case, the diffusion matrix A_r, the infection matrix F and the evolution matrix V are given by

$$A_r = \begin{pmatrix} 0 & 0 \\ 0 & \alpha^2 \end{pmatrix}, \quad F = \begin{pmatrix} 0 & \sigma S_f \\ \beta & 0 \end{pmatrix}, \quad V = \begin{pmatrix} (1-q) & 0 \\ 0 & (1-s) \end{pmatrix}.$$

The new subsystem is given by

$$\frac{\partial \tilde{z}}{\partial t} - A_r \frac{\partial^2 \tilde{z}}{\partial x^2} = (F - V)\tilde{z}. \tag{4.5}$$

Note that the matrix $M = F - V$ is an irreducible Metzler matrix, with $s(-V) < 0$. To study the large time behavior of the population modeled by system (4.5), we analyze the existence of a unique positive stationary solution using functions of the form $\tilde{z}(x,t) = e^{Mt}\varphi(x,t)$, which transforms the initial system into the system, $\frac{\partial \varphi}{\partial t} = A_r \frac{\partial^2 \varphi}{\partial x^2}$. Now, consider $\varphi(x,t) = T(t)w(x)$, we just have to find the principal eigenvalue of the elliptic eigenvalue problem

$$\begin{aligned} w''(x) + \gamma^2 w(x) &= 0 \\ w'(x)\big|_{\delta\Omega} &= 0. \end{aligned} \tag{4.6}$$

In [15] provides the existence of the principal eigenvalue of an elliptic eigenvalue problem in a bounded smooth domain, under Newmann boundary condition, where some diffusion coefficients may be zero. This eigenvalue is simple and the associated eigenfunction is positive in the smooth domain. In a similar way we solve the problem (4.6) being $\mu^* = -\gamma^{*2}$ this principal eigenvalue. From the comparation principle [13] the solution $\tilde{z}^*(x,t) = e^{(M+\mu^* A_r)t} v_{\mu^*} w^*(x)$ determines the behavior of the system around the disease-free equilibrium point. Then to study the asymptotically stability of system (4.5) we need studied the spectrum of matrix $M + \mu^* A_r$, that is, system (4.5) is asymptotically stable if and only is $s(M + \mu^* A_r) < 0$.

The conditions that must meet the parameters of the model to ensure its stability around the disease-free equilibrium point are set out in the following result.

Proposition 4.1 *Given system (4.5) the following statements are equivalent:*

(i) $s(M + \mu^* A_r) < 0$.
(ii) $\frac{\sigma\beta N}{(1-q)(1-s+(\gamma^*\alpha)^2)} < 1$.

Proof From the Routh-Hurwitz stability test, see [7], $s(M + \mu^* A_r) < 0$ if and only if all the coefficients of the polynomial given by $|\lambda I - (M + \mu^* A_r)| = 0$ are positives. By a simple calculation it is easy to check that all the coefficients are positives if and only if $(1-q)(1-s+(\gamma^*\alpha)^2) - \sigma\beta N > 0$ and taking into account the interpretation of the parameters involved in this expression, $(1-q) > 0$, $(1-s) > 0$ then $1 - s + (\gamma^*\alpha)^2) > 0$. Thus, all the coefficients are positives if and only if $\frac{\sigma\beta S_f}{(1-q)(1-s+(\gamma^*\alpha)^2)} < 1$. □

On the other hand, it is known that the basic reproductive number of the epidemiological process, R_0, is a measure or indicator to know whether the disease will disappear. Recall that this parameter is the expected number of secondary cases produced by a infective individual, see [5, 10]. If $R_0 < 1$ the disease tends to disappear around the disease-free equilibrium point and otherwise it remains. In the next section, we propose an expression of this parameter for the considered reaction-diffusion epidemic model.

4.3 The Basic Reproduction Number

Consider system (4.5) rewritten as

$$\frac{\partial \tilde{z}}{\partial t} - A_r \frac{\partial^2 \tilde{z}}{\partial x^2} = (-V)\tilde{z} + F\tilde{z},$$

with $s(-V + \mu^* A_r) < 0$, that is, system (4.5) is asymptotically stable in absence of infection. From definition of the basic reproduction number we analyze the accumulated secondary infective individuals from a primary infective individual. Hence, we observe that

$$(V - \mu^* A_r)^{-1} = \int_0^\infty e^{(-V+\mu^* A_r)t} dt.$$

Thus, given an initial state, this matrix $(V - \mu^* A)^{-1}$ represents its expected transition throughout his life. When we consider the infection problem given by system (4.5), the matrix $F(V - \mu^* A)^{-1}$ generates the secondary infected individuals and secondary contaminants from the primary state. So, the basic reproduction number is the spectral radius of the matrix $F(V - \mu^* A)^{-1}$. That is,

$$R_0^D = \rho(F(V - \mu^* A_r)^{-1}).$$

In [15] R_0 is established for reaction-diffusion epidemic models where some diffusion coefficients may be zero. Applying the results given in [15] to our case it follows that $R_0 = R_0^D$.

Proposition 4.2 *Given system (4.5) the following statements are equivalent:*

(i) $s(F - (V - \mu^* A_r)) < 0$.
(ii) $R_0^D < 1$.

Proof It is easy to check that $F - (V - \mu^* A_r)$ is a Metzler matrix and $(V - \mu^* A_r)$ is nonsingular with $(V - \mu^* A_r)^{-1} \geq 0$, then $F(V - \mu^* A_r)^{-1} \geq 0$. Let $\mathscr{A} = -F + (V - \mu^* A_r)$ be then

(i)→ (ii). If $F - (V - \mu^* A_r)$ is also Hurwitz then $\mathscr{A}$ is a nonsingular M-matrix, [2]. Moreover $\mathscr{A} = (V - \mu^* A_r) - F$ where $(V - \mu^* A_r)$ and F are non-negative matrices with $(V - \mu^* A_r)^{-1} \geq 0$, then $F(V - \mu^* A_r)^{-1} \geq 0$. This implies that $\rho(F(V - \mu^* A_r)^{-1}) < 1$, [2].

(ii)→ (i). If $\rho(F(V - \mu^* A_r)^{-1}) < 1$ then matrix $(I - F(V - \mu^* A_r)^{-1})$ is a nonsingular M-matrix, and, since $\mathscr{A} = (I - F(V - \mu^* A_r)^{-1})(V - \mu^* A_r)$, $\mathscr{A}^{-1} =$

$(V - \mu^* A_r)^{-1}(I - F(V - \mu^* A_r)^{-1})^{-1} \geq 0$. Therefore, $\mathscr{A}$ is a nonsingular M-matrix what implies that $s(-\mathscr{A}) < 0$, [2]. □

Remark 4.1 For the model represented by system (4.5) the basic reproduction number is given by

$$R_0^D = \sqrt{\frac{\sigma\beta N}{(1-q)(1-s+(\gamma^*\alpha)^2)}}.$$

From Propositions 4.1 and 4.2, the following result is directly obtained.

Corollary 4.1 *The system (4.5) is asymptotically stable to P_f if and only if $R_0^D < 1$.*

Remark 4.2 Given the model represented by system (4.5) if $\alpha = 0$ we can obtain the basic reproduction number using the next-generation matrix [5]. So,

$$R_0 = \sqrt{\frac{\sigma\beta N}{(1-q)(1-s)}}.$$

Note that $R_0^D < R_0$ since $(R_0^D)^2 = \vartheta R_0^2$ with

$$\vartheta = \frac{1}{1 + \frac{(\gamma^*\alpha)^2}{1-s}} < 1.$$

From the above it follows that, if the system is asymptotically stable without diffusion also is asymptotically stable with diffusion. Furthermore, if the system without diffusion is not asymptotically stable it could find a diffusion coefficient such that the new diffusion system is asymptotically stable.

4.4 A Discrete-Time Model

The main aim of this section is to obtain a discrete-model associated with system (4.4). For discretizing the partial differential equation we replace the partial derivatives in (4.4) by difference quotients. For that, we consider both time and space discretizations on a uniform grid with grid parameter $h = \Delta x$, where h is the distance between two neighboured nodes of the grid and we discretize the interval $(0, T)$ by an one dimensional grid with step size $k = \Delta t$. We denote by $\hat{z}_{i,j} = \hat{z}(i\Delta x, j\Delta t)$, $i = 1, \ldots, M, j > 0$.

In this case, we use the forward-looking difference operator in time to approximate the first-order derivative in the following way,

$$\frac{\partial \hat{z}}{\partial t} \approx \frac{\hat{z}_{i,j+1} - \hat{z}_{i,j}}{k},$$

and the Euler scheme backward in space should be used in our model to obtain a discrete-model associated to system (4.4), so the second-order derivative is approximated by

$$\frac{\partial^2 \hat{z}}{\partial x^2} \approx \frac{\hat{z}_{i+1,j+1} - 2\hat{z}_{i,j+1} + \hat{z}_{i-1,j+1}}{h^2},$$

with initial condition $\hat{z}_{i,0} = f(x_i),\ i = 1, \ldots, M.$

The discretized problem can be written as

$$\frac{\hat{z}_{i,j+1} - \hat{z}_{i,j}}{k} - A\frac{\hat{z}_{i+1,j+1} - 2\hat{z}_{i,j+1} + \hat{z}_{i-1,j+1}}{h^2} = B\hat{z}_{i,j+1}, \tag{4.7}$$

After some algebraic manipulations, we obtain

$$E\hat{z}_{i+1,j+1} = (I + 2E - kB)\hat{z}_{i,j+1} - \hat{z}_{i,j} - E\hat{z}_{i-1,j+1},$$

where $E = \frac{k}{h^2}A.$

As in the previous section we focus on the subsystem involving infected individuals and the contaminant. That is

$$E_r\tilde{z}_{i+1,j+1} = (I + 2E_r - kM)\tilde{z}_{i,j+1} - \tilde{z}_{i,j} - E_r\tilde{z}_{i-1,j+1},$$

where $E_r = \frac{k}{h^2}A_r$. So, we focus on the solutions of the polynomial

$$\left|w_1 w_2 E_r - w_2(I + 2E_r - kM) + E_r w_1^{-1} w_2 + I\right| = 0,$$

in particular we analyze the solutions of

$$\left|E_r(1 - w_1^{-1})^2 - w_1^{-1}(I(1 - w_2^{-1}) - kM)\right| = 0. \tag{4.8}$$

Changing $\lambda_i = 1 - w_i^{-1},\ i = 1, 2,$ we have $\left|E_r\lambda_1^2 - (1 - \lambda_1)(I\lambda_2 - kM)\right| = 0.$ Note that, we are interested in the property of stability, then we need information of λ_2. So, we analyze the solutions of $\left|I\lambda_2 - (E_r\frac{\lambda_1^2}{(1-\lambda_1)} + kM)\right| = 0$, that is the eigenvalues of the matrix $\Lambda = E_r\frac{\lambda_1^2}{(1-\lambda_1)} + kM$. Note that, if the spectral bound $s(\Lambda) < 0$ then $|w_2| < 1$ for all w_2 solution of (4.8).

Proposition 4.3 *Consider the matrix $\Lambda = E_r\frac{\lambda_1^2}{(1-\lambda_1)} + kM$, being λ_1 a real number. Then all its eigenvalues are real. Moreover, denoting by $\kappa = \alpha^2\frac{\lambda_1^2}{(1-\lambda_1)h^2}$ and $R_\kappa = \sqrt{\frac{\sigma\beta N}{(1-q)(1-s-\kappa)}}$ the following statements hold*

(i) if $\kappa + s + q < 2$ and $R_\kappa < 1$ then its eigenvalues are less than zero.
(ii) if $\kappa + s + q > 2$ or $R_\kappa \geq 1$ then at least one of its eigenvalues is positive.

Proof The eigenvalues of Λ are the solutions of

$$P(\lambda) = \left| \begin{pmatrix} \lambda - k(q-1) & -k\sigma S_f \\ -k\beta & \lambda - k(s-1+\kappa) \end{pmatrix} \right| = 0.$$

Note that, if $\lambda_1 > 1$ then $\kappa < 0$ and if $\lambda_1 < 1$ then $\kappa > 0$. The discriminant of both roots can be written as $(\kappa + (s-q))^2 + 4\sigma\beta Sf > 0$ so both are real numbers. From Routh-Hurwitz stability criterion, to study the sign of the roots must analyze the sign of the coefficients of $P(\lambda)$. So, $\kappa + s + q - 2 < 0$ and $(q-1)(\kappa + s - 1) - \sigma\beta N > 0$. Thus, if $\kappa + s + q > 2$ and $\frac{\sigma\beta N}{(1-q)(1-s-\kappa)} < 1$ all roots of $P(\lambda)$ have part real negative. Otherwise at least one of its roots will be positive. □

References

1. Beaumont, C., Burie, J., Ducrot, A., Zongo, P.: Propagation of salmonella within an industrial hen house. SIAM J. Appl. Math. **72**(4), 1113–1148 (2012)
2. Berman, A., Plemmons, R.J.: Nonnegative Matrices in Mathematical Sciences. Academic Press, New York (1979)
3. Capasso, V.: Mathematical Structures of Epidemic Systems. Lecture Notes in Biomathematics, vol. 97. Springer, Heidelberg (1993)
4. Capasso, V., Wilson, R.E.: Analysis of a reaction-diffusion system modeling man-environment-man epidemics. SIAM J. Appl. Math. **57**(2), 327–346 (1997)
5. Diekmann, O., Heesterbeek, J.A.P., Metz, J.A.J.: On the definition and the computation of the basic reproduction ratio R_0 in models for infectious diseases in heterogeneus populations. J. Math. Biol. **28**, 365–382 (1990)
6. Farina, L., Rinaldi, S.: Positive Linear Systems, Theory and Applications. Pure and Applied Mathematics. Wiley (2000)
7. Gantmacher, F.R.: Applications of the Theory of Matrices. Interscience Publishers, New York (1959)
8. Kaczorek, T.: Linear Control System, vol. 2. Wiley, New York (1991)
9. Keeling, M.J., Rohani, P.: Modeling Infectious Diseases in Humans and Animals. Priceton University Press, Princeton and Oxford (2008)
10. Li, C.K., Schneider, H.: Applications of Perron-Frobenius theory to population dynamics. J. Math. Biol. **44**, 450–462 (2002)
11. Pazy, A.: Semigroups of Linear Operators and Applications to Partial Differential Equations. Springer, New York (1983)
12. Prévost, K., Beaumont, C., Magal, P.: Asymptotic behavior in a Salmonella infection model. Math. Modell. Nat. Phenomena Epidemiology **2**(1), 1–22 (2006)
13. Smith, H.L.: Monotone Dynamical system: an introduction to the theory of Competitive and Cooperative Systems, Math. Surveys Monogr. **41**, American Mathematical Society, Providence (1995)
14. Varga, R.S.: Matrix Iterative Analysis. Springer, Heidelberg (2000)
15. Wang, W., Zhao, X.Q.: Basic reproduction numbers for reaction-diffusion epidemic models. SIAM J. Appl. Dyn. Syst **11**, 1652–1673 (2012)

Part II
Positive Systems with Delay and Disturbance

Chapter 5
On Feedback Transformation and Integral Input-to-State Stability in Designing Robust Interval Observers for Control Systems

Thach Ngoc Dinh and Hiroshi Ito

Abstract The problem of designing interval observers is addressed for output feedback control of a class of nonlinear systems in this chapter. The framework of integral input-to-state stability is exploited to drive the estimated intervals and the state variables to the origin asymptotically when disturbances converge to zero. Moreover interval observers are tuned with feedback gain. A reduced-order interval observer is proposed, and the flexibility offered by gains in designing observer is related to the existence of reduced-order interval observers. Comparative simulations are given to illustrate the theoretical results.

Keywords Interval observers · Reduced-order observers · Nonlinear systems · Output feedback control · Guaranteed state estimation.

5.1 Introduction

Interval observers generate upper bounds and lower bounds of state variables of dynamical systems at each time instant based on given information about bounds of unknown disturbances and of unknown initial conditions [6]. The bounds give intervals where the state variables are sure to stay during transient periods in which classical observers do not provide any guarantee. The usefulness of interval estimates is evident for monitoring purposes when large disturbances or uncertainties are present [1]. A typical mechanism to allow the construction of such interval observers is to let the estimation errors be governed by positive systems. Some examples of extensive studies on design of interval observers have been reported in [4, 5, 8–14] (see also references therein).

T.N. Dinh is an International Research Fellow of the Japan Society for the Promotion of Science.

T.N. Dinh
Université de Valenciennes et du Hainaut-Cambrésis, Le Mont Houy,
59313 Valenciennes Cedex 9, France
e-mail: ngocthach.dinh@univ-valenciennes.fr

H. Ito (✉)
Kyushu Institute of Technology, 680-4 Kawazu, Iizuka, 820-8502 Fukuoka, Japan
e-mail: hiroshi@ces.kyutech.ac.jp

F. Cacace et al. (eds.), *Positive Systems*, Lecture Notes in Control
and Information Sciences 471, DOI 10.1007/978-3-319-54211-9_5

Recently, an interval observer was proposed in [3] for nonlinear control systems which are affine in unmeasured state variables, and it was investigated further in [7] to provide design guidelines for guaranteeing the length of estimated intervals to converge to zero for converging disturbances and guaranteeing (integral) input-to-state stability ((i)ISS) of the entire controlled system. The iISS approach developed in [7] has allowed one to deal with a larger class of nonlinearities than the original approach [3]. This chapter continues investigating the iISS framework and introduces a modification by incorporating feedback gain into the observer for control systems. The modification, in addition to state transformation of the error systems, offers flexibility in obtaining positive systems leading to tighter interval estimates and swifter convergence of the interval length and state variables of the plant to zero. This chapter also proposes a reduced-order interval observer aiming at swifter behavior of the estimates and the plant state with less control effort. It also discusses how the positivity of error systems allows the existence of a full-order observer to imply the existence of a reduced-order observer. Comparative simulations are given to illustrate these ideas.

In this chapter, the set of real numbers is denoted by $\mathbb{R}$. The set of non-negative real numbers is denoted by $\mathbb{R}_{\geq 0}$. The symbol $|\cdot|$ denotes Euclidean norm of vectors. Inequalities must be understood *component-wise*, i.e., for $a = [a_1, ..., a_n]^\top \in \mathbb{R}^n$ and $b = [b_1, ..., b_n]^\top \in \mathbb{R}^n$, $a \leq b$ if and only if, for all $i \in \{1, ..., n\}$, $a_i \leq b_i$. For a square matrix $Q \in \mathbb{R}^{n\times n}$, let $Q^+ \in \mathbb{R}^{n\times n}$ denote $Q^+ = \left(\max\{q_{i,j}, 0\}\right)_{i,j=1,1}^{n,n}$, where $Q = \left(q_{i,j}\right)_{i,j=1,1}^{n,n}$. Let $Q^- = Q^+ - Q$. This notation is limited to square matrices, and the superscripts $+$ and $-$ for other purposes are defined appropriately when they appear. A square matrix $Q \in \mathbb{R}^{n\times n}$ is said to be Metzler if each off-diagonal entry of this matrix is nonnegative. The symbol I denotes the identity matrix of appropriate dimension. For $\alpha, \beta : \mathbb{R}_{\geq 0} \to \mathbb{R}^n$, by $\alpha(s) \equiv \beta(s)$ we mean $\alpha(s) = \beta(s)$ for all $s \in \mathbb{R}_{\geq 0}$. A function $\alpha : \mathbb{R}_{\geq 0} \to \mathbb{R}_{\geq 0}$ is said to be positive definite and written as $\alpha \in \mathscr{P}$ if α is continuous and satisfies $\alpha(0) = 0$ and $\alpha(s) > 0$ for all $s \in (0, \infty)$. A function $\alpha \in \mathscr{P}$ is said to be of class $\mathscr{K}$ if α is strictly increasing. A class $\mathscr{K}$ function is said to be of class $\mathscr{K}_\infty$ if it is unbounded. A continuous function $\beta : \mathbb{R}_{\geq 0} \times \mathbb{R}_{\geq 0} \to \mathbb{R}_{\geq 0}$ is said to be of class $\mathscr{KL}$ if, for each fixed $t \in \mathbb{R}_{\geq 0}$, $\beta(\cdot, t)$ is of class $\mathscr{K}$ and, for each fixed $s > 0$, $\beta(s, \cdot)$ is strictly decreasing and $\lim_{t\to\infty} \beta(s, t) = 0$. Logical sum and logical product are denoted by $\vee$ and $\wedge$, respectively.

5.2 Setups and Objectives

Consider the system

$$\dot{x}(t) = A(y(t))x(t) + B(y(t))u(y(t), \hat{x}^+(t)) + \delta(t) \tag{5.1a}$$

$$y(t) = Cx(t) \tag{5.1b}$$

with time $t \in \mathbb{R}_{\geq 0}$, the state $x(t) \in \mathbb{R}^n$, the measurement output $y(t) \in \mathbb{R}^p$ and the initial condition $x(0) = x_0$, where the functions $A : \mathbb{R}^p \to \mathbb{R}^{n\times n}$ and $B : \mathbb{R}^p \to \mathbb{R}^{n\times q}$

are supposed to be locally Lipschitz, and $C \in \mathbb{R}^{p\times n}$ is a constant matrix. The term $u(y(t), \hat{x}^+(t)) \in \mathbb{R}^q$ is the control input indicating output feedback, and the function $u : \mathbb{R}^p \times \mathbb{R}^n \to \mathbb{R}^q$ is supposed to be locally Lipschitz. The signal $\hat{x}^+(t) \in \mathbb{R}^n$ denotes an estimate of $x(t)$, which has yet to be defined. The disturbance vector $\delta : \mathbb{R}_{\geq 0} \to \mathbb{R}^n$ is supposed to be piecewise continuous. It is stressed that $x(t)$ is not measured. Instead, the output $y(t)$ is available as a measurement for all $t \in \mathbb{R}_{\geq 0}$. Assume that the vectors $x_0^-, x_0^+ \in \mathbb{R}^n$ and piecewise continuous functions $\delta^+, \delta^- : \mathbb{R}_{\geq 0} \to \mathbb{R}^n$ satisfying

$$x_0^- \leq x_0 \leq x_0^+ \tag{5.2}$$

$$\delta^-(t) \leq \delta(t) \leq \delta^+(t), \quad \forall t \in \mathbb{R}_{\geq 0} \tag{5.3}$$

are known, while $x(0) = x_0$ and $\delta(t)$ are not known. The design problem to be addressed in this chapter is mainly to achieve two objectives simultaneously. One is to drive $x(t)$ to the origin asymptotically for an arbitrary initial condition satisfying (5.2) by output feedback control when $\delta(t)$ converges to zero. The other is to estimate an envelope $x^-(t), x^+(t) \in \mathbb{R}_{\geq 0}$ such that the framer property

$$x^-(t) \leq x(t) \leq x^+(t), \quad \forall t \in \mathbb{R}_{\geq 0} \tag{5.4}$$

holds in the presence of any piecewise continuous disturbance $\delta(t)$ satisfying (5.3). The former is for the purpose of control, and the latter is for monitoring. Other important features of the simultaneous control and monitoring problem are described mathematically in Sect. 5.4.

5.3 Observer Candidates

5.3.1 Full-Order Interval Observer

Divide the control input u into a direct output feedback term and the remainder as

$$u(y, \hat{x}^+) = K(y)y + u_a(y, \hat{x}^+). \tag{5.5}$$

The locally Lipschitz function function $K : \mathbb{R}^p \to \mathbb{R}^{q\times p}$ can be given arbitrarily since $K(y)y$ can be absorbed by the locally Lipschitz function $u_a : \mathbb{R}^p \times \mathbb{R}^n \to \mathbb{R}^q$. Define an observer candidate as

$$\dot{\hat{x}}^+ =(A(y) + B(y)K(y)C)\hat{x}^+ + B(y)u_a + H(y)[C\hat{x}^+ - y] + S[R^+\delta^+ - R^-\delta^-] \tag{5.6a}$$

$$\dot{\hat{x}}^- =(A(y) + B(y)K(y)C)\hat{x}^- + B(y)u_a + H(y)[C\hat{x}^- - y] + S[R^+\delta^- - R^-\delta^+] \tag{5.6b}$$

with the initial condition defined by

$$\hat{x}^+(0) = \hat{x}_0^+ := S[R^+ x_0^+ - R^- x_0^-] \tag{5.7a}$$
$$\hat{x}^-(0) = \hat{x}_0^- := S[R^+ x_0^- - R^- x_0^+] \tag{5.7b}$$

and the output equation

$$x^+ = S^+ R\hat{x}^+ - S^- R\hat{x}^-, \; x^- = S^+ R\hat{x}^- - S^- R\hat{x}^+, \tag{5.8}$$

where $S = R^{-1}$. The invertible matrix $R \in \mathbb{R}^{n\times n}$, the locally Lipschitz functions $H : \mathbb{R}^p \to \mathbb{R}^{n\times p}$ and $K : \mathbb{R}^p \to \mathbb{R}^{q\times p}$ are design parameters. The observer candidate (5.6) includes the one proposed in [3] as a special case given by $K = 0$. For $K = 0$, sufficient conditions for achieving (5.4) and the nominal convergence ($(x(t), x^+(t), x^-(t) \to 0$ as $t \to \infty$ for $\delta(t) \equiv 0$) are given in [3]. The convergence by the observer with $K = 0$ was made robust to allow $\delta(t) \not\equiv 0$ in [7]. Inspired by the result in [7], this chapter introduces the following two assumptions as guidelines for selecting K and H for (5.6).

Assumption 5.1 The matrix

$$\Gamma(y) = R[A(y) + B(y)K(y)C + H(y)C]R^{-1} \tag{5.9}$$

is Metzler for each fixed $y \in \mathbb{R}^p$.

Assumption 5.2 There exist a C^1 function $V : \mathbb{R}^n \to \mathbb{R}_{\geq 0}$, continuous functions $\underline{\nu}$, $\overline{\nu} \in \mathcal{K}_\infty$, $\omega \in \mathcal{P}$ and $\eta^+, \eta^- \in \mathcal{K}$ such that $\underline{\nu}(|\xi|) \leq V(\xi) \leq \overline{\nu}(|\xi|)$ and

$$\begin{aligned} &\frac{\partial V}{\partial \xi}(\xi)\left\{[A(y) + B(y)K(y)C + H(y)C]\xi + S[R^+\rho^+ + R^-\rho^-]\right\} \\ &\quad \leq -\omega(|\xi|) + \eta^+(|\rho^+|) + \eta^-(|\rho^-|) \end{aligned} \tag{5.10}$$

hold for all $\xi \in \mathbb{R}^n$, $y \in \mathbb{R}^p$, $\rho^+ \in \mathbb{R}^n$ and $\rho^- \in \mathbb{R}^n$.

The former assumption aims at securing the framer property (5.4). The latter assumption guarantees the convergence of $x^+(t) - x^-(t)$ to zero even in the presence of disturbance $\delta(t) \not\equiv 0$ by requiring the error systems of $\hat{x}^+ - x$ and $\hat{x}^- - x$ corresponding to (5.6a) and (5.6b) to be integral input-to-state stable (iISS) with respect to $\rho^+ := \delta^+ - \delta$ and $\rho^- := \delta - \delta^-$, respectively. Based on the idea of separating feedback design from the observer design, the following assumption is introduced as guidelines for selecting the feedback input u.

Assumption 5.3 There exist a positive definite radially unbounded C^1 function $U : \mathbb{R}^n \to \mathbb{R}_{\geq 0}$, continuous functions $\mu \in \mathcal{P}$ and $\gamma, \zeta \in \mathcal{K}$ such that

$$\frac{\partial U}{\partial x}(x)[A(Cx)x + B(Cx)u(Cx, x + d) + \delta] \leq -\mu(|x|) + \gamma(|d|) + \zeta(|\delta|) \tag{5.11}$$

holds for all $x \in \mathbb{R}^n$, $d \in \mathbb{R}^n$ and $\delta \in \mathbb{R}^n$.

This assumption requires the closed-loop system with the fictitious state feedback u using x instead of $\hat{x}^+$ to be iISS with respect to the estimation error $d = \hat{x}^+ - x$ and the disturbance δ.

5.3.2 *Reduced-Order Interval Observer*

Consider the following partition of the state vector x:

$$x = \begin{bmatrix} x_m \\ x_{\overline{m}} \end{bmatrix} \begin{matrix} \} \, p \text{ components} \\ \} \, n - p \text{ components.} \end{matrix} \tag{5.12}$$

Accordingly, A, B, δ and x_0 are partitioned as

$$A(y) = \begin{bmatrix} A_{m,m}(y) & A_{m,\overline{m}}(y) \\ A_{\overline{m},m}(y) & A_{\overline{m},\overline{m}}(y) \end{bmatrix}, \; B(y) = \begin{bmatrix} B_m(y) \\ B_{\overline{m}}(y) \end{bmatrix}, \; \delta = \begin{bmatrix} \delta_m \\ \delta_{\overline{m}} \end{bmatrix}, \; x_0 = \begin{bmatrix} x_{m,0} \\ x_{\overline{m},0} \end{bmatrix} \tag{5.13}$$

and it is assumed that

$$C = [I \quad 0] \in \mathbb{R}^{p \times n} \tag{5.14}$$

holds. Since the component vector $x_m(t) \in \mathbb{R}^p$ is measured, one needs to estimate the remainder $x_{\overline{m}}(t) \in \mathbb{R}^{n-p}$. Let $\hat{w}_{\overline{m}}(t)$ denote such an estimate which has yet to be defined. Then the output feedback control law based on the estimation can be represented by $u(y, \hat{w}_{\overline{m}})$ instead of $u(y, \hat{x}^+)$. For a constant matrix $G \in \mathbb{R}^{(n-p)\times p}$ to be chosen later, let $\hat{w}_{\overline{m}}$ be called an estimate of $x_{\overline{m}}$ by defining $\hat{w}_{\overline{m}} = \hat{x}_{\overline{m}}^+ - Gy$ and generating $\hat{x}_{\overline{m}}^+(t)$ appropriately. Then we have

$$u(y, \hat{w}_{\overline{m}}) = u(y, \hat{x}_{\overline{m}}^+ - Gy). \tag{5.15}$$

To construct a reduced-order observer, we replace (5.3) with

$$x_{\overline{m},0}^- \le x_{\overline{m},0} \le x_{\overline{m},0}^+, \tag{5.16}$$

$$\delta_m^-(t) \le G\delta_m(t) \le \delta_m^+(t), \quad \forall t \in \mathbb{R}_{\ge 0}, \tag{5.17}$$

$$\delta_{\overline{m}}^-(t) \le \delta_{\overline{m}}(t) \le \delta_{\overline{m}}^+(t), \quad \forall t \in \mathbb{R}_{\ge 0}, \tag{5.18}$$

where the vectors $x_{\overline{m},0}^-, x_{\overline{m},0}^+ \in \mathbb{R}^{n-p}$ and piecewise continuous functions $\delta_m^+, \delta_m^- : \mathbb{R}_{\ge 0} \to \mathbb{R}^p$, $\delta_{\overline{m}}^+, \delta_{\overline{m}}^- : \mathbb{R}_{\ge 0} \to \mathbb{R}^{n-p}$ are assumed to be known and satisfy

$$G = 0 \Rightarrow \delta_m^-(t) \equiv \delta_m^+(t) \equiv 0. \tag{5.19}$$

The bounds δ_m^- and δ_m^+ are meaningless unless (5.19) holds.

Define a reduced-order observer candidate as

$$\dot{\hat{x}}_{\overline{m}}^{+} = \left[A_{\overline{m},\overline{m}}(y) + GA_{m,\overline{m}}(y)\right]\hat{x}_{\overline{m}}^{+} + \left[A_{\overline{m},m}(y) - A_{\overline{m},\overline{m}}(y)G - GA_{m,\overline{m}}(y)G + GA_{m,m}(y)\right]y + \left[B_{\overline{m}}(y) + GB_m(y)\right]u + S_{\overline{m}}[R_{\overline{m}}^{+}\delta_{\overline{m}}^{+} - R_{\overline{m}}^{-}\delta_{\overline{m}}^{-}] + S_{\overline{m}}[R_{\overline{m}}^{+}\delta_{m}^{+} - R_{\overline{m}}^{-}\delta_{m}^{-}] \tag{5.20a}$$

$$\dot{\hat{x}}_{\overline{m}}^{-} = \left[A_{\overline{m},\overline{m}}(y) + GA_{m,\overline{m}}(y)\right]\hat{x}_{\overline{m}}^{-} + \left[A_{\overline{m},m}(y) - A_{\overline{m},\overline{m}}(y)G - GA_{m,\overline{m}}(y)G + GA_{m,m}(y)\right]y + \left[B_{\overline{m}}(y) + GB_m(y)\right]u + S_{\overline{m}}[R_{\overline{m}}^{+}\delta_{\overline{m}}^{-} - R_{\overline{m}}^{-}\delta_{\overline{m}}^{+}] + S_{\overline{m}}[R_{\overline{m}}^{+}\delta_{m}^{-} - R_{\overline{m}}^{-}\delta_{m}^{+}] \tag{5.20b}$$

with

$$\hat{x}_{\overline{m}}^{+}(0) = \hat{x}_{\overline{m},0}^{+} := S_{\overline{m}}[R_{\overline{m}}^{+}x_{\overline{m},0}^{+} - R_{\overline{m}}^{-}x_{\overline{m},0}^{-}] + Gy(0) \tag{5.21a}$$

$$\hat{x}_{\overline{m}}^{-}(0) = \hat{x}_{\overline{m},0}^{-} := S_{\overline{m}}[R_{\overline{m}}^{+}x_{\overline{m},0}^{-} - R_{\overline{m}}^{-}x_{\overline{m},0}^{+}] + Gy(0) \tag{5.21b}$$

and

$$x_{\overline{m}}^{+} = S_{\overline{m}}^{+}R_{\overline{m}}\hat{x}_{\overline{m}}^{+} - S_{\overline{m}}^{-}R_{\overline{m}}\hat{x}_{\overline{m}}^{-} - Gy \tag{5.22a}$$

$$x_{\overline{m}}^{-} = S_{\overline{m}}^{+}R_{\overline{m}}\hat{x}_{\overline{m}}^{-} - S_{\overline{m}}^{-}R_{\overline{m}}\hat{x}_{\overline{m}}^{+} - Gy \tag{5.22b}$$

$$x^{+} = \begin{bmatrix} y \\ x_{\overline{m}}^{+} \end{bmatrix}, \quad x^{-} = \begin{bmatrix} y \\ x_{\overline{m}}^{-} \end{bmatrix}. \tag{5.22c}$$

Here, $S_{\overline{m}} = R_{\overline{m}}^{-1}$. The invertible matrix $R_{\overline{m}} \in \mathbb{R}^{(n-p)\times(n-p)}$ is a design parameter. For the reduced-order observer, this chapter proposes the following assumptions as guidelines to select the gain G and the control input u.

Assumption 5.4 The matrix

$$\Gamma_{\overline{m}}(y) = R_{\overline{m}}\left[A_{\overline{m},\overline{m}}(y) + GA_{m,\overline{m}}(y)\right]R_{\overline{m}}^{-1} \tag{5.23}$$

is Metzler for each fixed $y \in \mathbb{R}^p$.

Assumption 5.5 There exist a C^1 function $V : \mathbb{R}^{n-p} \to \mathbb{R}_{\geq 0}$, continuous functions $\underline{\nu}, \overline{\nu} \in \mathcal{K}_\infty$, $\omega \in \mathcal{P}$ and $\eta^+, \eta^- \in \mathcal{K}$ such that $\underline{\nu}(|\xi|) \leq V(\xi) \leq \overline{\nu}(|\xi|)$ and

$$\frac{\partial V}{\partial \xi}(\xi)\left\{\left[A_{\overline{m},\overline{m}}(y) + GA_{m,\overline{m}}(y)\right]\xi + S_{\overline{m}}[R_{\overline{m}}^{+}\rho_{\overline{m}}^{+} + R_{\overline{m}}^{-}\rho_{\overline{m}}^{-}] + S_{\overline{m}}[R_{\overline{m}}^{+}\rho_{m}^{+} + R_{\overline{m}}^{-}\rho_{m}^{-}]\right\} \leq -\omega(|\xi|) + \eta^{+}(|\rho^{+}|) + \eta^{-}(|\rho^{-}|) \tag{5.24}$$

hold for all $\xi \in \mathbb{R}^{n-p}$, $y \in \mathbb{R}^p$, $\rho^+ = [\rho_m^{+\top}, \rho_{\overline{m}}^{+\top}]^\top \in \mathbb{R}^{p+(n-p)}$ and $\rho^- = [\rho_m^{-\top}, \rho_{\overline{m}}^{-\top}]^\top \in \mathbb{R}^{p+(n-p)}$.

Assumption 5.6 There exist a positive definite radially unbounded C^1 function $U : \mathbb{R}^n \to \mathbb{R}_{\geq 0}$, continuous functions $\mu \in \mathcal{P}$ and $\gamma, \zeta \in \mathcal{K}$ such that

$$\frac{\partial U}{\partial x}(x)[A(x_m)x + B(x_m)u(x_m, x_{\overline{m}} + d_{\overline{m}}) + \delta] \le -\mu(|x|) + \gamma(|d_{\overline{m}}|) + \zeta(|\delta|) \tag{5.25}$$

holds with (5.12) for all $x \in \mathbb{R}^n$, $d_{\overline{m}} \in \mathbb{R}^{n-p}$ and $\delta \in \mathbb{R}^n$.

5.4 Guarantees

Define the following vectors:

$$\begin{array}{l}\eta = \eta^+ + \eta^-, \\ \delta^\pm = \delta^+ - \delta^-,\end{array} \quad X = \begin{bmatrix} x \\ \hat{x}^+ \\ \hat{x}^- \end{bmatrix}, \quad \Delta = \begin{bmatrix} \delta \\ \delta^+ \\ \delta^- \end{bmatrix}, \quad \hat{z} = \begin{bmatrix} \hat{x}^+ - x \\ x^+ - x^- \end{bmatrix}, \quad \hat{\rho} = \begin{bmatrix} \delta^+ - \delta \\ \delta^- - \delta \end{bmatrix}.$$

Since the assumptions in Sect. 5.3 are imposed separately on the observer mechanism (5.6) and the feedback mechanism $u(\cdot, \cdot)$, the following two theorems provide conditions under which their coupling results in desired boundedness and convergence for control and monitoring.

Theorem 5.1 *Suppose that Assumptions 5.1, 5.2 and 5.3 are satisfied with $\mu \in \mathcal{K}$. Then in the case of $\delta(t) \equiv \delta^+(t) \equiv \delta^-(t) \equiv 0$, for any x_0 satisfying* (5.2)*, the unique solution $X(t)$ to (5.1) and (5.6) satisfies* (5.4) *and* $\lim_{t\to\infty} |x^+(t) - x^-(t)| = 0$*, and moreover, $X = 0$ is globally asymptotically stable. If*

$$\omega \in \mathcal{K}_\infty \vee \left[\omega \in \mathcal{K} \wedge \left\{\gamma \notin \mathcal{K}_\infty \vee \lim_{s\to\infty} \omega(s) > \sup_{t\in\mathbb{R}_{\ge 0}} \eta(\sqrt{2}|\delta^\pm(t)|)\right\}\right] \tag{5.26}$$

holds, there exist $\hat{\theta} \in \mathcal{KL}$, $\hat{\psi} \in \mathcal{K}$ and $\hat{\chi} \in \mathcal{K}_\infty$ such that

$$\hat{\chi}\left(|\hat{z}(t)|\right) \le \hat{\theta}(|\hat{z}(0)|, t) + \int_0^t \hat{\psi}(|\hat{\rho}(\tau)|)d\tau, \ \forall t \in \mathbb{R}_{\ge 0}. \tag{5.27}$$

$$\int_0^\infty \hat{\psi}(|\hat{\rho}(\tau)|)d\tau < \infty \Rightarrow \lim_{t\to\infty} |\hat{z}(t)| = 0 \tag{5.28}$$

hold for any x_0 and δ satisfying (5.2) *and* (5.3)*, and moreover, the closed-loop system consisting of (5.1) and (5.6) is iISS with respect to the input Δ and the state X. If*

$$\mu \in \mathcal{K}_\infty \wedge \omega \in \mathcal{K}_\infty \tag{5.29}$$

holds, there exist $\hat{\theta} \in \mathcal{KL}$ and $\hat{\phi} \in \mathcal{K}$ such that

$$|\hat{z}(t)| \leq \hat{\theta}(|\hat{z}(0)|, t) + \hat{\phi}\left(\sup_{\tau\in[0,t]} |\hat{\rho}(\tau)|\right), \ \forall t \in \mathbb{R}_{\geq 0} \tag{5.30}$$

$$\lim_{t\to\infty} |\hat{\rho}(t)| = 0 \ \Rightarrow \ \lim_{t\to\infty} |\hat{z}(t)| = 0 \tag{5.31}$$

hold for any x_0 and δ satisfying (5.2) and (5.3), and moreover, the closed-loop system is ISS with respect to Δ and X.

Theorem 5.2 *The claims in Theorem 5.1 hold true even if $\mu \in \mathcal{K}$ and (5.26) are replaced by*

$$\int_0^1 \frac{\gamma \circ \underline{\nu}^{-1}(s)}{\omega \circ \overline{\nu}^{-1}(s)} ds < \infty \tag{5.32}$$

$$\omega \in \mathcal{K} \ \wedge \ \left\{\exists c > 0,\ \exists k \geq 1,\ \forall s \in \mathbb{R}_{\geq 0},\ c\gamma \circ \underline{\nu}^{-1}(s) \leq [\omega \circ \overline{\nu}^{-1}(s)]^k\right\}, \tag{5.33}$$

respectively.

The proofs are omitted due to the space limitation. The above theorems can be verified by following the arguments developed in [7]. Modification of the arguments also proves that Theorems 5.1 and 5.2 hold true for the reduced-order observer candidate (5.20) by replacing Assumptions 5.1, 5.2 and 5.3 with Assumptions 5.4, 5.5 and 5.6, respectively, and redefining

$$X = \begin{bmatrix} x \\ \hat{x}^+_{\overline{m}} - Gy \\ \hat{x}^-_{\overline{m}} - Gy \end{bmatrix}, \ \hat{z} = \begin{bmatrix} \hat{x}^+_{\overline{m}} - x_{\overline{m}} - Gy \\ x^+_{\overline{m}} - x^-_{\overline{m}} \end{bmatrix}.$$

5.5 Difference Between Observers

5.5.1 Utility of H and K, and Difference

Property (5.11) is independent of the state transformation R and the gains $H(y)$ and $K(y)$. The state transformation R contributes to only (5.9), while the gain $H(y)$ contributes to (5.9) and (5.10) and has the same effect as $B(y)K(y)$. The observer (5.6) varies with the choice of $K(y)$ for a given and fixed u. Thus, $K(y)$ offers freedom to change the behavior of the interval estimates $x^+(t)$ and $x^-(t)$ within the aforementioned guarantees. This change in estimates influences the behavior of $x(t)$ of the plant. The standard Luenberger observer also admits $K(y)$ influencing the closed-loop response. However, the freedom is not much appreciated since the standard observer aims at only closed-loop stability and convergence and it is not built for monitoring. In contrast, interval observers provide estimates in the transient phase and the freedom of $K(y)$ matters. Notice that for a given and fixed feedback control law u, the choice of $H(y)$ does not influence u_a in the observer (5.6), while the choice of $K(y)$ does. This flexibility of $K(y)$ in addition to $H(y)$ can be utilized to construct a bundle of interval observers for generating a tighter estimate, as done for instance in [2].

5.5.2 *Benefits of Reduced-Order Design*

In the case of partial measurement (5.14), the reduced-order interval observer (5.20) lets the exact measurement x_m be used instead of estimating intervals for x_m. Since the reduced-order observer is free from dynamics estimating the measured part x_m, its closed loop can be expected to have relatively swifter response with less control effort than the control loop based on the full-order estimates.

To illustrate another advantage of the reduced-order observer, consider the simplest choice $G = 0$ in (5.20). Suppose that Assumption 5.1 is achieved with

$$R = \begin{bmatrix} R_m & 0 \\ 0 & R_{\overline{m}} \end{bmatrix}, \quad R_m \in \mathbb{R}^{p\times p}. \tag{5.34}$$

Then we have

$$R(BKC + HC)R^{-1} = [R(BK + H)R_m^{-1}\ 0] \tag{5.35}$$

Thus, to render $\Gamma(y)$ Metzler, the observer gain $H(y)$ and the feedback gain $K(y)$ modify the first p columns which correspond to the measurable part x_m of x, Therefore,

$$R[A(y) + B(y)K(y)C + H(y)C]R^{-1} \text{ is Metzler} \Rightarrow R_{\overline{m}}A_{\overline{mm}}(y)R_{\overline{m}}^{-1} \text{ is Metzler.} \tag{5.36}$$

holds for each fixed $y \in \mathbb{R}^p$ since every principal minor of a Metzler matrix is Metzler. The modification of A within the limited freedom of (5.35) is unnecessary if a reduced-order interval observer is constructed. The reduced-order design is concerned with only the part $R_{\overline{m}}A_{\overline{mm}}(y)R_{\overline{m}}^{-1}$ which can be influenced by neither $K(y)$ nor $H(y)$ of the full-order observer design. In this way, the reduced-order design allows us to get rid of the unnecessarily "Metzlerization" in the partial measurement case (5.14). In addition, the matrix G in the reduced-order design provides another degree of freedom to modify $R_{\overline{m}}A_{\overline{mm}}(y)R_{\overline{m}}^{-1}$ for the "Metzlerization".

Now, we pay attention to Assumption 5.2. The next proposition demonstrates that in many cases, attainability of (5.9) and (5.10) for the full-order interval observer (5.6) implies the existence of a reduced-order interval observer unless the state transformation R is fully exploited.

Proposition 5.1 *Suppose that* (5.14) *holds and* $A_{\overline{mm}}(y)$ *is independent of* y. *If Assumptions 5.1 and 5.2 are satisfied with a non-singular matrix* $R \in \mathbb{R}^{n\times n}$ *of the form* (5.34) *and a quadratic function* $V : \mathbb{R}^n \to \mathbb{R}_{\geq 0}$. *then Assumptions 5.4 and 5.5 hold with* $G = 0$.

Exploiting $G \neq 0$ can yield a better (larger) ω in (5.24). Furthermore, $A_{\overline{mm}}(y)$ is allowed to depend on y in Proposition 5.1 if the quadratic function $V : \mathbb{R}^n \to \mathbb{R}_{\geq 0}$ is chosen as a quadratic form of a block-diagonal matrix. In the partial measurement case (5.14), producing a Metzler matrix Γ within the freedom of (5.35) imposes severe constraints on the choice of H and K in obtaining a better (larger) ω in (5.10) for the full-order observer (5.6).

Finally, it should be stressed that the above discussions on benefits of the reduced-order observer are not precise when R is not block diagonal. The use of non-diagonal R is crucial for allowing $H(y)$ and $K(y)$ to offer more flexibility than the reduced-order design.

5.6 Comparative Simulations

To illustrate the design flexibility introduced in this chapter, we borrow the following plant from [7]:

$$\begin{bmatrix} \dot{x}_1 \\ \dot{x}_2 \end{bmatrix} = \begin{bmatrix} 0 & -x_1^2 - \frac{1}{2} \\ 0 & -2x_1^2 - \frac{1}{2} \end{bmatrix} \begin{bmatrix} x_1 \\ x_2 \end{bmatrix} + \begin{bmatrix} \frac{x_2}{2} + u_1 + \delta_1 \\ -\frac{x_2}{2} + u_2 + \delta_2 \end{bmatrix} \tag{5.37a}$$

$$y = x_1. \tag{5.37b}$$

Fix the feedback control input as

$$u(y, \hat{x}_2) = \frac{1}{2} \begin{bmatrix} -4y^3 + \hat{x}_2 \\ -\hat{x}_2 \end{bmatrix}, \tag{5.38}$$

where $\hat{x}_2$ denotes an estimate of x_2. The full-order interval observer in Sect. 5.3.1 employs $\hat{x}_2 = \hat{x}_2^+$, while the reduced-order interval observer in Sect. 5.3.2 employs $\hat{x}_2 = \hat{w}_2^+$. Let $U(x) = x^\top x$. As verified in [7], (5.11) and (5.25) are satisfied with $\mu(s) = \frac{1}{4}\min\{s^4, s^2\}$, $\gamma(s) = \max\left\{\frac{3}{2}s^{\frac{4}{3}}, s^2\right\}$, $\zeta(s) = \max\left\{3s^{\frac{4}{3}}, 2s^2\right\}$. Let

$$H(y) = \begin{bmatrix} -2y^2 - 3/4 \\ -1/2 \end{bmatrix}. \tag{5.39}$$

For the choice $K = 0$, (5.10) is satisfied for $V(\xi) = \xi^\top \xi$ with $\omega(s) = s^2/60$, $\eta^+(s) = 10s^2$ and $\eta^-(s) = 13s^2$. For

$$K(y) = \begin{bmatrix} -2y^2 - 1 \\ 0 \end{bmatrix} \tag{5.40}$$

property (5.10) is achieved by letting $\omega(s) = 2s^2/5$. The matrix $\Gamma(y)$ for both $K = 0$ and (5.40) is Metzler with (5.34) and $R_m = 1$, $R_{\overline{m}} = -1/2$. Thus, Assumptions 5.1, 5.2, 5.3 and (5.29) in Theorem 5.1 are satisfied. Since the diagonal matrix R and the quadratic function $V(\xi) = \xi^\top \xi$ led to the above two full-order observers (5.6) with $K = 0$, and (5.6) with (5.40), the discussion in Sect. 5.5.2 indicates that a reduced-order observer can be constructed. Define the reduced-order interval observer as (5.20). For any $G \geq 0$, Assumptions 5.4, 5.5, 5.6 and (5.29) are satisfied. For simulations, we use $x_0 = [5, -5]^\top$, $x_0^+ = [10, 0]^\top$, $x_0^- = [0, -10]^\top$ and

$$\delta(t) = \begin{bmatrix} \mathrm{sgn}(\sin(t)) \min\left\{|\sin(t)|, 5/t^2\right\} \\ \mathrm{sgn}(\cos(t)) \min\left\{|\cos(t)|, 5/t^2\right\} \end{bmatrix}. \tag{5.41}$$

Pick δ^+ by replacing $\sin(t)$ and $\cos(t)$ in (5.41) with 1. Use -1 instead for δ^-. The simulation results shown in Figs. 5.1, 5.2 and 5.3 verify that in all the three designs, the framer property (5.4) is achieved, and the estimated intervals and the plant state converge to the origin. Figures 5.1 and 5.2 show that the choice (5.40) in the observer (5.6) provides a tighter estimate than $K = 0$. Since the control law (5.38) uses the measured component $y = x_1$ instead of its estimate in the full-order designs, the behavior of x with the reduced-order observer (5.20) for $G = 0$ is almost identical with that of the full-order observers (The plots are omitted). For the reduced order observer (5.20) with $G = 2$, Fig. 5.3 not only verifies the achievement of the framer property and the convergence of the estimates and the plant state, but also shows that the change from $G = 0$ to $G = 2$ resulted in the slightly swifter convergence of the interval estimate and x to zero in Fig. 5.3.

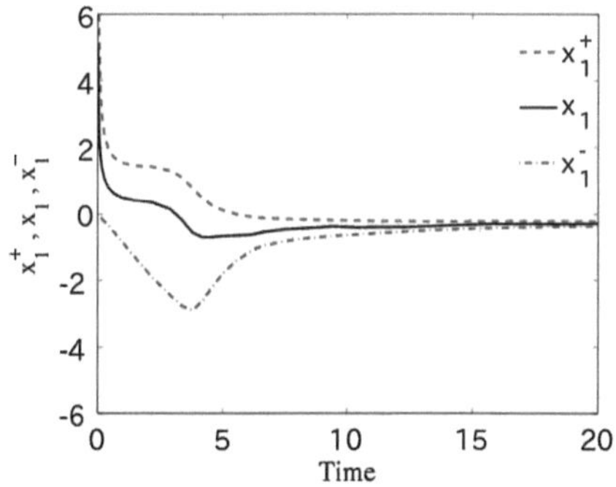

(a) $x_1^-(t), x_1(t), x_1^+(t)$

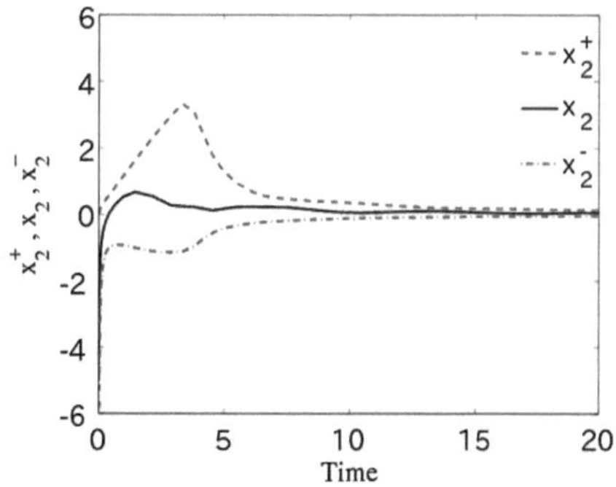

(b) $x_2^-(t), x_2(t), x_2^+(t)$

Fig. 5.1 Closed-loop response for (5.6) with $K = 0$ in the presence of (5.41)

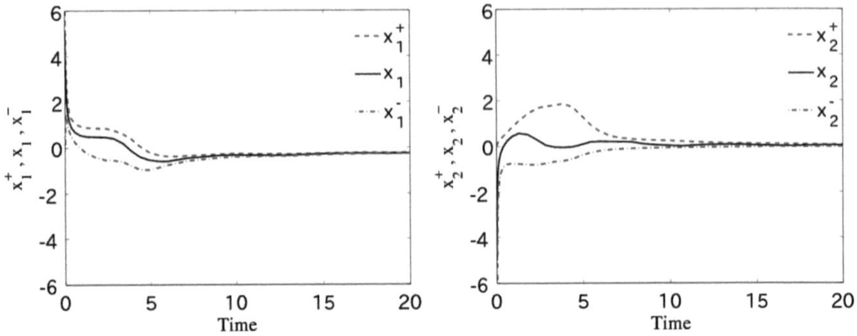

(a) $x_1^-(t), x_1(t), x_1^+(t)$ **(b)** $x_2^-(t), x_2(t), x_2^+(t)$

Fig. 5.2 Closed-loop response for (5.6) with K as in (5.40) in the presence of (5.41)

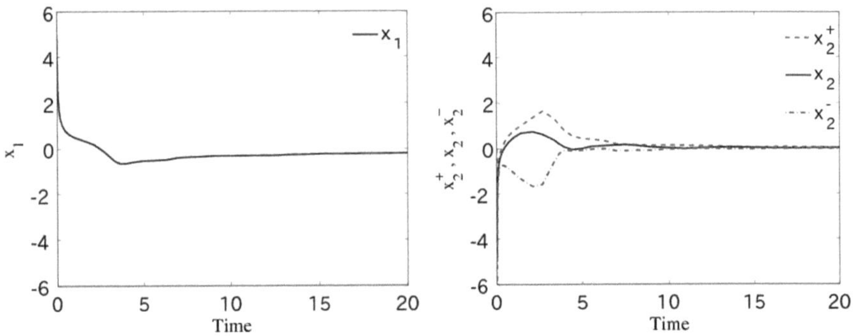

(a) $x_1(t)$ (b) $x_2^-(t), x_2(t), x_2^+(t)$

Fig. 5.3 Closed-loop response for (5.20) with $G = 2$ in the presence of (5.41)

5.7 Conclusions

This chapter has presented an iISS approach to interval observer design for output feedback control of nonlinear systems to guarantee the convergence of the estimated interval length to zero in the presence of converging disturbances. A modification has been proposed by incorporating feedback gain into the interval observer presented in the preceding study [7]. The simple modification offers flexibility to obtain better transient behavior of estimated intervals without altering the observer gain and the control law. For possible improvement of performance for control and estimation, this chapter has also proposed a reduced-order interval observer to avoid estimating measured variables. As a unique consequence of the interval observer design based on Metzler matrices, it has been shown that the existence of a full-order observer implies the existence of a reduced-order observer unless state transformation is fully exploited.

References

1. Alcaraz-Gonzalez, V., Harmand, J., Rapaport, A., Steyer, J.P., Gonzalez-Alvarez, V., Pelayo-Ortiz, C.: Software sensors for highly uncertain WWTPs: a new approach based on interval observers. Water Res. **36**(10), 2515–2524 (2002)
2. Bernard, O., Gouzé, J.L.: Closed loop observers bundle for uncertain biotechnical models. J. Process Control **14**(7), 765–774 (2004)
3. Dinh, T.N., Mazenc, F., Niculescu, S.-I.: Interval observer composed of observers for nonlinear systems. In: Proceedings of the 13th European Control Conference, pp. 660–665 (2014)
4. Efimov, D., Perruquetti, W., Richard, J.P.: Interval estimation for uncertain systems with time-varying delays. Int. J. Control **86**(10), 1777–1787 (2013)
5. Efimov, D., Raissi, T., Chebotarev, S., Zolghadri, A.: Interval state observer for nonlinear time varying systems. Automatica **49**(1), 200–205 (2013)
6. Gouzé, J.L., Rapaport, A., Hadj-Sadok, M.Z.: Interval observers for uncertain biological systems. Ecol. Modell. **133**(1–2), 45–56 (2000)
7. Ito, H., Dinh, T.N.: Interval observers for nonlinear systems with appropriate output feedback. In: Proceedings of the 2nd SICE International Symposium on Control Systems, pp. 9–14 (2016)
8. Mazenc, F., Bernard, O.: Interval observers for linear time-invariant systems with disturbances. Automatica **47**(1), 140–147 (2011)
9. Mazenc, F., Dinh, T.N.: Construction of interval observers for continuous-time systems with discrete measurements. Automatica **50**(10), 2555–2560 (2014)
10. Mazenc, F., Dinh, T.N., Niculescu, S.-I.: Robust interval observers and stabilization design for discrete-time systems with input and output. Automatica **49**(11), 3490–3497 (2013)
11. Mazenc, F., Niculescu, S.-I., Bernard, O.: Exponentially stable interval observers for linear systems with delay. SIAM J. Control Optim. **50**(1), 286–305 (2012)
12. Polyakova, A., Efimov, D., Perruquetti, W.: Output stabilization of time-varying input delay systems using interval observation technique. Automatica **49**(11), 3402–3410 (2013)
13. Raissi, T., Efimov, D., Zolghadri, A.: Interval state estimation for a class of nonlinear systems. IEEE Trans. Autom. Control **57**(1), 260–265 (2012)
14. Raissi, T., Ramdani, N., Candau, Y.: Bounded error moving horizon state estimation for non-linear continuous time systems: application to a bioprocess system. J. Process Control **15**(5), 537–545 (2005)

Chapter 6
Stability Analysis of Neutral Type Time-Delay Positive Systems

Yoshio Ebihara, Naoya Nishio and Tomomichi Hagiwara

Abstract This chapter is concerned with asymptotic stability analysis of neutral type time-delay positive systems (TDPSs). We focus on a neutral type TDPS represented by a feedback system constructed from a finite-dimensional LTI positive system and the pure delay, and give a necessary and sufficient condition for the stability. In the case where we deal with a retarded type TDPS, i.e., if the direct-feedthrough term of the finite-dimensional LTI positive system is zero, it is well known that the retarded type TDPS is stable if and only if its delay-free finite-dimensional counterpart is stable. In the case of neutral type TDPS, i.e., if the direct-feedthrough term is nonzero, however, we clarify that the neutral type TDPS is stable if and only if its delay-free finite-dimensional counterpart is stable and the direct-feedthrough term is Schur stable. Namely, we need additional condition on the direct-feedthrough term.

Keywords Asymptotic stability · Time-delay positive systems · Neutral type

6.1 Introduction

The theory of finite-dimensional linear time-invariant positive systems (FDLTIPSs) is deeply rooted in the theory of nonnegative matrices [3, 13], and celebrated Perron-Frobenius theorem [13] has played a central role in analysis and synthesis. Recently, positive system theory has gained renewed interest from the viewpoint of convex optimization, and excellent papers have been published along this direction, see, e.g., those by Rantzer [18, 19], Shorten et al. [9, 16, 21], Tanaka and Langbort [22], Blanchini et al. [4], Briat [5], and Najison [17]. We also emphasize that the study on consensus problems of multi-agent positive systems is a promising direction, and this issue is treated actively by Valcher and Misra [23] and Ebihara et al. [7]. On the other hand, study on the analysis and synthesis of time-delay positive systems (TDPSs)

Y. Ebihara (✉) · N. Nishio · T. Hagiwara
Department of Electrical Engineering, Kyoto University, Kyotodaigaku-Katsura, Nishikyo-ku, Kyoto 615-8510, Japan
e-mail: ebihara@kuee.kyoto-u.ac.jp

F. Cacace et al. (eds.), *Positive Systems*, Lecture Notes in Control and Information Sciences 471, DOI 10.1007/978-3-319-54211-9_6

has also been active, and fruitful results have been obtained, e.g., by Haddad and Chellaboina [10], Ait Rami et al. [1], and Shen and Lam [20]. In particular, Haddad and Chellaboina [10] showed a prominent result verifying that a *retarded type* TDPS is stable if and only if its delay-free finite-dimensional counterpart is stable. However, to the best of the author's knowledge, existing studies on TDPSs are restricted to retarded type TDPSs, and those for *neutral type* TDPSs are surprisingly scarce. This is probably due to the fact that the definition of solutions for neutral type TDSs is rather difficult [2] irrespective of positivity.

Even though the delay-differential equations (DDEs) of the form $\dot{q}(t) = Jq(t) + K\dot{q}(t-h) + Lq(t-h)$ together with the description of the initial condition as in $q(t) = \phi(t)\,(-h \le t < 0)$ and $q(0) = \xi$ are widely accepted in representing TDSs of constant delay length $h > 0$ [2, 12], in this study we focus on the time-delay feedback system (TDFS) representation shown in Fig. 6.1. In this figure, G is an FDLTI system described by $\dot{x}(t) = Ax(t) + Bw(t)$ and $z(t) = Cx(t) + Dw(t)$, while H is the pure delay of delay length $h > 0$ and thus $w(t) = z(t-h)$ $(t \ge h)$. In the following we denote by $G \star H$ the TDFS shown in Fig. 6.1. If $K \neq 0$ then the DDE shown above represents a neutral type TDS [2], and as already noted, the definition of solutions for neutral type TDSs is involved. This issue is deeply studied by Hagiwara and Kobayashi [11], and the authors provided proper definitions of the solutions (depending upon the discontinuity of the initial function ϕ) and proved their existence and uniqueness by way of conversion techniques from DDE form to TDFS form. The results there suggest that the ability of TDFS form in describing TDSs is higher than the conventional DDE form, in the sense that the properly defined solutions in DDE form can always be represented by signals in TDFS form provided that the FDLTI system G is properly constructed and the initial condition (i.e., $w(t)$ $(0 \le t < h)$ and $x(0)$) is suitably determined. Among several solutions defined there, we adopt the continuous concatenated solution (CCS) as a solution for the neutral type DDE since this solution has the natural continuity property. To summarize, in this chapter, we consider the stability of CCSs of neutral type TDPSs based on the TDFS form. In particular, if we follow the conversion from DDE to TDFS shown in [11], we see that the direct-feedthrough term D of the FDLTI system G coincides with the coefficient K in the DDE form. Therefore, we justifiably focus on TDSs given by TDFS form with $D \neq 0$. On the basis of this preliminary result, we next provide a proper definition for the positivity of TDSs given by TDFS form. This definition leads us to the sound conclusion that a TDS is positive if and only if the FDLTI system G is positive (in the sense of FD systems). Then, we move on to the main

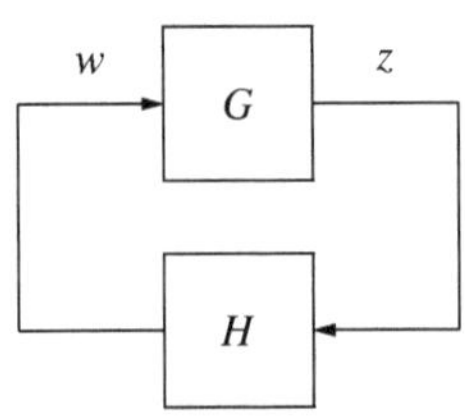

Fig. 6.1 Time-Delay Feedback System

issue of this study, i.e., asymptotic stability analysis of neutral type TDPSs. We first provide an explicit definition of the asymptotic stability in terms of 1 norm of $x_t = x(t)$ and $L_1[0,h)$ norm of $w_t = w_t(\theta) = w(t+\theta)$ $(0 \le \theta < h)$. Then, as the main result of this study, we show that TDPS $G \star H$ is stable if and only if D is Schur stable and $A + B(I - D)^{-1}C$ is Hurwitz stable. This result implies that (i) the stability of $G \star H$ does not depend on the length of the delay h; (ii) $G \star H$ is unstable whenever D is not Schur stable. We provide a rigorous proof for this main result, and in particular, we prove the sufficiency part by concretely constructing a linear Lyapunov functional with respect to $x_t = x(t)$ and $w_t(\theta) = w(t+\theta)$ $(0 \le \theta < h)$.

We use the following notation. We denote by $\mathbb{R}$, $\mathbb{R}_+$, and $\mathbb{R}_{++}$ the set of real, nonnegative, and strictly positive numbers, respectively. The set of natural numbers is denoted by $\mathbb{N}$. For given two matrices A and B of the same size, we write $A > B$ $(A \ge B)$ if $A_{ij} > B_{ij}$ $(A_{ij} \ge B_{ij})$ holds for all (i, j), where A_{ij} stands for the (i, j)-entry of A. We define $\mathbb{R}^n_+ := \{x \in \mathbb{R}^n : x \ge 0\}$ and $\mathbb{R}^n_{++} := \{x \in \mathbb{R}^n : x > 0\}$. We also define $\mathbb{R}^{n\times m}_+$ and $\mathbb{R}^{n\times m}_{++}$ with obvious modifications. For $x \in \mathbb{R}^n$, we denote its 1-norm by $\|x\|$, i.e., $\|x\| := \sum_{i=1}^n \|x_i\|$. Finally, in relation to the definition of CCS and positivity for TDSs in TDFS form, we introduce the following function spaces

$$\begin{aligned}
\mathscr{C}^m_{[0,h)} &:= \{f : f(\theta) \in \mathbb{R}^m,\ f \text{ is continuous over } [0,h)\}, \\
\mathscr{K}^m_h &:= \left\{ f \in \mathscr{C}^m_{[0,h)} : \lim_{\theta \to h-0} f(\theta) \text{ exists} \right\}, \\
\mathscr{K}^m_{h+} &:= \left\{ f \in \mathscr{K}^m_h : f(\theta) \ge 0\ (\forall \theta \in [0,h)) \right\}.
\end{aligned} \tag{6.1}$$

The $L_1[0,h)$ norm for $f \in \mathscr{K}^m_h$ is well-defined by $\|f\| := \int_0^h \|f(\theta)\| d\theta$.

6.2 Representation of Linear Time-Invariant (LTI) Time-Delay Systems (TDSs)

6.2.1 Delay-Differential Equations (DDEs)

In the literature, linear time-invariant (LTI) time-delay systems (TDSs) represented by the following delay-differential equation (DDE) are studied extensively [2, 12].

$$\dot{q}(t) = Jq(t) + K\dot{q}(t-h) + Lq(t-h), \quad J, K, L \in \mathbb{R}^{n\times n}. \tag{6.2}$$

Here, $h > 0$ stands for the delay length. TDSs represented by the DDE (6.2) with $K = 0$ are historically referred to as *retarded type* TDSs, while TDSs represented by the DDE (6.2) with $K \ne 0$ are referred to as *neutral type* TDSs [2, 12].

The "solution" of (6.2) is determined under the initial condition

$$q(t) = \phi(t)\ (t \in [-h, 0)), \quad q(0) = \xi \tag{6.3}$$

where $\phi(t)$ is usually assumed to be (continuous and) continuously differentiable on the closed interval $[-h, 0]$. However, it is rather difficult to define the concept of solutions due to the appearance of indifferentiability (or even discontinuity) especially in the neutral case. This issue is deeply studied by Hagiwara and Kobayashi [11], and proper definitions of solutions are given. Among them, in this chapter, we adopt the continuous concatenated solution (CCS) defined there.

6.2.2 *Continuous Concatenated Solutions (CCSs)*

Let us introduce the definition of CCSs for the DDE (6.2) equipped with the initial condition (6.3).

Definition 6.1 [11] Suppose $\phi(t)$ in (6.3) is bounded, continuously differentiable on $[-h, 0)$, and has the limit $\lim_{t\to 0-0} \phi(t)$. Then, $q(t)$ $(t \geq -h)$ is said to be a continuous concatenated solution (CCS) of the DDE (6.2) under the initial condition (6.3) if (i) it is continuous for $t \geq 0$ and (ii) it is differentiable and satisfies (6.2) for $t \geq 0$ except possibly for time instants $t = kh$ $(k \in \mathbb{N})$.

As the definition says, a CCS is continuous over $t \geq 0$ but not necessarily differentiable at $t = kh$ $(k \in \mathbb{N})$. On the other hand, a stronger solution that is differentiable over $t \geq -h$ is referred to as a regular solution in [11]. However, for the existence of such a regular solution, it is obviously necessary that $\phi(0) = \xi$ and $\dot{\phi}(-0) = J\phi(0) + K\dot{\phi}(-h) + L\phi(-h)$. The latter requirement is rather stringent, and hence it is reasonable to introduce somehow relaxed solutions. Among them the CCS is believed to be a natural one since it possesses continuity property that is essentially required in describing the behavior of physical (real-world) systems.

6.2.3 *Time-Delay Feedback Systems (TDFSs)*

In the community of control theory, it is common to describe LTITDSs in time-delay feedback system (TDFS) form shown in Fig. 6.1. In Fig. 6.1, G stands for an FDLTI system described by

$$G: \begin{cases} \dot{x}(t) = Ax(t) + Bw(t), \\ z(t) = Cx(t) + Dw(t), \end{cases} \quad A \in \mathbb{R}^{n\times n},\ B \in \mathbb{R}^{n\times m},\ C \in \mathbb{R}^{m\times n},\ D \in \mathbb{R}^{m\times m}. \tag{6.4}$$

On the other hand, H is the pure delay of constant delay length $h > 0$ and thus

$$H : w(t) = z(t - h). \tag{6.5}$$

The behavior of TDS $G \star H$ can be determined by the initial condition given for

$$w(t)\ (t \in [0, h)), \quad x(0). \tag{6.6}$$

Once we can describe a given TDS in TDFS form, we can apply fully matured control-oriented techniques such as (scaled) small-gain theorem for its stability analysis. Moreover, recently, a more advanced and rigorous monodromy operator approach has been build for the stability analysis of TDSs in TDFS form, see [15] and references cited there in. This is the motivation of [11] to seek for a conversion technique from DDE form to TDFS form so that the monodromy operator approach can also be applied to TDSs in DDE form. More precisely, the issue in [11] is how to determine the FDLTI system G (or say, the matrices A, B, C, and D) and the initial condition (6.6) from given DDE (or say, the matrices J, K, and L) and given initial condition (6.3) so that the (continuous concatenated) solution of the DDE can be represented as a signal in TDS $G \star H$. The results in [11] that are relevant to this issue are quickly reviewed in the next subsection.

6.2.4 Conversion from DDE to TDFS

The next result shows that the CCS of the DDE (6.2) can always be represented by the state x of the FDLTI system G in the TDS $G \star H$. In the following we write

$$w_t = w_t(\theta) = w(t + \theta)\ (\theta \in [0, h)),\ z_t = z_t(\theta) = z(t + \theta)\ (\theta \in [0, h)). \tag{6.7}$$

Proposition 6.1 *[11] Suppose $\phi(t)$ in* (6.3) *is bounded, continuously differentiable on $[-h, 0)$, and has the limit $\lim_{t \to 0-0} \phi(t)$. Then, the DDE* (6.2) *has a unique CCS $q(t)$, and it coincides, over $t \geq 0$, with $x(t)$ resulting from $G \star H$ with A, B, C, and D given by*

$$A = J, \quad B = I, \quad C = L + KJ, \quad D = K \tag{6.8}$$

and with the initial condition

$$w_0(\theta) = K\dot{\phi}(\theta - h) + L\phi(\theta - h)\ (\theta \in [0, h)), \quad x(0) = \xi. \tag{6.9}$$

As clearly shown in (6.8), the direct-feedthrough term D of the FDLTI system G coincides with the coefficient K in the DDE form. Since we are mainly interested in the neutral type TDSs and hence $K \neq 0$ in (6.2), we justifiably focus on TDSs given by TDFS form with $D \neq 0$ in the following.

Since from now on we focus on TDSs in TDFS form, it is of great benefit in removing ambiguity if we clarify the behavior of the solution $x(t)$ we employ in Proposition 6.1. To this end, we note that the initial function w_0 in (6.9) satisfies $w_0 \in \mathscr{K}_h^m$, which is confirmed by the assumption imposed on ϕ. With this in mind, we assume $w_0 \in \mathscr{K}_h^m$ in the initial condition (6.6) and first focus on the behavior of TDS $G \star H$ over $t \in [0, h)$. Then, from the variation of constant formula, we see that

$$x(t) = e^{At}x(0) + \int_0^t e^{A(t-\tau)}Bw(\tau)d\tau \quad (0 \leq t < h) \tag{6.10}$$

holds and hence $x(t)$ is uniquely determined and continuous over $t \in [0, h)$. In particular, since $w_0 \in \mathscr{K}_h^m$ (or more precisely since $\lim_{t\to h-0} w(t)$ exists from the definition of $\mathscr{K}_h^m$), we see that $\lim_{t\to h-0} x(t)$ exists, and from the continuity requirement on x we can let $x(h) := \lim_{t\to h-0} x(t)$. On the other hand, from

$$z(t) = Cx(t) + Dw(t) \ (0 \leq t < h), \tag{6.11}$$

we see that the important property $z_0 \in \mathscr{K}_h^m$ holds. To summarize, for the next time interval $t \in [h, 2h)$, we know that $x(h)$ is determined and w_h has exactly the same property with $w_0 \in \mathscr{K}_h^m$ since $w(t) = z(t-h)$ $(h \leq t < 2h)$ and $z_0 \in \mathscr{K}_h^m$. Therefore by repeating the same arguments, we see that continuous solution $x(t)$ exists over $t \in [0, 2h)$, and by repeating the same arguments recursively (or say, by concatenating the solutions determined over $[kh, (k+1)h)$ repeatedly), we can conclude that continuous solution $x(t)$ exists over $t \geq 0$. We note that $x(t)$ thus constructed is continuous but might not be differentiable for time instants $t = kh$ $(k \in \mathbb{N})$. This is the continuous solution we employ for TDS $G \star H$ given in TDFS form. To ensure the existence of such continuous solutions, we assume $w_0 \in \mathscr{K}_h^m$ throughout the rest of the chapter.

6.3 Neutral Type Time-Delay Positive Systems (TDPSs)

The goal of this section is to provide a proper definition of positivity of neutral-type TDS in TDFS form. To this end, we first quickly review basic results for FDLTI positive systems.

6.3.1 Basics for Finite-Dimensional LTI Positive Systems (FDLTIPSs)

In this subsection, we gather basic definitions and fundamental results for FDLTI positive systems.

Definition 6.2 (Metzler Matrix)[8] A matrix $A \in \mathbb{R}^{n\times n}$ is said to be *Metzler* if its off-diagonal entries are all nonnegative, i.e., $A_{ij} \geq 0$ $(i \neq j)$.

In the following, we denote by $\mathbb{M}^{n\times n}$ ($\mathbb{H}^{n\times n}$) the set of the Metzler (Hurwitz stable) matrices of size n. Under these notations, the next lemmas hold.

Lemma 6.1 *[8, 14, 16] For given $A \in \mathbb{M}^{n\times n}$, the following conditions are equivalent.*

(i) The matrix A is Hurwitz stable, i.e., $A \in \mathbb{H}^{n\times n}$.
(ii) The matrix A is nonsingular and $A^{-1} \leq 0$.
(iii) There exists $h \in \mathbb{R}^n_{++}$ such that $h^T A < 0$.
(iv) For any $g \in \mathbb{R}^n_+ \setminus \{0\}$, the vector Ag has at least one strictly negative entry.

Lemma 6.2 *[6, 7] For given $P \in \mathbb{M}^{n_1\times n_1}$, $Q \in \mathbb{R}_+^{n_1\times n_2}$, $R \in \mathbb{R}_+^{n_2\times n_1}$, and $S \in \mathbb{M}^{n_2\times n_2}$, the following conditions are equivalent.*

(i) $\Pi := \begin{bmatrix} P & Q \\ R & S \end{bmatrix} \in \mathbb{H}^{(n_1+n_2)\times(n_1+n_2)}$.
(ii) $P \in \mathbb{H}^{n_1\times n_1}$, $S - RP^{-1}Q \in \mathbb{H}^{n_2\times n_2}$.
(iii) $S \in \mathbb{H}^{n_2\times n_2}$, $P - QS^{-1}R \in \mathbb{H}^{n_1\times n_1}$.

To recall the definition of FDLTI positive systems, let us consider the FDLTI system G given by (6.4) (note that there is no need for G to be square for the definition of positivity). The definition of positivity and a basic result are given in the following.

Definition 6.3 [8] The FDLTI system (6.4) is said to be *positive* if its state and output are both nonnegative for any nonnegative initial state and nonnegative input.

Proposition 6.2 *[8] The FDLTI system* (6.4) *is positive if and only if*

$$A \in \mathbb{M}^{n\times n},\ B \in \mathbb{R}_+^{n\times m},\ C \in \mathbb{R}_+^{m\times n},\ D \in \mathbb{R}_+^{m\times m}. \tag{6.12}$$

6.3.2 Positivity of TDSs in TDFS Form

We are now ready to give the definition of positivity for TDSs in TDFS form. Note that, for the definition of positivity, we naturally replace the initial condition $w_0 \in \mathscr{K}_h^m$ with $w_0 \in \mathscr{K}_{h+}^m$.

Definition 6.4 A TDS in TDFS form $G \star H$ constructed from (6.4) and (6.5) is said to be positive if $x(t) \geq 0$ and $w(t) \geq 0$ ($\forall t \geq 0$) hold for any $x(0) \in \mathbb{R}_+^n$ and $w_0 \in \mathscr{K}_{h+}$.

We then give a necessary and sufficient condition for a TDS $G \star H$ to be positive.

Theorem 6.1 *A TDS in TDFS form $G \star H$ constructed from (6.4) and (6.5) is positive in the sense of Definition 6.4 if and only if FDLTI system G is positive.*

Proof of Theorem 6.1 The proof for the necessity of the positivity of G is exactly the same as that of Proposition 6.2 [8] and hence omitted here. To prove the sufficiency, suppose $x(0) \in \mathbb{R}_+^n$ and $w_0 \in \mathscr{K}_{h+}$. Then, from the positivity of G and (6.10) and (6.11), we see that for the first time interval $t \in [0, h)$ the positivity $x(t) \geq 0$ ($\forall t \in [0, h)$) and $z_0 \in \mathscr{K}_{h+}$ hold. In particular, from the discussion around (6.10), $\lim_{t \to h-0} x(t)$ exists and hence $x(h) = \lim_{t \to h-0} x(t) \geq 0$ holds. To summarize, for the next time interval $t \in [h, 2h)$, we know that $x(h) \in \mathbb{R}_+^n$ is determined and w_h has exactly the same property with $w_0 \in \mathscr{K}_{h+}^m$ since $w(t) = z(t-h)$ $(h \leq t < 2h)$ and $z_0 \in \mathscr{K}_{h+}^m$. Therefore by repeating the same arguments recursively, we can conclude that $x(t) \geq 0$ and $w(t) \geq 0$ ($\forall t \geq 0$) hold. ∎

6.4 Stability Analysis of TDPSs

We now move on to the main issue of this study, i.e., asymptotic stability analysis of TDPS $G \star H$ in TDFS form. We fist provide the definition of asymptotic stability for general (not necessarily positive) TDSs in TDFS form. For the consistency with the notation given by (6.7), we define $x_t := x(t)$ in the FDLTI system G given by (6.4).

Definition 6.5 A TDS in TDFS form $G \star H$ constructed from (6.4) and (6.5) is said to be asymptotically stable if $\|x_t\| + \|w_t\| \to 0$ $(t \to \infty)$ for any $x_0 \in \mathbb{R}^n$ and $w_0 \in \mathscr{K}_h$.

In the following, "asymptotic stability" is abbreviated as "stability" just for brevity. The next theorem, which provides a necessary and sufficient condition for the stability of TDPSs in TDFS form, is the main result of this chapter.

Theorem 6.2 *A TDPS in TDFS form $G \star H$ constructed from FDLTIPS G given by* (6.4) *and* (6.12) *and the pure delay H given by* (6.5) *is stable if and only if $D - I \in \mathbb{H}^{m \times m}$ and $A_{\mathrm{cl}} = A + B(I-D)^{-1}C \in \mathbb{H}^{n \times n}$.*

An immediate fact that follows this theorem is that the stability of $G \star H$ is independent of the delay length $h > 0$. Other important implications of Theorem 6.2 will be given after its detailed and rigorous proof.

Remark 6.1 Since TDS $G \star H$ is linear, we can rephrase Definition 6.5 in the way that $G \star H$ is said to be stable if $\|x_t\| + \|w_t\| \to 0$ $(t \to \infty)$ for any $x_0 \in \mathbb{R}_+^n$ and

$w_0 \in \mathscr{K}_{h+}^m$. Namely, we can restrict the initial condition to be "positive orthant." In particular, as long as TDPS $G \star H$ is concerned, $x_0 \in \mathbb{R}_+^n$ and $w_0 \in \mathscr{K}_{h+}^m$ ensures $x(t) \geq 0$ and $w(t) \geq 0$ for all $t \geq 0$ from the definition of TDPS (see Definition 6.4). This positivity property is the key to validate Theorem 6.2 as we see in the following.

Proof of Theorem 6.2 In the proof we assume that $x_0 \in \mathbb{R}_+^n$ and $w_0 \in \mathscr{K}_{h+}$ on the basis of the reasoning given in Remark 6.1.

Sufficiency: Suppose $D - I \in \mathbb{H}^{m\times m}$ and $A_{\rm cl} = A + B(I - D)^{-1}C \in \mathbb{H}^{n\times n}$. Then it is clear that $D - I \in \mathbb{M}^{m\times m} \cap \mathbb{H}^{m\times m}$ and therefore from (ii) of Lemma 6.1 we see $(D - I)^{-1} \leq 0$. It follows that $B(I - D)^{-1}C \in \mathbb{R}_+^{n\times n}$ and hence $A_{\rm cl} \in \mathbb{M}^{n\times n} \cap \mathbb{H}^{n\times n}$. Since $D - I \in \mathbb{M}^{m\times m} \cap \mathbb{H}^{m\times m}$ and $A_{\rm cl} \in \mathbb{M}^{n\times n} \cap \mathbb{H}^{n\times n}$ as just proved, we see from Lemma 6.2 that

$$\begin{bmatrix} A & B \\ C & D - I \end{bmatrix} \in \mathbb{H}^{(n+m)\times(n+m)}.$$

Then, again from Lemma 6.2, we have $A \in \mathbb{M}^{n\times n} \cap \mathbb{H}^{n\times n}$ and $\Psi - I \in \mathbb{M}^{m\times m} \cap \mathbb{H}^{m\times m}$ hold where

$$\Psi := -CA^{-1}B + D. \tag{6.13}$$

Since $A_{\rm cl} \in \mathbb{M}^{n\times n} \cap \mathbb{H}^{n\times n}$ and $\Psi - I \in \mathbb{M}^{m\times m} \cap \mathbb{H}^{m\times m}$, we see from (iii) of Lemma 6.1 that there exist $p_1 \in \mathbb{R}_{++}^n$ and $p_2 \in \mathbb{R}_{++}^m$ such that

$$p_1^T A_{\rm cl} < 0, \quad p_2^T(\Psi - I) < 0. \tag{6.14}$$

By using such $p_1 \in \mathbb{R}_{++}^n$ and $p_2 \in \mathbb{R}_{++}^m$, define

$$r_x := p_1 - A^{-T}C^T p_2 \in \mathbb{R}^n, \quad r_w := p_2 - (D - I)^{-T}B^T p_1 \in \mathbb{R}^m. \tag{6.15}$$

Then, we readily see that $r_x \in \mathbb{R}_{++}^n$ and $r_w \in \mathbb{R}_{++}^m$. By using $r_x \in \mathbb{R}_{++}^n$ and $r_w \in \mathbb{R}_{++}^m$ given above, let us define the Lyapunov functional as in

$$V(x_t, w_t) := r_x^T x_t + r_w^T \int_0^h w_t(\theta)d\theta. \tag{6.16}$$

Here, since $r_x \in \mathbb{R}_{++}^n$ and $r_w \in \mathbb{R}_{++}^m$, and since both $x(t)$ and $w(t)$ are nonnegative for any $t \geq 0$ for any $x_0 \in \mathbb{R}_+^n$ and $w_0 \in \mathscr{K}_{h+}$, we see that the following relationship holds: $V(x_t, w_t) = 0 \iff \|x_t\| + \|w_t\| = 0$. Therefore to prove the stability of $G \star H$ it suffices to show $V(x_t, w_t) \to 0$ $(t \to \infty)$ for each fixed $x_0 \in \mathbb{R}_+^n$ and $w_0 \in \mathscr{K}_{h+}$.

To this end, let us consider the time-derivative of $V(x_t, w_t)$ along the trajectory of $G \star H$. Then, we readily obtain

$$
\begin{aligned}
\frac{dV(x_t, w_t)}{dt} &= r_x^T \dot{x}(t) + r_w^T (w(t+h) - w(t)) = r_x^T (Ax(t) + Bw(t)) + r_w^T (z(t) - w(t)) \\
&= r_x^T (Ax(t) + Bw(t)) + r_w^T \{Cx(t) + (D - I)w(t)\} \\
&= (r_x^T A + r_w^T C)x(t) + \{r_x^T B + r_w^T (D - I)\}w(t) \\
&= \{p_1^T A - p_2^T C + p_2^T C - p_1^T B(D - I)^{-1} C\}x(t) \\
&\qquad + \{p_1^T B - p_2^T C A^{-1} B + p_2^T (D - I) - p_1^T B\}w(t) \\
&= p_1^T A_{\mathrm{cl}} x(t) + p_2^T (\Psi - I)w(t) \quad (kh < t < (k+1)h,\ k = 0, 1, \ldots)
\end{aligned} \tag{6.17}
$$

where we used (6.15) and (6.13). In the above calculation we do not evaluate the time-derivative of $V(x_t, w_t)$ at $t = kh$ $(k = 0, 1, \ldots)$ since, as we have already mentioned, $x_t = x(t)$ is not differentiable at these time instants in general.

In (6.17), since both $x(t)$ and $w(t)$ are nonnegative for any $t \geq 0$, and since (6.14) holds, we see that $dV(x_t, w_t)/dt \leq 0$ holds except for $t = kh$ $(k = 0, 1, \ldots)$. With this fact and the fact that $V(x_t, w_t)$ is continuous for any $t \geq 0$, we can conclude that $V(x_t, w_t)$ is non-increasing over $t \geq 0$. It follows that $V(x_t, w_t) \leq V(x_0, w_0)$ $(\forall t \geq 0)$. In addition, since $V(x_t, w_t) \geq 0$ $(\forall t \geq 0)$ from the definition of $V(x_t, w_t)$ given by (6.16), there exists $V_\infty \geq 0$ such that $V_\infty = \lim_{t \to \infty} V(x_t, w_t)$. To summarize the above arguments, it remains to prove that $V_\infty = 0$.

To prove $V_\infty = 0$ by contradiction, suppose $V_\infty > 0$. We first note from (6.16) that

$$
V(x_t, w_t) \leq \alpha(\|x_t\| + \|w_t\|) \quad (\forall t \geq 0) \tag{6.18}
$$

holds where $\alpha := \max\{\|r_x\|, \|r_w\|\} > 0$. Since $V(x_t, w_t)$ is non-increasing as just proved, we see from (6.18) that

$$
\frac{V_\infty}{\alpha} \leq \|x_t\| + \|w_t\| \quad (\forall t \geq 0) \tag{6.19}
$$

On the other hand, if we take the (improper) integral of the left-hand side of (6.17) over $[kh, (k+1)h]$ $(k = 0, 1, 2, \ldots)$, we have

$$
\begin{aligned}
&V(x_{(k+1)h}, w_{(k+1)h}) - V(x_{kh}, w_{kh}) \\
&= p_1^T A_{\mathrm{cl}} \int_0^h x(kh + \theta)d\theta + p_2^T (\Psi - I) \int_0^h w(kh + \theta)d\theta.
\end{aligned} \tag{6.20}
$$

From the first equation of (6.4), the term $x(kh + \theta)$ satisfies

$$
x(kh + \theta) = e^{A\theta} x(kh) + \int_0^\theta e^{A(\theta - \tau)} Bw(kh + \tau)d\tau \geq e^{A\theta} x(kh) \ (0 \leq \theta \leq h) \tag{6.21}
$$

where we used the fact that $w(t) \geq 0$ $(\forall t \geq 0)$ and $e^{At} \in \mathbb{R}_+^n$ $(\forall t \geq 0)$ to verify the above inequality. It follows that

$$\int_0^h x(kh+\theta)d\theta \geq \int_0^h e^{A\theta}x(kh)d\theta = -A^{-1}(I - e^{Ah})x(kh) \geq 0. \tag{6.22}$$

From this inequality and (6.20) and noting $p_1^T A_{\mathrm{cl}} < 0$, we obtain

$$\begin{aligned} & V(x_{(k+1)h}, w_{(k+1)h}) - V(x_{kh}, w_{kh}) \\ & \leq p_1^T A_{\mathrm{cl}}\{-A^{-1}(I - e^{Ah})\}x(kh) + p_2^T(\Psi - I)\int_0^h w(kh+\theta)d\theta \\ & = -v_x^T x(kh) - v_w^T \int_0^h w(kh+\theta)d\theta \end{aligned} \tag{6.23}$$

where

$$v_x := -\left(-A^{-1}(I - e^{Ah})\right)^T A_{\mathrm{cl}}^T p_1 \in \mathbb{R}_{++}^n, \quad v_w := -(\Psi - I)^T p_2 \in \mathbb{R}_{++}^m. \tag{6.24}$$

The fact that $v_w \in \mathbb{R}_{++}^m$ readily follows from (6.14). The proof for $v_x \in \mathbb{R}_{++}^n$ is given in the appendix section. From (6.23), we obtain

$$V(x_{(k+1)h}, w_{(k+1)h}) - V(x_{kh}, w_{kh}) \leq -\beta(\|x_{kh}\| + \|w_{kh}\|) \tag{6.25}$$

where $\beta := \min\{\min(v_x), \min(v_w)\} > 0$. It follows from the two inequalities (6.19) and (6.25) that $V(x_{(k+1)h}, w_{(k+1)h}) - V(x_{kh}, w_{kh}) \leq -\beta V_\infty/\alpha$. By applying this inequality recursively, we have $V(x_{(k+1)h}, w_{(k+1)h}) - V(x_0, w_0) \leq -k\beta V_\infty/\alpha$. In this inequality, since $V(x_0, w_0)$ takes a finite value depending upon the initial condition $x_0 \in \mathbb{R}_+^n$ and $w_0 \in \mathscr{K}_{h+}^m$, since $\alpha > 0$, $\beta > 0$, and since $V_\infty > 0$ from the underlying assumption for contradiction, we arrive at the conclusion that $V(x_{(k+1)h}, w_{(k+1)h}) < 0$ for k large enough. This contradicts $V(x_t, w_t) \geq 0$ $(\forall t \geq 0)$. Therefore $V_\infty = 0$ and the proof is completed.

Necessity: To prove the necessity by contradiction, we consider the following two cases: (A1) $D - I \notin \mathbb{H}^m$; (A2) $D - I \in \mathbb{H}^m$ and $A_{\mathrm{cl}} \notin \mathbb{H}^m$. By showing that $G \star H$ is not stable for both cases, we can complete the proof. To this end, we first consider the case (A1). Then, it is apparent that $\rho(D) \geq 1$. In addition, since $D \in \mathbb{R}_+^m$, we see from Perron-Frobenius Theorem [13] that there exists $v \in \mathbb{R}_+^m \setminus \{0\}$ such that $Dv = \rho(D)v$. On the other hand, as for the signal w in $G \star H$, we readily obtain

$$\begin{aligned} w(kh+\theta) &= z((k-1)h+\theta) = \{Cx((k-1)h+\theta) + Dw((k-1)h+\theta)\} \\ &\geq Dw((k-1)h+\theta) \cdots \geq D^k w_0(\theta) \quad (0 \leq \theta < h) \end{aligned} \tag{6.26}$$

where we used the fact that both $x(t)$ and $w(t)$ are nonnegative for any $t \geq 0$ and $C \geq 0,\ D \geq 0$. If we let $w_0(\theta) = v\ (0 \leq \theta < h)$, we have $w(kh+\theta) \geq (D)^k v = \rho(D)^k v\quad (0 \leq \theta < h)$. Since $\rho(D) \geq 1$ as above, this inequality implies $\|w_{kh}\| \geq \rho(D)^k h\|v\| \geq h\|v\| > 0$. From the definition of stability of $G \star H$ given in Definition 6.5, this clearly show that $G \star H$ is not stable.

We next consider the case (A2). Then, we see from (the transposed version of) (iv) of Lemma 6.1 that there exists $p_1 \in \mathbb{R}_+^n \setminus \{0\}$ such that $p_1^T A_{\rm cl} \geq 0$. With such p_1 and $p_2 = 0 \in \mathbb{R}^m$, let us define the linear functional $V(x_t, w_t)$ by (6.16). Then, for each fixed $x_0 \in \mathbb{R}_+^n$ and $w_0 \in \mathscr{K}_{h+}$, we have from (6.17) that $dV(x_t, w_t)/dt = p_1^T A_{\rm cl} x(t) \geq 0\quad (kh < t < (k+1)h,\ k = 0, 1, \ldots)$. From this inequality and again from the continuity of $V(x_t, w_t)$, we can conclude that $V(x_t, w_t)$ is non-decreasing over $t \geq 0$. It follows that $V(x_t, w_t) \geq V(x_0, w_0)\ (\forall t \geq 0)$. Here if we let $x(0) = p_1$ and $w_0(\theta) = 0\ (0 \leq \theta < h)$, it is clear that $V(x_t, w_t) \geq V(x_0, w_0) = p_1^T p_1 > 0\ (\forall t \geq 0)$. This clearly show that $G \star H$ is not stable and hence the proof is completed. ■

We conclude this section by providing several important remarks about the implication of the main result, Theorem 6.2.

Remark 6.2 (i) As we have shown at the beginning of the proof for sufficiency, $D - I \in \mathbb{H}^m$ and $A_{\rm cl} \in \mathbb{H}^{n\times n}$ requires $A \in \mathbb{H}^{n\times n}$. It follows that $G \star H$ is stable *only if* the finite-dimensional part G is (internally) stable.

(ii) Since $D \in \mathbb{R}_+^{m\times m}$, the condition $D - I \in \mathbb{H}^m$ can be restated equivalently as $\rho(D) < 1$. It follows that $G \star H$ is stable *only if* the direct-feedthrough term D of the finite-dimensional part G is Schur stable.

(iii) For the retarded type TDPS, i.e., if $D = 0$, it is obvious that the stability condition in Theorem 6.2 reduces to the well-known condition $A - BC \in \mathbb{H}^{n\times n}$ shown in [10]. In other words, Theorem 6.2 includes this well-known result as a special case.

6.5 Conclusion

In this chapter we dealt with asymptotic stability analysis of neutral-type time-delay positive systems given in feedback system form between a finite-dimensional LTI positive system G and the pure delay. We first introduce the continuous concatenated solution as a proper solution of neutral-type time-delay systems, and on the basis of this preliminary result, we define positivity and asymptotic stability. As the main result, we showed a solid result verifying that a neutral type time-delay positive system is asymptotically stable if and only if its delay-free finite-dimensional counterpart is asymptotically stable and the direct-feedthrough term of G is Schur stable.

Acknowledgements This work was supported by JSPS KAKENHI Grant Number 25420436.

Appendix

6.5.1 *Proof of (6.24)*

Let us define $S : \mathbb{R}_+ \to \mathbb{R}^{n\times n}$ by $S(t) = -A^{-1}(I - e^{At})$. Then, v_x can be rewritten as $v_x = -S(h)^T A_{\rm cl}^T p_1$. Since $-A_{\rm cl}^T p_1 \in \mathbb{R}^n_{++}$ from (6.14), it suffices to prove $S(h) \geq 0$ and $S(h)_{ii} > 0$ $(i = 1, \ldots, n)$. The first inequality obviously holds since $S(0) = 0$ and $dS(t)/dt = e^{At} \geq 0$ $(\forall t \geq 0)$. In particular, from the Taylor series expansion $dS(t)/dt = e^{At} = I + At + \frac{1}{2}(At)^2 \cdots$, it is obvious that there exists $t' > 0$ such that $(dS(t)/dt)_{ii} > 0$ $(\forall t \in [0, t'], i = 1, \ldots, n)$. Since $S(0) = 0, dS(t)/dt \geq 0$ $(\forall t \geq 0)$, and $(dS(t)/dt)_{ii} > 0$ $(\forall t \in [0, t'],\ i = 1, \ldots, n)$, we can conclude that $S(h)_{ii} > 0$ $(i = 1, \ldots, n)$. This completes the proof.

References

1. Ait Rami, M., Jordan, A.J., Schonlein, M., Schonlein, M.: Estimation of linear positive systems with unknown time-varying delays. Eur. J. Control **19**(3), 179–187 (2013)
2. Bellman, R., Cooke, K.L.: Differential Difference Equations. Academic Press, New York (1963)
3. Berman, A., Plemmons, R.J.: Nonnegative Matrices in the Mathematical Sciences. Academic Press, New York (1979)
4. Blanchini, F., Colaneri, P., Valcher, M.E.: Co-positive Lyapunov functions for the stabilization of positive switched systems. IEEE Trans. Autom. Control **57**(12), 3038–3050 (2012)
5. Briat, C.: Robust stability and stabilization of uncertain linear positive systems via integral linear constraints: L_1-gain and L_∞-gain characterization. Int. J. Robust Nonlinear Control **23**(17), 1932–1954 (2013)
6. Ebihara, Y., Peaucelle, D., Arzelier, D.: L_1 gain analysis of linear positive systems and its application. In: Proceedings Conference on Decision and Control, pp. 4029–4034 (2011)
7. Ebihara, Y., Peaucelle, D., Arzelier, D.: Analysis and synthesis of interconnected positive systems. IEEE Trans. Autom. Control **62**(2), 652–667 (2017)
8. Farina, L., Rinaldi, S.: Positive Linear Systems: Theory and Applications. Wiley (2000)
9. Gurvits, L., Shorten, R., Mason, O.: On the stability of switched positive linear systems. IEEE Trans. Autom. Control **52**(6), 1099–1103 (2007)
10. Haddad, W.M., Chellaboina, V.: Stability theory for nonnegative and compartmental dynamical systems with time delay. Syst. Control Lett. **51**(5), 355–361 (2004)
11. Hagiwara, T., Kobayashi, M.: Concatenated solutions of delay-differential equations and their representation with time-delay feedback systems. Int. J. Control **84**(6), 1126–1139 (2011)
12. Hale, J.K.: Theory of Functional Differential Equations. Springer, New York (1977)
13. Horn, R.A., Johnson, C.A.: Topics in Matrix Analysis. Cambridge University Press, New York (1991)
14. Kaczorek, T.: Positive 1D and 2D Systems. Springer, London (2001)
15. Kim, J.H., Hagiwara, T., Hirata, K.: Spectrum of monodromy operator for a time-delay system with application to stability analysis. IEEE Trans. Autom. Control **60**(12), 3385–3390 (2015)
16. Mason, O., Shorten, R.: On linear copositive Lyapunov function and the stability of switched positive linear systems. IEEE Trans. Autom. Control **52**(7), 1346–1349 (2007)
17. Najson, F.: On the Kalman-Yakubovich-Popov lemma for discrete-time positive linear systems: A novel simple proof and some related results. Int. J. Control **86**(10), 1813–1823 (2013)
18. Rantzer, A.: Scalable control of positive systems. Eur. J. Control **24**(1), 72–80 (2015)

19. Rantzer, A.: On the kalman-yakubovich-popov lemma for positive systems. IEEE Trans. Autom. Control **61**(5), 1346–1349 (2016)
20. Shen, J., Lam, J.: L_∞-gain analysis for positive systems with distributed delays. Automatica **50**(2), 547–551 (2014)
21. Shorten, R., Mason, O., King, C.: An alternative proof of the Barker, Berman, Plemmons (BBP) result on diagonal stability and extensions. Linear Algebra Appl. **430**, 34–40 (2009)
22. Tanaka, T., Langbort, C.: The bounded real lemma for internally positive systems and H_∞ structured static state feedback. IEEE Trans. Autom. Control **56**(9), 2218–2223 (2011)
23. Valcher, M.E., Misra, P.: On the stabilizability and consensus of positive homogeneous multi-agent dynamical systems. IEEE Trans. Autom. Control **59**(7), 1936–1941 (2014)

Chapter 7
Internally Positive Representations and Stability Analysis of Linear Delay Systems with Multiple Time-Varying Delays

Francesco Conte, Vittorio De Iuliis and Costanzo Manes

Abstract This chapter introduces the Internally Positive Representation of linear systems with multiple time-varying state delays. The technique, previously introduced for the undelayed case, aims at building a positive representation of systems whose dynamics is, in general, indefinite in sign. As a consequence, stability criteria for positive time-delay systems can be exploited to analyze the stability of the original system. As a result, an easy-to-check sufficient condition for the delay-independent stability is provided, that is compared with some well known conditions available in the literature.

Keywords Positive delay systems · Time-varying delays · Internally positive representation (IPR) · Stability analysis

7.1 Introduction

Positive linear systems have been extensively studied in the last decades due to their well known properties and applications [6, 18]. More recently, several works on positive linear time-delay systems appeared in the literature, some of them providing insightful stability results [1, 12, 15–17, 19, 23]. To exploit the properties of positive systems also for not necessarily positive systems, an useful tool has recently been developed in the linear undelayed case: the Internally Positive Representation (IPR). The technique, introduced in the discrete-time framework in [4, 8, 9] and in the continuous-time one in [2, 3], aims at constructing internally positive representations of systems whose dynamics is indefinite in sign. The method presented in

F. Conte
Dipartimento di Ingegneria Navale, Elettrica, Elettronica e delle Telecomunicazioni, Università degli Studi di Genova, Via All'Opera Pia, 11A, 16145 Genova, GE, Italy
e-mail: fr.conte@unige.it

V. De Iuliis (✉) · C. Manes
Dipartimento di Ingegneria e Scienze Dell'Informazione, e Matematica, Università degli Studi dell'Aquila, Via Vetoio, 67100 Coppito, AQ, Italy
e-mail: vittorio.deiuliis@graduate.univaq.it

C. Manes
e-mail: costanzo.manes@univaq.it

F. Cacace et al. (eds.), *Positive Systems*, Lecture Notes in Control and Information Sciences 471, DOI 10.1007/978-3-319-54211-9_7

[2], although very easy and straightforward, can produce in some cases an unstable positive system even if the original system is stable. Later works on the IPR focused on this issue, showing how to construct IPRs whose stability properties are equivalent to those of the original system [3].

As is typical in Systems and Control, one usually tries to extend to the more general case what is well known in the particular one: to this end, the main part of this chapter focuses on the extension of the IPR construction method to linear continuous time-delay systems, in the general case of multiple time-varying delays. Then, a stability analysis follows, leading to the conclusion that only delay systems that are stable for any set of delays, constant or time-varying, can admit a stable IPR. As a result, an easy-to-check sufficient condition for the delay-independent stability of the original system is provided, whose efficacy with respect to other similar sufficient conditions available in the literature is tested by numerical examples.

This chapter is organized as follows: in Sect. 7.2, the Internally Positive Representation for linear systems with multiple time-varying delays is introduced. Section 7.3 reports a discussion on the stability properties of IPRs and presents the new stability condition. In Sect. 7.4 the condition is compared with similar existing results, and in Sect. 7.5 an illustrative example is reported. Conclusions follow.

Notations. $\mathbb{R}_+$ is the set of nonnegative real numbers. $\mathbb{C}^-$ and $\mathbb{C}^+$ are the open left-half and right-half complex planes, respectively. $\mathbb{R}^n_+$ is the nonnegative orthant of $\mathbb{R}^n$. $\mathbb{R}^{m\times n}_+$ is the cone of positive $m \times n$ matrices. I_n is the $n \times n$ identity matrix. $\Re(z)$ and $\Im(z)$ are the real and imaginary parts of a complex number z, respectively. $\mathscr{C}([a, b], \mathbb{R}^n)$ denotes the Banach space of all continuous functions on $[a, b]$ with values in $\mathbb{R}^n$, endowed with the uniform convergence norm $\|\cdot\|_\infty$. $A \in \mathbb{R}^{n\times n}$ is said to be *Metzler* if all its off-diagonal elements are nonnegative. $d(A)$ denotes the diagonal matrix extracted from A. $\sigma(A)$ and $\alpha(A)$ denote the spectrum and the spectral abscissa of A, respectively. A is said to be stable or *Hurwitz* if $\sigma(A) \subset \mathbb{C}^-$ or, equivalently, if $\alpha(A) < 0$. $\mathscr{L}_1^p$ and $\mathscr{L}_{1,+}^p$ are the sets of locally integrable functions with values in $\mathbb{R}^p$ and $\mathbb{R}^p_+$, respectively. Finally, $\underline{m} = \{1, 2, \ldots, m\}$ and $\underline{m}_0 = \{0, 1, \ldots, m\}$.

7.2 Internally Positive Representation of Delay Systems

7.2.1 Internally Positive Delay Systems

Let $S = \big\{\{A_k\}_0^m, B, C, D\big\}_{n,p,q}$ denote a continuous-time delay system, with possibly time-varying delays, having the following form

$$\begin{aligned} \dot{x}(t) &= A_0 x(t) + \sum_{k=1}^{m} A_k x(t - \delta_k(t)) + Bu(t), \qquad t \geq t_0, \\ y(t) &= Cx(t) + Du(t), \\ x(t) &= \phi(t - t_0), \qquad t \in [t_0 - \delta, t_0], \end{aligned} \tag{7.1}$$

where $u(t) \in \mathbb{R}^p$ is the input, with $u \in \mathscr{L}_1^p$, $y(t) \in \mathbb{R}^q$ is the output, $x(t) \in \mathbb{R}^n$ is the system variable and $\phi \in \mathscr{C}([-\delta, 0], \mathbb{R}^n)$ is a *pre-shape* function (initial state). $\delta_k : \mathbb{R} \to \mathbb{R}_+$ are time-delays, which are bounded continuous functions

$$0 \le \delta_k(t) \le \delta, \quad \forall t \ge t_0. \tag{7.2}$$

$B \in \mathbb{R}^{n\times p}$, $C \in \mathbb{R}^{q\times n}$, $D \in \mathbb{R}^{q\times p}$, and $A_k \in \mathbb{R}^{n\times n}$, for $k \in \underline{m}_0$. It is well known that the delay differential equation in (7.1) admits a unique solution satisfying a given initial condition ϕ (see e.g. [13]). Throughout the chapter, the solution $x(t)$ and the corresponding output trajectory $y(t)$ associated to a system S will be denoted as

$$\big(x(t), y(t)\big) = \Phi_S\big(t, t_0, \phi, u\big). \tag{7.3}$$

Following [14, 17], an *internally positive* linear delay system is defined as follows.

Definition 7.1 A delay system $S = \big\{\{A_k\}_0^m, B, C, D\big\}_{n,p,q}$ is said to be internally positive if

$$\left\{ \begin{matrix} \phi \in \mathscr{C}([-\delta, 0], \mathbb{R}^n_+) \\ u \in \mathscr{L}^p_{1,+} \end{matrix} \right\} \;\Rightarrow\; \left\{ \begin{matrix} x(t) \in \mathbb{R}^n_+, \\ y(t) \in \mathbb{R}^q_+, \end{matrix} \;\; \forall t \ge t_0 \right\}. \tag{7.4}$$

Stated informally, S is internally positive if nonnegative initial states and input functions produce nonnegative state and output trajectories. The following result gives necessary and sufficient conditions to fulfill Definition 7.1 (see [12, 23]).

Lemma 7.1 *A delay system $S = \{\{A_k\}_0^m, B, C, D\}_{n,p,q}$ is internally positive if and only if A_0 is Metzler and B, C, D and A_k, for $k \in \underline{m}$, are nonnegative.*

7.2.2 Positive Representation of Vectors and Matrices

Given a matrix (or vector) $M \in \mathbb{R}^{m\times n}$, the symbols M^+, M^- denote the componentwise *positive* and *negative* parts of M, while $|M|$ stands for its componentwise absolute value. It follows that $M = M^+ - M^-$ and $|M| = M^+ + M^-$.

Let $\Delta_n = [I_n \;\; -I_n] \in \mathbb{R}^{n\times 2n}$. The definitions reported below are taken from [4, 9].

Definition 7.2 A positive representation of a vector $x \in \mathbb{R}^n$ is any vector $\tilde{x} \in \mathbb{R}^{2n}_+$ such that

$$x = \Delta_n \tilde{x}. \tag{7.5}$$

The min-positive representation of a vector $x \in \mathbb{R}^n$ is the positive vector $\pi(x) \in \mathbb{R}^{2n}_+$ defined as

$$\pi(x) = \begin{bmatrix} x^+ \\ x^- \end{bmatrix}. \tag{7.6}$$

The min-positive representation of a matrix $M \in \mathbb{R}^{m\times n}$ is the positive matrix $\Pi(M) \in \mathbb{R}_+^{2m\times 2n}$ defined as

$$\Pi(M) = \begin{bmatrix} M^+ & M^- \\ M^- & M^+ \end{bmatrix} \tag{7.7}$$

while the min-Metzler representation of a matrix $A \in \mathbb{R}^{n\times n}$ is the Metzler matrix $\Gamma(A) \in \mathbb{R}^{2n\times 2n}$ defined as

$$\Gamma(A) = \begin{bmatrix} d(A) + (A - d(A))^+ & (A - d(A))^- \\ (A - d(A))^- & d(A) + (A - d(A))^+ \end{bmatrix}. \tag{7.8}$$

Of course, if $d(A) \in \mathbb{R}_+^{n\times n}$ then $\Gamma(A) = \Pi(A)$. Moreover, for any $x \in \mathbb{R}^n$ and matrices $M \in \mathbb{R}^{m\times n}$, $A \in \mathbb{R}^{n\times n}$ the following properties hold true:

(a) $x = \Delta_n \pi(x)$;
(b) $\Delta_m \Pi(M) = M\Delta_n$, so that $\Delta_m \Pi(M)\pi(x) = Mx$;
(c) $\Delta_n \Gamma(A) = A\Delta_n$, so that $\Delta_n \Gamma(A)\pi(x) = Ax$.

7.2.3 Internally Positive Representations

The concept of Internally Positive Representation (IPR) of an arbitrary system has been introduced in [4, 8, 9], for discrete-time systems, and in [2, 3] for continuous-time systems. The IPR construction presented in [2] can be extended to the case of time-varying delays systems by the following definition.

Definition 7.3 An *Internally Positive Representation* (IPR) of a delay system $S = \big\{\{A_k\}_0^m, B, C, D\big\}_{n,p,q}$ is an internally positive system $\widetilde{S} = \big\{\{\widetilde{A}_k\}_0^m, \widetilde{B}, \widetilde{C}, \widetilde{D}\big\}_{\tilde{n},\tilde{p},\tilde{q}}$ together with four transformations $\{T_X^f, T_X^b, T_U, T_Y\}$,

$$T_X^f : \mathbb{R}^n \mapsto \mathbb{R}_+^{\tilde{n}}, \quad T_X^b : \mathbb{R}_+^{\tilde{n}} \mapsto \mathbb{R}^n, \quad T_U : \mathbb{R}^p \mapsto \mathbb{R}_+^{\tilde{p}}, \quad T_Y : \mathbb{R}_+^{\tilde{q}} \mapsto \mathbb{R}^q, \tag{7.9}$$

such that $\forall t_0 \in \mathbb{R}, \forall\big(\phi, u\big) \in \mathscr{C}([-\delta, 0], \mathbb{R}^n) \times \mathscr{L}_1^p$, the following implication holds:

$$\left\{ \begin{array}{ll} \tilde{\phi}(\tau) = T_X^f\big(\phi(\tau)\big), & \forall \tau \in [-\delta, 0] \\ \tilde{u}(t) = T_U\big(u(t)\big), & \forall t \geq t_0 \end{array} \right\} \implies \left\{ \begin{array}{l} x(t) = T_X^b\big(\tilde{x}(t)\big), \\ y(t) = T_Y\big(\tilde{y}(t)\big), \end{array} \ \forall t \geq t_0 \right\} \tag{7.10}$$

where

$$\begin{aligned} \big(x(t), y(t)\big) &= \Phi_S\big(t, t_0, \phi, u\big), \\ \big(\tilde{x}(t), \tilde{y}(t)\big) &= \Phi_{\tilde{S}}\big(t, t_0, \tilde{\phi}, \tilde{u}\big). \end{aligned}$$

T_X^f and T_X^b in (7.9) are the *forward* and *backward* state transformations of the IPR, respectively, while T_U and T_Y are the input and output transformations, respectively. The implication (7.10) means that **if** the (nonnegative) pre-shape function $\tilde{\phi}$ of the IPR is computed as the forward state transformation T_X^f of the pre-shape function ϕ of the *original* system, **and** the (nonnegative) input $\tilde{u}$ to the IPR is computed as the input transformation T_U of the input u to the *original* system, **then** the state trajectory of the *original* system is given by the backward transformation T_X^b of the (nonnegative) state $\tilde{x}$ of the IPR, **and** the output trajectory y of the *original* system is given by the output transformation T_Y of the (nonnegative) output $\tilde{y}$ of the IPR. For consistency, the *backward* map T_X^b must be a left-inverse of the *forward* map T_X^f, i.e. $x = T_X^b\big(T_X^f(x)\big), \forall x \in \mathbb{R}^n$.

The following theorem provides a method for the IPR construction of arbitrary time-varying delays systems.

Theorem 7.1 *Consider a delay system S as in* (7.1), *with* $S = \big\{\{A_k\}_0^m, B, C, D\big\}_{n,p,q}$. *An internally positive system* $\bar{S} = \big\{\{\bar{A}_k\}_0^m, \bar{B}, \bar{C}, \bar{D}\big\}_{2n,2p,2q}$, *with*

$$\bar{A}_0 = \Gamma(A_0), \quad \bar{B} = \Pi(B), \quad \bar{C} = \Pi(C), \quad \bar{D} = \Pi(D), \quad \bar{A}_k = \Pi(A_k), \quad k \in \underline{m}, \tag{7.11}$$

together with the four transformations

$$\bar{x} = T_X^f(x) = \pi(x), \qquad x = T_X^b(\bar{x}) = \Delta_n \bar{x}, \tag{7.12}$$

$$\bar{u} = T_U(u) = \pi(u), \qquad y = T_Y(\bar{y}) = \Delta_q \bar{y}, \tag{7.13}$$

is an IPR of S.

Proof First of all, since $\bar{A}_0$ is Metzler and $\bar{B}, \bar{C}, \bar{D}$, and $\bar{A}_k, k \in \underline{m}$, are all nonnegative, from Lemma 7.1 it follows that system $\bar{S}$ is internally positive. For any pre-shape function $\phi \in \mathscr{C}([-\delta, 0], \mathbb{R}^n)$, let $\bar{x}(t)$ and $\bar{y}(t)$ denote the state and output trajectories

$$\big(\bar{x}(t), \bar{y}(t)\big) = \Phi_{\bar{S}}\big(t, t_0, \bar{\phi}, \bar{u}\big) \tag{7.14}$$

where $\bar{\phi}(\tau) = T_X^f(\phi(\tau)) = \pi(\phi(\tau)), \forall \tau \in [-\delta, 0]$ and $\bar{u}(t) = T_U(u(t)) = \pi(u(t))$, $\forall t \geq t_0$. Thus, (7.14) solves the system

$$\begin{aligned}
\dot{\bar{x}}(t) &= \bar{A}_0\bar{x}(t) + \sum_{k=1}^{m} \bar{A}_k\bar{x}(t - \delta_k(t)) + \bar{B}\bar{u}(t), \qquad t \geq t_0 \\
\bar{y}(t) &= \bar{C}\bar{x}(t) + \bar{D}\bar{u}(t), \\
\bar{x}(t) &= \bar{\phi}(t - t_0), \qquad t \in [t_0 - \delta, t_0].
\end{aligned} \tag{7.15}$$

Consider now the vectors

$$z(t) = T_X^b(\bar{x}(t)) = \Delta_n \bar{x}(t), \tag{7.16}$$

$$v(t) = T_Y(\bar{y}(t)) = \Delta_q \bar{y}(t). \tag{7.17}$$

The theorem is proved by showing that $x(t) = z(t)$ and $y(t) = v(t)$ for all $t \geq t_0$. Using properties (b) and (c), given in Sect. 7.2.2, and (7.16), it results that, for $t \geq t_0$,

$$\dot{z}(t) = \Delta_n \dot{\bar{x}}(t) = \Delta_n \bar{A}_0 \bar{x}(t) + \sum_{k=1}^{m} \Delta_n \bar{A}_k \bar{x}(t - \delta_k(t)) + \Delta_n \bar{B} \pi\big(u(t)\big)$$
$$= A_0 z(t) + \sum_{k=1}^{m} A_k z(t - \delta_k(t)) + Bu(t). \tag{7.18}$$

and for $t \in [t_0 - \delta, t_0]$

$$z(t) = \Delta_n \bar{x}(t) = \Delta_n \bar{\phi}(t - t_0) = \Delta_n \pi\big(\phi(t - t_0)\big) = \phi(t - t_0), \tag{7.19}$$

and

$$v(t) = \Delta_q \bar{y}(t) = \Delta_q \bar{C} \bar{x}(t) + \Delta_q \bar{D} \pi\big(u(t)\big) = Cz(t) + Du(t), \qquad t \geq t_0. \tag{7.20}$$

Note that $\big(z(t), v(t)\big)$ obey the same equations of (7.1), with the same initial condition. From the uniqueness of the solution we get $\big(z(t), v(t)\big) = \big(x(t), y(t)\big)$, and this concludes the proof. □

Remark 7.1 If $A_k = 0$ for all $k \in \underline{m}$ the IPR proposed in Theorem 7.1 coincides with the *normal-form* IPR proposed in [2] (Theorem 7.4) for the delay-free case.

7.3 Stability Analysis

In this section we investigate the relationships between the stability of a delay system and of its IPR. A quite obvious consequence of the boundedness of the state transformations $T_X^f(\cdot)$ and $T_X^b(\cdot)$ in (7.12) is that if an IPR of a system is stable, then the original system is stable as well. As we will see, the converse is not always true.

Throughout this chapter we will use a standard nomenclature about stability. The trivial solution $x(t) \equiv 0$ of a delay system of the type (7.1) is said to be stable if any solution $x(t)$ for all $t \geq t_0$ satisfies a bound of the type $\|x(t)\| \leq k\|\phi\|_\infty$, for some $k > 0$. If in addition $\lim_{t\to\infty} \|x(t)\| = 0$, the trivial solution is asymptotically stable. If there exist $k > 0$ and $\eta > 0$ such that $\|x(t)\| \leq k\, e^{-\eta t}\|\phi\|_\infty$, the trivial solution is said to be exponentially stable.

A delay system as in (7.1) is said to be *stable* if the trivial solution is asymptotically stable. It is worth recalling that the stability of a delay system of the type (7.1) depends on the nature of delays (see e.g. [7, 11]): one can have stability for a given set or for any set of constant delays, for commensurate constant delays, for time-varying delays, within a given bound or without a specific bound, fast or slowly varying, etc.

For reasons that will soon be clear, in this chapter we are mainly concerned with stability for any set of constant or time-varying delays without a specific bound (delay-independent stability).

7.3.1 Stable IPRs of Delay-Free Systems

For the case of delay-free systems ($A_k = 0$, $k \in \underline{m}$) in [2] it has been shown that the IPR construction method there presented when applied to stable systems in some cases may produce unstable IPRs. Indeed, the spectrum of $\bar{A}_0 = \Gamma(A_0)$ properly contains the spectrum of A_0, and the additional eigenvalues can be unstable. However, a change of coordinates on the original system can generally affect the stability of the IPR, and this fact can be exploited to obtain stable IPRs. In [2] it has been proved that such a change of coordinates exists if $\sigma(A_0)$ belongs to the sector of $\mathbb{C}^-$ characterized by $\Re(z) + |\Im(z)| < 0$. In [3], the IPR construction method of [2] has been suitably extended so that stable IPRs can be constructed for any stable system, without any limitation on the location of the eigenvalues of A_0 within $\mathbb{C}^-$.

7.3.2 Stability of Positive Delay-Systems

The IPR produced by the method in Theorem 7.1 is by construction a linear positive delay system. For this reason we recall below the stability conditions for such a class of systems. Consider a system of the type (7.1) which is internally positive (i.e., A_0 is Metzler and A_k, $k \in \underline{m}$, are nonnegative, Lemma 7.1). In [12] it has been proved that, when the delays δ_k are constant, a necessary and sufficient stability condition is that there exist p and r in $\mathbb{R}^n$ such that

$$\left(\sum_{k=0}^{m} A_k\right)^T p + r = 0 \qquad p > 0, \quad r > 0. \tag{7.21}$$

Note that, being $\sum_{k=0}^{m} A_k$ a Metzler matrix, condition (7.21) is equivalent to $\sum_{k=0}^{m} A_k$ Hurwitz, i.e.

$$\alpha\left(\sum_{k=0}^{m} A_k\right) < 0. \tag{7.22}$$

Another interesting equivalent condition (see [6]), that does not require the explicit computation of eigenvalues (condition (7.22)) or solving a linear problem (condition (7.21)) is that all the leading principal minors of the matrix

$$M = -\sum_{k=0}^{m} A_k$$

are positive, i.e. $M_i > 0$ for $i = 1, ..., n$, where M_i is the determinant of the matrix obtained removing the last $n - i$ rows and columns from M. Note that all these equivalent conditions do not depend on the size of the delays. In [1] and in [17] it has been proved that (7.22) is necessary and sufficient for stability even in the case of time-varying delays $\delta_k(t)$, without limitation on the size of the delays and of their derivatives. Ngoc in [23] proved a similar condition also for the case of distributed delays.

Remark 7.2 It should be remarked that condition (7.22) is necessary and sufficient for the delay-independent stability of a positive delay-system, while it is only necessary for the stability of general (not necessarily positive) systems, being required for the stability of the associated delay-free system.

To summarize, we have the following:

Proposition 7.1 *If a system S as in* (7.1)*, with A_0 Metzler and B, C, D, A_k, for $k \in \underline{m}$, nonnegative, is stable for a given set of constant delays δ_k, then it is also delay-independent stable, i.e. stable for any arbitrary set of constant or time-varying delays.*

Liu and Lam [16] showed that if a positive delay system is stable for all continuous and bounded delays, then the trivial solution is exponentially stable for all continuous and bounded delays. On the other hand, if the delays are continuous but unbounded, the trivial solution may be asymptotically stable but not exponentially stable.

7.3.3 Stable IPRs of Delay Systems

Consider the equations (7.15) of the IPR given in Theorem 7.1. We have the following:

Theorem 7.4 *If a delay system S as in* (7.1) *admits a stable IPR, then necessarily S is delay-independent stable.*

Proof As discussed in the previous paragraph, since the IPR is a positive delay system, a necessary and sufficient condition for its stability is that the Metzler matrix $\sum_{k=0}^{m} \bar{A}_k$ is Hurwitz, and this in turn implies that the IPR is delay-independent stable. The boundedness of the state transformations $T_X^f(\cdot)$ and $T_X^b(\cdot)$ defined in (7.12) trivially implies the delay-independent stability of the original system. □

Stated in another way, Theorem 7.4 claims that only delay systems that are delay-independent stable admit stable Internally Positive Representations.

Theorem 7.4 suggests the following sufficient condition of delay-independent stability for not necessarily positive delay systems.

Theorem 7.5 *Consider a delay system S as in* (7.1). *If*

$$\alpha\left(\Gamma(A_0) + \sum_{k=1}^{m} \Pi(A_k)\right) < 0, \tag{7.23}$$

then S is delay-independent stable.

Proof Note first that the Metzler matrix in (7.23) coincides with $\sum_{k=0}^{m} \bar{A}_k$, where $\bar{A}_k$ are the matrices of the IPR of Theorem 7.1. Thus, if condition (7.23) is satisfied, then the IPR of S is stable, and thanks to Theorem 7.4 the original system S is delay-independent stable. □

Remark 7.3 As pointed out in Sect. 7.3.2, checking condition (7.23) does not require the explicit computation of the eigenvalues of the Metzler matrix $\Gamma(A_0) + \sum_{k=1}^{m} \Pi(A_k)$. Indeed, an easy equivalent condition only requires to check that all the leading principal minors of $M = -\left(\Gamma(A_0) + \sum_{k=1}^{m} \Pi(A_k)\right)$ are positive.

7.4 Comparison with Similar Conditions of Delay-Independent Stability

Many stability conditions exist for delay systems with multiple delays, based on different techniques: frequency sweeping [5], spectral analysis [20], Linear Matrix Inequalities [7, 10] and others (see [11]). These results refer to different cases such as commensurate or incommensurate delays, constant or time-varying delays, slowly or fast varying delays. Many stability tests rely on numerical computations and some have a not negligible computational complexity (particularly the necessary and sufficient ones). Coming to delay-independent stability, in [21] and [22], for the case of single and constant delay, the following sufficient condition for delay-independent stability has been given

$$\mu_p(A_0) + \|A_1\|_p < 0 \tag{7.24}$$

where $\mu_p(A)$ is the logarithmic norm (or measure) of matrix A induced by the operator norm $\|A\|_p$, defined as:

$$\mu_p(A) = \lim_{\varepsilon \to 0} \frac{\|I + \varepsilon A\|_p - 1}{\varepsilon}.$$

The expression of $\mu_p(\cdot)$ can easily be computed for $p = 1, 2, \infty$:

$$\mu_1(A) = \max_{j=1...n}\left(a_{ii} + \sum_{i=1,\ i\neq j}^{n} |a_{ij}|\right),$$

$$\mu_2(A) = \frac{1}{2}\lambda_{max}(A^T + A),$$

$$\mu_\infty(A) = \max_{i=1...n}\left(a_{ii} + \sum_{j=1,\ j\neq i}^{n} |a_{ij}|\right).$$

The extended condition:

$$\mu_p(A_0) + \sum_{k=1}^{m} \|A_k\|_p < 0 \tag{7.25}$$

has been shown [26] to be sufficient for the stability of systems with multiple commensurate delays, although only for the case of $p = 2$. In [25] and [24] the same condition has been proven sufficient, for any p, also in the case of non commensurate and time-varying delays of any size, and therefore is a sufficient condition of delay-independent stability of the system.

As a matter of fact, it is rather easy to find delay-independent stable systems which satisfy condition (7.23) given in Theorem 7.5 and do not satisfy condition (7.25): an example is reported in Sect. 7.5. Further investigations are needed to compare the conservativeness of the new condition with respect to the classical one.

7.5 Example

Consider the problem of verifying the delay-independent asymptotic stability of a system $S = \big\{\{A_0, A_1, A_2\}, B, C, D\big\}_{3,p,q}$ with:

$$A_0 = \begin{bmatrix} -25 & -5 & -14 \\ 0 & -19 & 0.1 \\ 0.7 & 1.2 & -16 \end{bmatrix}, \quad A_1 = \begin{bmatrix} -1.5 & -0.4 & 0 \\ 0.5 & -2.9 & 1 \\ -1.5 & 0.5 & -3.4 \end{bmatrix}, \quad A_2 = \begin{bmatrix} -7 & 2 & 6.8 \\ 1.8 & -1.6 & -2.1 \\ 0.5 & 1.6 & -3.3 \end{bmatrix}$$

Since S is not an internally positive system, (7.22) is only a necessary condition for its delay-independent stability (see Remark 7.2). We have that:

$$\alpha\left(\sum_{k=0}^{m} A_k\right) = -23.131 < 0$$

and therefore condition (7.22) is satisfied. Hence we can check the proposed sufficient condition (7.23), verifying that all the leading principal minors of the matrix

$$M = -\big(\Gamma(A_0) + \Pi(A_1) + \Pi(A_2)\big)$$

are positive (Remark 7.3). We get:

$$M_1 = 25, \quad M_2 = 470.4, \quad M_3 = 7.2 \cdot 10^3, \quad M_4 = 1.4 \cdot 10^5,$$
$$M_5 = 2.3 \cdot 10^6, \quad M_6 = \det(M) = 1.5 \cdot 10^7$$

and this is sufficient to conclude that the system is delay-independent stable.

Actually, the exact value of condition (7.23) is:

$$\alpha\Bigg(\Gamma(A_0) + \Pi(A_1) + \Pi(A_2)\Bigg) = -2.436.$$

It is not possible to achieve the same conclusion on the stability of the system applying the classical sufficient condition (7.25), since:

$$\mu_1(A_0) + \|A_1\|_1 + \|A_2\|_1 = 14.700 > 0,$$
$$\mu_2(A_0) + \|A_1\|_2 + \|A_2\|_2 = 2.896 > 0,$$
$$\mu_\infty(A_0) + \|A_1\|_\infty + \|A_2\|_\infty = 15.200 > 0,$$

and therefore the condition is not satisfied at least for $p = 1, 2, \infty$.

To sum up, for the system in this example the criterion (7.25) fails to assess the stability, which has been proved using the proposed condition (7.23).

Figure 7.1 depicts some examples of time evolution of $\log(\|x(t)\|)$ obtained with $u(t) = 0$ for $t \in [0, 200]$ and different constant values of the two delays. In all cases, the plotted quantity decreases linearly, thus confirming the asymptotic stability, which in the case of constant delays is also exponential.

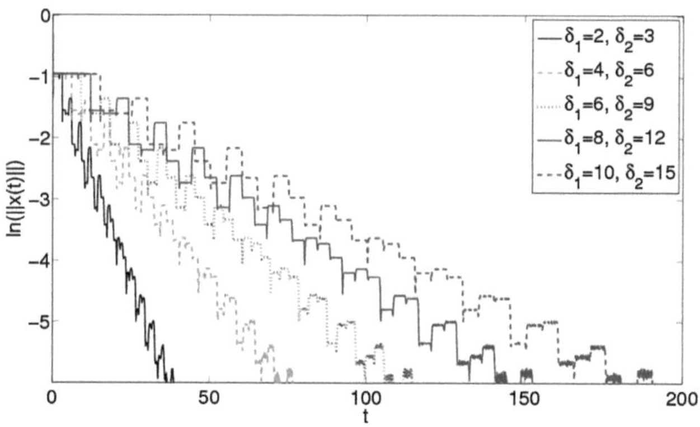

Fig. 7.1 Plot of $\log(\|\mathbf{x(t)}\|)$ with different constant delays values

7.6 Conclusions and Future Work

In this chapter the Internally Positive Representation of linear delay systems with multiple delays, possibly time-varying, has been introduced, and its consequences on the study of the stability of the original system have been investigated, leading to an easy-to-check sufficient condition whose efficacy with respect to the delay-independent stability tests provided in [21, 24, 25] has been tested by means of numerical examples. Future work will be devoted to further stability analysis and to the extension of the IPR technique to other classes of delay systems.

Acknowledgements We would like to thank Alfredo Germani and Filippo Cacace for their encouragement and helpful suggestions in doing this work.

References

1. Ait Rami, M.: Positive Systems: Proceedings of the third Multidisciplinary International Symposium on Positive Systems: Theory and Applications (POSTA 2009) Valencia, Spain, September 2-4, 2009, pp. 205–215. Springer, Berlin (2009)
2. Cacace, F., Farina, L., Germani, A., Manes, C.: Internally positive representation of a class of continuous time systems. IEEE Trans. Autom. Control **57**(12), 3158–3163 (2012)
3. Cacace, F., Germani, A., Manes, C.: Stable internally positive representations of continuous time systems. IEEE Trans. Autom. Control **59**(4), 1048–1053 (2014)
4. Cacace, F., Germani, A., Manes, C., Setola, R.: A new approach to the internally positive representation of linear MIMO systems. IEEE Trans. Autom. Control **57**(1), 119–134 (2012)
5. Chen, J., Latchman, H.A.: Frequency sweeping tests for stability independent of delay. IEEE Trans. Autom. Control **40**(9), 1640–1645 (1995)
6. Farina, L., Rinaldi, S.: Positive Linear Systems: Theory and Applications, vol. 50. Wiley (2011)
7. Fridman, E.: Introduction to Time-Delay Systems. Birkhuser Basel (2014)
8. Germani, A., Manes, C., Palumbo, P.: State space representation of a class of MIMO systems via positive systems. In: 2007 46th IEEE Conference on Decision and Control, pp. 476–481. IEEE (2007)
9. Germani, A., Manes, C., Palumbo, P.: Representation of a class of MIMO systems via internally positive realization. Eur. J. Control **16**(3), 291–304 (2010)
10. Gu, K., Kharitonov, V.L., Chen, J.: Stability of Time-Delay Systems. Birkhuser Basel (2003)
11. Gu, K., Niculescu, S.I.: Survey on recent results in the stability and control of time-delay systems. J. Dyn. Syst. Meas. Control **125**(2), 158–165 (2003)
12. Haddad, W.M., Chellaboina, V.: Stability theory for nonnegative and compartmental dynamical systems with time delay. Syst. Control Lett. **51**(5), 355–361 (2004)
13. Hale, J., Lunel, S.M.V.: Introduction to Functional Differential Equations. Springer, New York (1993)
14. Kaczorek, T.: Positive 1D and 2D Systems. Springer, London, UK (2001)
15. Kaczorek, T.: Stability of positive continuous-time linear systems with delays. In: 2009 European Control Conference (ECC), pp. 1610–1613. IEEE (2009)
16. Liu, X., Lam, J.: Relationships between asymptotic stability and exponential stability of positive delay systems. Int. J. Gen. Syst. **42**(2), 224–238 (2013)
17. Liu, X., Yu, W., Wang, L.: Stability analysis for continuous-time positive systems with time-varying delays. IEEE Trans. Autom. Control **55**(4), 1024–1028 (2010)
18. Luenberger, D.: Introduction to Dynamic Systems: Theory, Models, and Applications. Wiley (1979)

19. Mazenc, F., Malisoff, M.: Stability analysis for time-varying systems with delay using linear Lyapunov functionals and a positive systems approach. IEEE Trans. Autom. Control **61**(3), 771–776 (2016)
20. Michiels, W., Niculescu, S.I.: Stability, Control, and Computation for Time-Delay Systems: An Eigenvalue-Based Approach, vol. 27. Siam (2014)
21. Mori, T., Fukuma, N., Kuwahara, M.: Simple stability criteria for single and composite linear systems with time delays. Int. J. Control **34**(6), 1175–1184 (1981)
22. Mori, T., Fukuma, N., Kuwahara, M.: On an estimate of the decay rate for stable linear delay systems. Int. J. Control **36**(1), 95–97 (1982)
23. Ngoc, P.H.A.: Stability of positive differential systems with delay. IEEE Trans. Autom. Control **58**(1), 203–209 (2013)
24. Schoen, G.M.: Stability and stabilization of time-delay systems. Ph.D. thesis, Swiss Federal Institute of Technology, Zurich, Switzerland (1995). http://e-collection.library.ethz.ch
25. Schoen, G.M., Geering, H.P.: On stability of time delay systems. In: Proceedings of the 31st Annual Allerton Conference on Communications Control and Computing, pp. 1058–1060. Monticello, Il (1993)
26. Wang, S.S., Lee, C.H., Hung, T.H.: New stability anlysis of system with multiple time delays. In: American Control Conference (ACC 1991), pp. 1703–1704 (1991)

Part III
Switched and Fractional Positive Systems

Chapter 8
On Robust Pseudo State Estimation of Fractional Order Systems

Tarek Raïssi and Mohamed Aoun

Abstract The goal of this chapter is to design robust observers for fractional dynamic continuous-time linear systems described by pseudo state space representation. The fractional observer is guaranteed to compute a domain enclosing all the system pseudo states that are consistent with the model, the disturbances and the measurement noise realizations. Uncertainties on the initial pseudo state and noises are propagated in a reliable way to estimate the bounds of the fractional pseudo state. Only the bounds of the uncertainties are used and no additional assumptions about their stationarity or ergodicity are taken into account. A fractional observer is firstly built for a particular case where the estimation error can be designed to be positive. Then, the general case is investigated through changes of coordinates. Some numerical simulations illustrate the proposed methodology.

Keywords Fractional systems · Interval observers · Robust estimation

8.1 Introduction

Fractional differentiation is an extension of classical integer differentiation to deal with non-integer (fractional) orders. It was defined in the 19th century by Riemann and Liouville, see for instance [13]. First applications on automatic control are cited since 1945 by Bode [9] and subsequently in [24, 34, 45].

Nowadays, fractional calculus is widely used in many engineering fields [5, 12, 16, 18, 22, 23, 32, 33, 35, 39]. It is a powerful mathematical tool for the description of long memory and hereditary properties of various materials and processes. For

T. Raïssi (✉)
Conservatoire National des Arts et Métiers, Cedric-Lab, 292, Rue St-Martin, 75141 Paris, France
e-mail: tarek.raissi@cnam.fr

M. Aoun
Research Laboratory Modeling, Analysis and Control Systems, National Engineering School of Gabes, Street Omar Ibn El Khattab - Zerig, 6029 Gabes, Tunisia
e-mail: mohamed.aoun@gmail.com

F. Cacace et al. (eds.), *Positive Systems*, Lecture Notes in Control and Information Sciences 471, DOI 10.1007/978-3-319-54211-9_8

instance, in electrochemistry, diffusion processes of charges in acid batteries is governed by Randles models with an inherent fractional 0.5 derivatives of order [38]. Supercapacitor, which are highly energy device storage, are modeled with fractional integrator [30, 46]. In thermal diffusion, it is shown that the solution of the heat equation of a semi-infinite homogeneous medium depends on 0.5 order derivative [7]. Diffusion phenomena in semi-infinite planar, spherical and cylindrical media deals with a multiple of 0.5 differentiation order [31]. Experimental results prove that fractional models are appropriate to represent vibrations on viscoelastic materials [41]. The electromagnetic fields in dielectric media is described by a model with fractional differentiation [6, 43].

Some methods to estimate the pseudo states of fractional systems have been developed in the literature. For instance, fractional Kalman filters have been proposed for both discrete and continuous-time systems [1, 4, 21, 42]. Luenberger-based fractional observers have been also investigated in [14].

The main drawback of these techniques design is the difficulty to take into account uncertainties (unknown parameters or/and external disturbances). In the presence of uncertainty, design of conventional observers/filters, converging to the ideal value of the pseudo state is difficult to achieve. In such context, interval observers can be considered as an alternative. The latters do not permit to compute only an approximation but the set of all admissible values is characterized at each time instant. The width of the estimated domain should be proportional to the size of the uncertainties. With respect to conventional observers, the mid-value can be considered as a point estimation while the interval width is the uncertainty/deviation from such point value.

The theory of interval observers is well developed in the context of integer differentiation systems. In this chapter, such methodology is extended to fractional differentiation systems based on the theory of positive systems. It will be shown that, under some mild conditions, an interval observer can be developed for any linear fractional system subject to bounded noises and disturbances. To the best of our knowledge, it is the first time that interval observers are considered for this class of systems.

The chapter is organized as follows: some properties of fractional systems are recalled in Sect. 8.2. The main contribution is given in Sect. 8.3 where two observers are proposed for a particular case and also for general fractional linear systems. Finally, some numerical simulations are presented in Sect. 8.4 to illustrate this methodology.

8.2 Fractional Systems

Riemann and Liouville extended differentiation by using not only integer but also non-integer orders (fractional order). The γth fractional order differentiation of a continuous real function $f(t)$ is defined as [29]:

$$D^{\gamma} f(t) = \frac{1}{\Gamma(1-\gamma)} \left(\frac{d}{dt}\right)^{\lfloor \gamma+1 \rfloor} \int_0^t \frac{f(\tau)}{(t-\tau)^{\gamma}} d\tau \tag{8.1}$$

In the field of engineering sciences another definition of fractional differentiation has been proposed by Caputo [10]:

$$D^{\gamma} f(t) = \frac{1}{\Gamma(\lceil \gamma \rceil - \gamma)} \int_0^t \frac{f^{(\lceil \gamma \rceil)}(\tau)}{(t-\tau)^{1-\lceil \gamma \rceil + \gamma}} d\tau \tag{8.2}$$

where $f^{(\lceil \gamma \rceil)}(\tau)$ denotes the integer derivative at $(\lceil \gamma \rceil)$ of f.

The fractional differentiation can be numerically evaluated using the Grünwald approximation [3]:

$$D^{\gamma} f(t) \simeq \frac{1}{h^{\gamma}} \sum_{k=0}^{\infty} (-1)^k \binom{\gamma}{k} f(t-kh) \tag{8.3}$$

where h is a small real number and

$$\binom{\gamma}{k} = \frac{\Gamma(\gamma+1)}{k!\Gamma(\gamma-k+1)} \tag{8.4}$$

A continuous-time fractional linear system can be described with a fractional differential equation:

$$\sum_{i=0}^{n_y} a_i D^{\alpha_i} y(t) = \sum_{j=0}^{n_u} b_j D^{\beta_j} u(t) \tag{8.5}$$

where u and y denote respectively the system input and output. The fractional differentiation orders $\alpha_i, i = 0 \dots n_y$ and $\beta_i, i = 0 \dots n_u$ are positive real numbers. Generally they are assumed to be rational, thus a commensurate fractional differential equation can be obtained:

$$\sum_{l=0}^{n'_y} a'_l D^{l\nu} y(t) = \sum_{k=0}^{n'_u} b'_k D^{k\nu} u(t) \tag{8.6}$$

where the input and the output are differentiated to integer multiple of the commensurate order ν. From (8.6) the following representation can be deduced [36]:

$$\begin{cases} x^{\nu}(t) = Ax(t) + Bu(t) \\ y(t) = Cx(t) + Du(t) \end{cases} \tag{8.7}$$

where A, B, C and D are constant matrices with $A \in \mathbb{R}^{n \times n}$, $B \in \mathbb{R}^{n \times p}$, $C \in \mathbb{R}^{m \times n}$ and $D \in \mathbb{R}^{m \times p}$. For single input single output systems ($m = p = 1$), the vector x is called a pseudo state and x^{ν} denotes its fractional derivative at order ν, $0 < \nu \leq 1$.

The variable x in (8.7) is not rigourously a state similar to the integer differentiation context and it has been shown in [44] that the dimension of the actual state of fractional systems is infinite. Indeed, the knowledge of x in (8.7) at t and all input values u over an arbitrary interval $[t, t + \Delta t]$ is not sufficient to compute the state at $t + \Delta t$. However, the representation (8.7) is widely used when dealing with fractional systems since the pseudo state is sufficient for modelling, control and simulation purposes. Roughly speaking, in the following, (8.7) will be called fractional state space representation and x a state.

The system described by (8.7) is stable when all eigenvalues of A verify [2, 25]:

$$|arg(spec(A))| > \nu\frac{\pi}{2} \tag{8.8}$$

In the following, a matrix M is called stable if its eigenvalues satisfy the condition (8.8).

The observability of fractional systems has been discussed in several papers [8, 17, 26, 27, 40] and a necessary and sufficient rank condition similar to the case of integer systems has been given in [17]:

$$rank\begin{pmatrix} C \\ CA \\ \vdots \\ CA^{n-1} \end{pmatrix} = n \tag{8.9}$$

In the following and without any loss of generality, we will suppose that $D = 0$. A classical fractional observer structure for the estimation of x is given by:

$$\begin{cases} \hat{x}^{\nu}(t) = A\hat{x}(t) + Bu(t) + L\left(y(t) - \hat{y}(t)\right) \\ \hat{y}(t) = C\hat{x}(t) \end{cases} \tag{8.10}$$

where $\hat{x}$ denotes the estimated state and L is the observer gain. The estimation error is given by:

$$\tilde{x}^{\nu}(t) = \hat{x}^{\nu}(t) - x^{\nu}(t) = (A - LC)(\hat{x}(t) - x(t)) \tag{8.11}$$

To ensure the convergence of the estimation error, the system (8.11) should verify the stability condition (8.8). The observer gain L is chosen such that:

$$|arg(spec(A - LC))| > \nu\frac{\pi}{2} \tag{8.12}$$

Note that the observer (8.10) converges asymptotically provided that the system (8.7) is not subject to noises and disturbances. Otherwise, the results can be unreliable. In the following, a robust approach is proposed to compute not only an approximation of the state but an interval which is guaranteed to enclose all the values of x consistent with the assumptions on the noises and disturbances. To the best of our knowledge, it is the first tentative to investigate this methodology for fractional systems.

A dynamical system is called internally positive if starting from any nonnegative condition and for any nonnegative input, its state remains always positive [19]. Furthermore, a matrix $A \in \mathbb{R}^{n\times n}$ is called Metzler if all its off-diagonal entries are nonnegative, i.e. $A = \{a_{i,j}\},\ a_{i,j} \geq 0, \forall i \neq j$.

Lemma 8.1 [20] *The fractional system described by (8.7) with $\nu \leq 1$ and $x(0) \geq 0$*[1] *is internally positive if and only if A is Metzler and all coefficients of the matrices B and C are nonnegative.*

Lemma 8.2 [15] *Given a vector $\sigma(t) \in \mathbb{R}^n$ verifying $\underline{\sigma}(t) \leq \sigma \leq \overline{\sigma}(t)$ for two vectors $\underline{\sigma}(t), \overline{\sigma}(t) \in \mathbb{R}^n$. Then,*

$$M^{+}\underline{\sigma}(t) - M^{-}\overline{\sigma}(t) \leq M\sigma(t) \leq M^{+}\overline{\sigma}(t) - M^{-}\underline{\sigma}(t) \tag{8.13}$$

8.3 Main Results

Given a matrix $M \in \mathbb{R}^{m\times n}$ and define $M^{+} = max(0, M)$ and $M^{-} = M^{+} - M$. $|M| = M^{+} + M^{-}$ is the matrix of absolute values of all elements of M.

8.3.1 Design of a Fractional Interval Observer

Consider the noisy fractional system

$$\begin{cases} x^{\nu}(t) = Ax(t) + Bu(t) + Gw(t) \\ y(t) = Cx(t) + v(t) \end{cases} \tag{8.14}$$

with $\nu \leq 1$. The input $u(t)$ is known and A, B,C and G are constant matrices. $w(t)$ and $v(t)$ are some bounded disturbances and noises.

In the context of interval observers, the goal is to derive two trajectories $\underline{x}(t)$ and $\overline{x}(t)$ such that, starting from some initial conditions $\overline{x}_0 \leq x_0 \leq \underline{x}_0$, we have:

$$\overline{x}(t) \leq x(t) \leq \underline{x}(t), \quad \forall t \geq t_0$$

The following theorem gives a first result for the design of interval observers for (8.14).

Theorem 8.1 *Given the system (8.14) with the initial condition x_0 satisfying $\underline{x}_0 \leq x_0 \leq \overline{x}_0$ for $\underline{x}_0, \overline{x}_0 \in \mathbb{R}^n$. Assume that the noises and disturbances are bounded,*

---

[1]The order relations $<, \leq, >, \geq$ should be understood componentwise throughout this chapter.

i.e. $|v(t)| \le V$, $|w(t)| \le W$. *If there exists a gain L such that* $A - LC$ *is Metzler and* $|arg(spec(A - LC))| > \nu\frac{\pi}{2}$, *then, the system (8.15) is an interval observer for (8.14):*

$$\begin{cases} \underline{x}^{\nu}(t) = (A - LC)\,\underline{x}(t) + Bu(t) + \underline{b}(t),\ \underline{x}(t_0) = \underline{x}_0 \\ \overline{x}^{\nu}(t) = (A - LC)\,\overline{x}(t) + Bu(t) + \overline{b}(t),\ \overline{x}(t_0) = \overline{x}_0 \end{cases} \tag{8.15}$$

with

$$\begin{cases} \underline{b}(t) = -|G|W + Ly(t) - |L|V \\ \overline{b}(t) = |G|W + Ly(t) + |L|V \end{cases} \tag{8.16}$$

Proof Consider the observer error $\overline{e}_x = \overline{x} - x$. Based on (8.14) and (8.15), the dynamics of $\overline{e}_x$ is described by:

$$\begin{aligned} \overline{e}_x^{\nu}(t) &= (A - LC)\overline{x} + Bu(t) + |G|W + Ly + |L|V - (Ax(t) + Bu(t) + Gw(t)) \\ &= (A - LC)\overline{e}_x + (|L|V + Lv(t)) + (|G|W - Gw(t)) \end{aligned} \tag{8.17}$$

Since the gain L is designed such that $(A - LC)$ is Metzler and by construction $|L|V + Lv(t) \ge 0$, $|G|W - Gw(t) \ge 0$, then the dynamics of $\overline{e}_x$ is positive, i.e. $\overline{e}_x = \overline{x} - x \ge 0, \forall t \ge t_0$. In addition, it is assumed that the gain L is chosen such that $A - LC$ is stable (i.e. $|arg(spec(A - LC))| > \nu\frac{\pi}{2}$), thus the upper error $\overline{e}_x$ is stable. Similarly, the same methodology can be followed to prove that $\underline{e}_x = x - \underline{x} \ge 0, \forall t \ge t_0$ and that $\underline{e}_x$ is stable. To conclude, it has been proven that the observation errors are stable and that:

$$\underline{x}(t) \le x(t) \le \overline{x}(t), \quad \forall t \ge t_0. \tag{8.18}$$

□

Note here that the observability is a sufficient condition (however, the detectability is necessary and sufficient) for the existence of a gain L ensuring the stability of both $\underline{e}_x$ and $\underline{e}_x$. In practice, computing a gain L satisfying both conditions of Theorem 8.1 is not obvious and may be impossible in some cases. To overcome this problem, some changes of coordinates can be used to generalize the previous result.

8.3.2 General Case

Usually, it is not possible to find a gain L such that $A - LC$ is simultaneously Metzler and stable. Furthermore, the eigenvalues of the matrix $A - LC$ are preserved under a change of coordinates. In this section, we propose a procedure to overcome this concern by computing a gain L such that $A - LC$ is stable and a nonsingular transformation matrix $P \in \mathbb{R}^{n \times n}$ such that, in the new coordinates $z = Px$, the matrix $\Gamma = P(A - LC)P^{-1}$ is Metzler. The conditions of existence of such a real transformation matrix P has been established by the following lemma.

Lemma 8.3 [37] *Given the matrices $A \in \mathbb{R}^{n\times n}$, $R \in \mathbb{R}^{n\times n}$ and $C \in \mathbb{R}^{p\times n}$. If there is a gain $L \in \mathbb{R}^{n\times p}$ such that the matrices $A - LC$ and R have the same eigenvalues, then there exists a matrix $P \in \mathbb{R}^{n\times n}$ such that $R = P(A - LC)P^{-1}$ provided that the pairs $(A - LC, e_1)$ and (R, e_2) are observable for some $e_1 \in \mathbb{R}^n$, $e_2 \in \mathbb{R}^n$.*

This result was used in [37] to design interval observers for integer linear time invariant systems with a Metzler matrix R.

Furthermore, based on the Jordan form, it has been shown in [28] that it is usually possible to design a transformation $z = Px$ such that $A - LC$ is Metzler. When the eigenvalues of $A - LC$ are real, the matrix P is constant, otherwise, it is time-varying. A similar methodology has been developed in [11] where the complex-valued transformations are used.

In the following, given a gain L such that $A - LC$ is stable and consider a change of coordinates $z(t) = Px(t)$ such that $P(A - LC)P^{-1}$ is Metzler. The matrix P can be computed using the Lemma 8.3 or the Jordan form investigated in [11, 28]. Therefore, an interval observer structure for (8.7) in the coordinated z and x is given in the following theorem.

Theorem 8.2 *Given (8.14) with the initial condition x_0 satisfying $\underline{x}_0 \le x_0 \le \overline{x}_0$ for $\underline{x}_0, \overline{x}_0 \in \mathbb{R}^n$. Assume that the noises and disturbances are bounded, i.e. $|v(t)| \le V$, $|w(t)| \le W$. Suppose also that P is chosen following Lemma 8.3 and L such that the stability condition (8.23) is verified. Then, the system (8.19), initialized with (8.21), is an interval observer for (8.14) in the coordinates $z = Px$.*

$$\begin{cases} \underline{z}^{\nu}(t) = \Gamma \underline{z}(t) + PBu(t) + \underline{b}(t) \\ \overline{z}^{\nu}(t) = \Gamma \overline{z}(t) + PBu(t) + \overline{b}(t) \end{cases} \tag{8.19}$$

with

$$\Gamma = P\,(A - LC)\,M, \quad M = P^{-1} \tag{8.20}$$

$$\begin{cases} \underline{z}(0) = P^{+}\underline{x}_0 - P^{-}\overline{x}_0 \\ \overline{z}(0) = P^{+}\overline{x}_0 - P^{-}\underline{x}_0 \end{cases} \tag{8.21}$$

$$\begin{cases} \underline{b}(t) = -|PG|W + PLy(t) - |PL|V \\ \overline{b}(t) = |PG|W + PLy(t) + |PL|V \end{cases} \tag{8.22}$$

$$|arg(spec(\Gamma))| > \nu\frac{\pi}{2} \tag{8.23}$$

In addition, an interval estimation of (8.14), in the coordinates x, is given by (8.24):

$$\begin{cases} \underline{x}(t) = M^{+}\underline{z}(t) - M^{-}\overline{z}(t) \\ \overline{x}(t) = M^{+}\overline{z}(t) - M^{-}\underline{z}(t) \end{cases} \tag{8.24}$$

Proof The system (8.14) can be rewritten as:

$$z^{\nu}(t) = \Gamma z(t) + PBu(t) + PLCP^{-1}z(t) + PGw(t). \tag{8.25}$$

Furthermore, according to (8.19) the dynamics of $\overline{z}$ is given by:

$$\begin{aligned} \overline{z}^{\nu}(t) &= \Gamma\overline{z}(t) + PBu(t) + \overline{b}(t) \\ &= \Gamma\overline{z}(t) + PBu(t) + |PG|W + PLy(t) + |PL|V \\ &= \Gamma\overline{z}(t) + PBu(t) + |PG|W + PLCP^{-1}z(t) \\ &\quad + PLv(t) + |PL|V. \end{aligned} \tag{8.26}$$

Consider now the observer error $\overline{e}_z = \overline{z} - z$. Based on (8.25) and (8.26), the dynamics of $\overline{e}_z$ is described by:

$$\begin{aligned} \overline{e}_z^{\nu}(t) &= \Gamma\overline{z}_z(t) + PBu(t) + |PG|W + PLCP^{-1}z(t) \\ &\quad + PLv(t) + |PL|V \\ &\quad - (\Gamma z(t) + PBu(t) + PLCP^{-1}z + PGw(t) \\ &= \Gamma\overline{e}_z(t) + |PL|V + PLv(t) + |PG|W - PGw(t) \end{aligned} \tag{8.27}$$

Recall that the matrix $\Gamma = P(A - LC)P^{-1}$ is Metzler and by construction $|PL|V + PLv(t) \geq 0$, $|PG|W - PGw(t) \geq 0$, therefore the dynamics of $\overline{e}_z$ is positive, i.e. $\overline{e}_z = \overline{z} - z \geq 0, \forall t \geq t_0$.

In addition, it is assumed that the gain L is chosen such that $A - LC$ (and consequently Γ) is stable, thus the upper error $\overline{e}_z$ is stable.

Moreover, the same methodology can be followed to prove that $\underline{e}_z = z - \underline{z} \geq 0, \forall t \geq t_0$ and that $\underline{e}_z$ is stable.

Now, based on Lemma 8.2, it is trivial to show that:

$$\underline{x} = M^{+}\underline{z} - M^{-}\overline{z} \leq Mz = x \leq M^{+}\overline{z} - M^{-}\underline{z} = \overline{x}.$$

In addition, the stability of $x - \underline{x}$ and $\overline{x} - x$ are deduced from that of $\underline{e}_z$ and $\overline{e}_z$ since such property is preserved under changes of coordinates.

8.4 Numerical Simulations

8.4.1 Example 1

Given a system described by:

$$\begin{cases} x^{\nu}(t) = Ax(t) + Bu(t) \\ y(t) = Cx(t) + v(t) \end{cases} \tag{8.28}$$

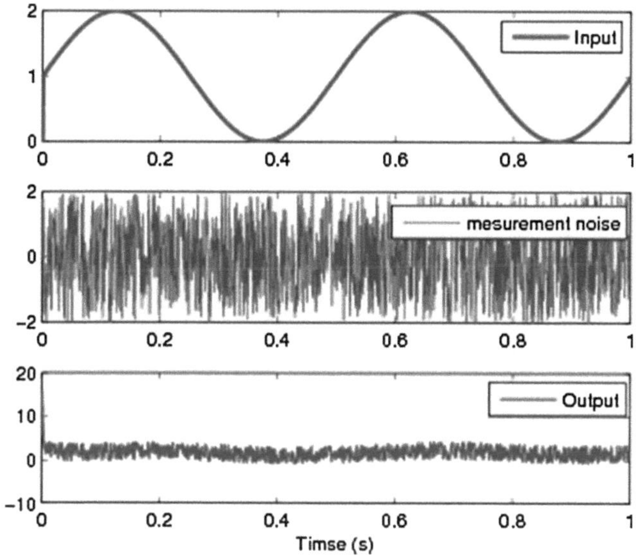

Fig. 8.1 Input, output and measurement noise for the system (8.28)

where:

$$A = \begin{bmatrix} -5 & 2 \\ 6 & -3 \end{bmatrix}, B = \begin{bmatrix} 1 \\ 1 \end{bmatrix}, C = \begin{bmatrix} 1 & 1 \end{bmatrix}$$

and the commensurate order is $\nu = 0.5$. $v(t)$ is a bounded noise such that $|v(t)| \leq V = 0.1$. The initial state is arbitrarily chosen as $(5, 10)^T$ and is supposed to be affected by 50% of uncertainty. Note that uncertainty on the initial state may model the insufficient information about the past of the system. The input and the output of the system are plotted on Fig. 8.1.

The pair (A, C) verifies the observability condition (8.9) and there exists a gain L verifying (8.12):

$$|arg(spec(A - LC))| > 0.5\frac{\pi}{2} \tag{8.29}$$

The gain $L = (0.12, 0.27)^T$ is used, it allows to the eigenvalues of $A - LC$ to be the same as those of A except the largest one which is multiplied by 4, i.e. $spec(A - LC) = \{-7.61, -1.58\}$. For the chosen gain L, the matrix

$$A - LC = \begin{bmatrix} -5.36 & 1.64 \\ 5.17 & -3.83 \end{bmatrix} \tag{8.30}$$

is Metzler. Therefore, the estimation error is positive and the fractional interval observer is designed according to (8.15). The actual state and its lower and upper bounds are plotted on Fig. 8.2. Clearly, the robustness is shown through this numerical example.

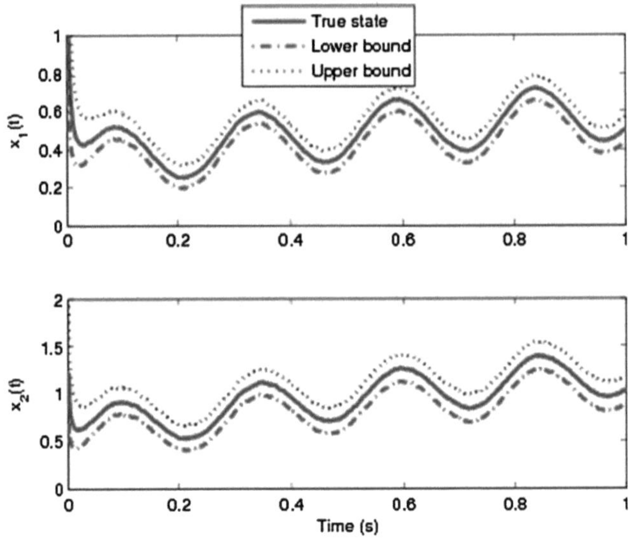

Fig. 8.2 The actual states and their lower and upper bounds for the system (8.28)

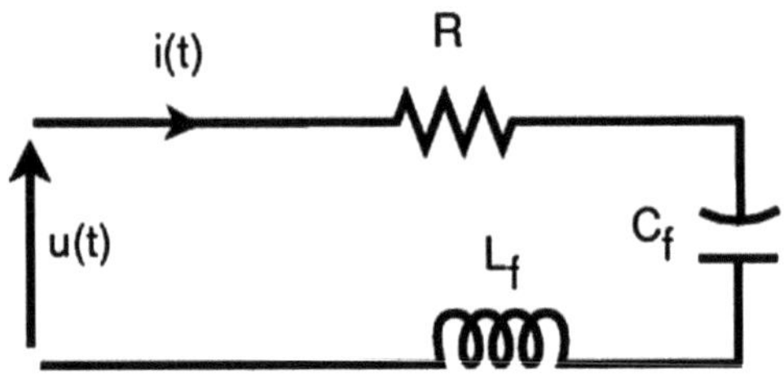

Fig. 8.3 Fractional electrical circuit

8.4.2 Example 2

Consider an electrical circuit to illustrate the design of a fractional interval observer in the general case. The fractional electrical circuit is given on Fig. 8.3 where R is the resistance, C_f is a fractional order supercapacitor and L_f is a fractional order inductance [20]. Analysing the circuit with the Kirchhoffs laws we obtain the fractional differential equations:

$$i(t) = C_f \frac{d^{\alpha} u_c(t)}{dt} \tag{8.31}$$

and

$$u(t) = Ri(t) + u_c(t) + L_f \frac{d^{\beta} i(t)}{dt} \tag{8.32}$$

Assuming that $\nu = \alpha = \beta$ and considering that only $u_c(t)$ is measured, the following fractional state space representation can be obtained:

$$\begin{cases} \begin{bmatrix} u_c(t) \\ i(t) \end{bmatrix}^{\nu} = \begin{bmatrix} 0 & 1/C_f \\ -1/L_f & -R/L_f \end{bmatrix} \begin{bmatrix} u_c(t) \\ i(t) \end{bmatrix} + \begin{bmatrix} 0 \\ 1/L_f \end{bmatrix} u(t) + w(t) \\ y(t) = u_c(t) + v(t) \end{cases} \tag{8.33}$$

where $w(t)$ and $v(t)$ are some unknown additive disturbances and noises. For simulation, the following numerical values are chosen:

$$R = 20\,\Omega \quad C_f = 600\,\mu\text{F},$$

$$L_f = 30\,\text{mH}, \nu = \alpha = \beta = 0.5$$

The observability rank condition (8.9) is verified and the gain

$$L = \begin{bmatrix} 503.3333 \\ -29.4667 \end{bmatrix}$$

is used. Thus, the closed-loop matrix is given by:

$$A - LC = \begin{bmatrix} -0.5033 & 1.6667 \\ -0.0039 & -0.6667 \end{bmatrix} 10^3 \tag{8.34}$$

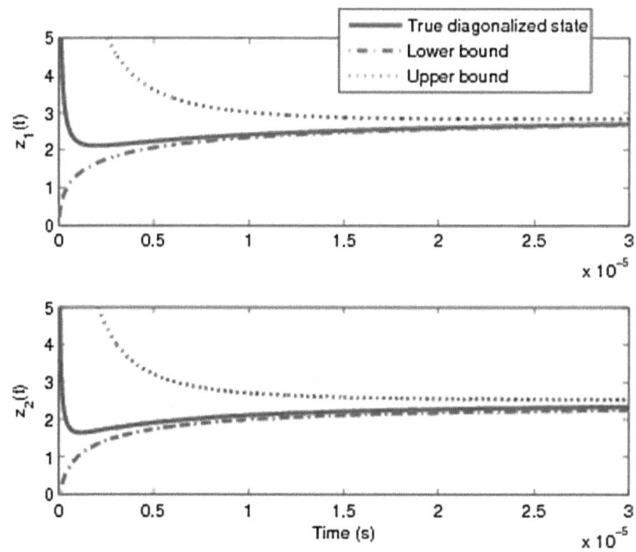

Fig. 8.4 Diagonalized states and their lower and upper bounds

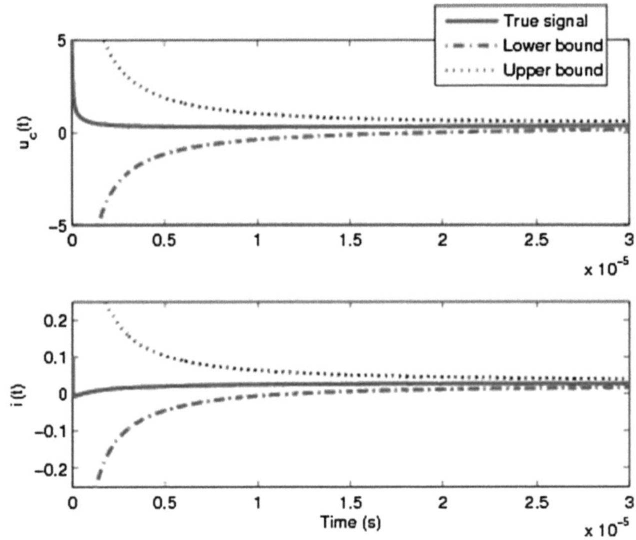

Fig. 8.5 u_c and $i(t)$ and their lower and upper bounds

Note that $(A - LC)$ is not Metzler. Therefore, the fractional interval observer of Theorem 8.1 cannot be applied. However, using a change of coordinates (a diagonalization of $A - LC$ in this case), the interval observer of Theorem 8.2 permits to estimate the lower and the upper bounds of the state in the coordinates z and also in the initial ones ($u_c(t)$ and $i(t)$).

Noises $w(t)$ and $v(t)$ are supposed to be bounded with $|W| = |V| = 0.1$. The initial state is chosen as $(u_c(0) = 2, i(0) = 0.5)^T$ and is supposed to be affected by large uncertainty, i.e.:

$$\underline{u}_c(0) = 0 \quad \overline{u}_c(0) = 20$$

$$\underline{i}(0) = 0 \quad \overline{i}(0) = 5$$

Applying Theorem 8.2, the estimated bounds of the state in the coordinates z are plotted on Fig. 8.4. Those of $u_c(t)$ and $i(t)$ are plotted on Fig. 8.5.

8.5 Conclusion

The design of interval observers for fractional differentiation systems is investigated in this work. Under some mild conditions (boundedness of noises and disturbances, observability), two Luenberger-based observers allow one to compute reliable bounds for the state values consistent with the bounds of the uncertainties. A first result is given to build interval observers when it is possible to design a gain L satisfying the

Metzler and stability properties of $A - LC$. In addition, by using a change of coordinates, a general result, which can be applied to any linear fractional differentiation system, is proposed. An extension of this approach to address the case of parameter uncertainties and time-varying systems will be the subject of further works.

Acknowledgements This work was developed within the "Research in Paris" project supported by the city of Paris.

References

1. Abdelhamid, M., Aoun, M., Najar, S., Abdelhamid, M.N.: Discrete fractional kalman filter. Intell. Control Syst. Signal Process. **2**, 520–525 (2009)
2. Aoun, M., Malti, R., Levron, E., Oustaloup, A.: Orthonormal basis functions for modeling continuous-time fractional systems. In: Elsevier editor, SySId, Rotterdam, Pays bas, 9. IFAC, Elsevier (2003)
3. Aoun, M., Malti, R., Levron, F., Oustaloup, A.: Numerical simulations of fractional systems: an overview of existing methods and improvements. Nonlinear Dyn. **38**(1–4), 117–131 (2004)
4. Aoun, M., Najar, S., Abdelhamid, M., Abdelkrim, M.N.: Continuous fractional kalman filter. In: 2012 9th International Multi-Conference on Systems, Signals and Devices (SSD), pp. 1–6, March 2012
5. Atanackovic, T.M., Pilipovic, S., Stankovic, B., Zorica, D.: Fractional Calculus with Applications in Mechanics: Vibrations and Diffusion Processes. ISTE, London (2014)
6. Baleanu, D., Golmankhaneh, A.K., Golmankhaneh, A.K., Baleanu, M.C.: Fractional electromagnetic equations using fractional forms. Int. J. Theoret. Phys. **48**(11), 3114–3123 (2009)
7. Battaglia, J.L., Le Lay, L., Batsale, J.C., Oustaloup, A., Cois, O.: Heat flux estimation through inverted non integer identification models. Int. J. Therm. Sci. **39**(3), 374–389 (2000)
8. Bettayeb, M., Djennoune, S.: A note on the controllability and the observability of fractional dynamical systems. Fract. Differ. Appl. **2**, 493–498 (2006)
9. Bode, H.W.: Network Analysis and Feedback Amplifier Design. New York (1945)
10. Caputo, M.: Linear models of dissipation whose q is almost frequency independent-ii. Geophys. J. Int. **13**(5), 529–539 (1967)
11. Combastel, C.: Stable interval observers in bbc for linear systems with time-varying input bounds. IEEE Trans. Autom. Control **58**(2), 481–487 (2013)
12. Das, S.: Functional Fractional Calculus for System Identification and Controls. Springer, Berlin (2008)
13. Dugowson, S.: Les différentielles métaphysiques: histoire et philosophie de la généralisation de l'ordre de dérivation. Ph.D. thesis, Université Paris XIII, France (1994)
14. Dzielinski, A., Sierociuk, D.: Observer for discrete fractional order state-space systems. Fract. Differ. Appl. **2**, 511–516 (2006)
15. Efimov, D., Raïssi, T., Chebotarev, S., Zolghadri, A.: Interval state observer for nonlinear time-varying systems. Automatica **49**(1), 200–205 (2013)
16. Gejji, V.: Fractional Calculus: Theory and Applications. Narosa Publishing House, New Delhi (2014)
17. Guo, T.L.: Controllability and observability of impulsive fractional linear time-invariant system. Comput. Math. Appl. **64**(10), 3171–3182 (2012)
18. Herrmann, R.: Fractional Calculus: An Introduction for Physicists. World Scientific, New Jersey (2014)
19. Kaczorek, T.: Fractional positive continuous-time linear systems and their reachability. Int. J. Appl. Math. Comput. Sci. **18**(2), 223–228 (2008)

20. Kaczorek, T., Rogowski, K.: Fractional linear systems and electrical circuits. In: Studies in Systems, Decision and Control. Springer (2014)
21. Koh, B.S., Junkins, J.L.: Kalman filter for linear fractional order systems. J. Guidance Control Dyn. **35**(6), 1816–1827 (2016)
22. Mainardi, F.: Fractional Calculus and Waves in Linear Viscoelasticity an Introduction to Mathematical Models. Imperial College Press, London Hackensack (2010)
23. Malinowska, A.: Introduction to the Fractional Calculus of Variations. Imperial College Press, London (2012)
24. Manabe, S.: The non-integer integral and its application to control systems. Jpn. Inst. Electr. Eng. **6**, 83–87 (1961)
25. Matignon, D.: Stability properties for generalized fractional differential systems. ESAIM Proc. Syst. Différ. Fract. Modèles Méth. Appl. **5**, 145–158 (1998)
26. Matignon, D., D'Andrea-Novel, B.: Observer-based controllers for fractional differential systems. IEEE Conf. Decis. Control **5**, 4967–4972 (1997)
27. Matignon, D., DAndrea-Novel, B.: Some results on controllability and observability of finite-dimensional fractional differential systems. Comput. Eng. Syst. Appl. **2**, 952–956 (1996)
28. Mazenc, F., Bernard, O.: Interval observers for linear time-invariant systems with disturbances. Automatica **47**(1), 140–147 (2011)
29. Miller, K.S., Ross, B.: An Introduction to the Fractional Calculus and Fractional Differential Equations. Wiley (1993)
30. Nelms, R.M., Cahela, D.R., Tatarchuk, B.J.: Modeling double-layer capacitor behavior using ladder circuits. IEEE Trans. Aerosp. Electron. Syst. **39**(2), 430–438 (2003)
31. Oldham, K., Spanier, J.: The Fractional Calculus: Theory and Applications of Differentiation and Integration to Arbitrary Order (1974)
32. Oldham, K.: The Fractional Calculus: Theory and Applications of Differentiation and Integration to Arbitrary Order. Dover Publications, Mineola (2006)
33. Ortigueira, M.: Fractional Calculus for Scientists and Engineers. Springer, Dordrecht (2011)
34. Oustaloup, A.: Linear feedback control systems of fractional order between 1 and 2. In: IEEE Symposium on Circuit and Systems, Chicago, USA (1981)
35. Oustaloup, A.: La Commande CRONE. Hermes, Paris (1991)
36. Oustaloup, A.: La dérivation non Entière: Théorie, Synthèse et Applications. Hermès, Paris (1995)
37. Raïssi, T., Efimov, D., Zolghadri, A.: Interval state estimation for a class of nonlinear systems. IEEE Trans. Autom. Control **57**(1), 260–265 (2012)
38. Sabatier, J., Aoun, M., Oustaloup, A., Grégoire, G., Ragot, F., Roy, P.: Fractional system identification for lead acid battery state of charge estimation. Signal Process. **86**(10), 2645–2657 (2006)
39. Sabatier, J., Lanusse, P., Melchior, P., Farges, C., Oustaloup, A.: Fractional order differentiation and robust control design: CRONE, H-infinity and motion control (Intelligent Systems, Control and Automation: Science and Engineering). Springer (2015)
40. Sabatier, J., Moze, M., Farges, C.: LMI stability conditions for fractional order systems. Comput. Math. Appl. **59**(5), 1594–1609 (2010)
41. Sasso, M., Palmieri, G., Amodio, D.: Application of fractional derivative models in linear viscoelastic problems. Mech. Time-Depend. Mater. **15**(4), 367–387 (2011)
42. Sierociuk, D., Dzielinski, A.: Fractional kalman filter algorithm for the states, parameters and order of fractional system estimation. Int. J. Appl. Math. Comput. Sci. **16**(1), 129 (2006)
43. Tarasov, V.E.: Fractional integro-differential equations for electromagnetic waves in dielectric media. Theoret. Math. Phys. **158**(3), 355–359 (2009)
44. Trigeassou, J.-C., Maamri, N., Sabatier, J., Oustaloup, A.: State variables and transients of fractional order differential systems. Comput. Math. Appl. **64**(10), 3117–3140 (2012)

45. Tustin, A., Allanson, J.T., Layton, J.M., Jakeways, R.J.: The design of systems for automatic control of the position of massive objects. In: Proceedings of Institution of Electrical Engineers, vol. 105-C, pp. 1–57 (1958)
46. Wang, Y., Hartley, T.T., Lorenzo, C.F., Adams, J.L., Carletta, J.E., Veillette, R.J.: Modeling ultracapacitors as fractional-order systems. In: Baleanu, D., Guvenc, Z.B., Machado, J.A.T. (eds.) New Trends in Nanotechnology and Fractional Calculus Applications, pp. 257–262. Springer, Netherlands (2010)

Chapter 9
Analysis of the Positivity and Stability of Fractional Discrete-Time Nonlinear Systems

Tadeusz Kaczorek

Abstract The positivity and asymptotic stability of the discrete-time nonlinear systems are addressed. Necessary and sufficient conditions for the positivity and sufficient conditions for the asymptotic stability of the nonlinear systems are established. The proposed stability tests are based on an extension of the Lyapunov method to the positive nonlinear systems. The effectiveness of the tests are demonstrated on examples.

Keywords Positivity · Stability · Fractional · Nonlinear · System

9.1 Introduction

A dynamical system is called positive if its trajectory starting from any nonnegative initial condition state remains forever in the positive orthant for all nonnegative inputs. An overview of state of the art in positive system theory is given in the monographs [8, 11] and in the papers [10, 12, 18, 21]. Models having positive behavior can be found in engineering, economics, social sciences, biology and medicine, etc.

The Lyapunov, Bohl and Perron exponents and stability of time-varying discrete-time linear systems have been investigated in [1–7]. The positive standard and descriptor systems and their stability have been analyzed in [10–12, 18, 21]. The positive linear systems with different fractional orders have been addressed in [12, 13] and the descriptor discrete-time linear systems in [9, 10]. Descriptor positive discrete-time and continuous-time nonlinear systems have been analyzed in [14, 19, 20] and the positivity and linearization of nonlinear discrete-time systems by state-feedbacks in [18]. The minimum energy control of positive linear systems has been addressed in [15–17]. The stability and robust stabilization of discrete-time switched systems have been analyzed in [23, 24].

T. Kaczorek (✉)
Faculty of Electrical Engineering, Bialystok University of Technology,
Bialystok, Poland
e-mail: kaczorek@isep.pw.edu.pl

F. Cacace et al. (eds.), *Positive Systems*, Lecture Notes in Control
and Information Sciences 471, DOI 10.1007/978-3-319-54211-9_9

In this chapter the positivity and asymptotic stability of the fractional discrete-time nonlinear systems will be investigated.

The chapter is organized as follows. In Sect. 9.2 the definitions and theorems concerning the positivity and stability of positive discrete-time and continuous-time linear systems are recalled. Necessary and sufficient conditions for the positivity of the fractional discrete-time nonlinear systems are established in Sect. 9.3. The asymptotic stability of the positive nonlinear systems is addressed in Sect. 9.4, where the sufficient conditions for the stability are proposed. Concluding remarks are given in Sect. 9.5.

The following notation will be used: $\mathbb{R}$—the set of real numbers, $\mathbb{R}^{n\times m}$—the set of $n \times m$ real matrices, $\mathbb{R}_+^{n\times m}$—the set of $n \times m$ matrices with nonnegative entries and $\mathbb{R}_+^n = \mathbb{R}_+^{n\times 1}$, $\mathbb{Z}_+$—the set of nonnegative integers, $\mathbb{M}_n$—the set of $n \times n$ Metzler matrices (with nonnegative off-diagonal entries), $\mathbb{I}_n$—the $n \times n$ identity matrix.

9.2 Positive Discrete-Time and Continuous-Time Linear Systems and Their Stability

Consider the discrete-time linear system

$$x_{i+1} = Ax_i + Bu_i, \quad i \in \mathbb{Z}_+ = \{0, 1, \ldots\}, \tag{9.1a}$$

$$y_i = Cx_i + Du_i, \tag{9.1b}$$

where $x_i \in \mathbb{R}^n$, $u_i \in \mathbb{R}^m$, $y_i \in \mathbb{R}^p$ are the state, input and output vectors and $A \in \mathbb{R}^{n\times n}$, $B \in \mathbb{R}^{n\times m}$, $C \in \mathbb{R}^{p\times n}$, $D \in \mathbb{R}^{p\times m}$.

Definition 9.1 [8, 11] The discrete-time linear system (9.1) is called (internally) positive if $x_i \in \mathbb{R}_+^n$, $y_i \in \mathbb{R}_+^p$, $i \in \mathbb{Z}_+$ for any initial conditions $x_0 \in \mathbb{R}_+^n$ and all inputs $u_i \in \mathbb{R}_+^m$, $i \in \mathbb{Z}_+$.

Theorem 9.1 [8, 11] *The discrete-time linear system (9.1) is positive if and only if*

$$A \in \mathbb{R}_+^{n\times n}, \quad B \in \mathbb{R}_+^{n\times m}, \quad C \in \mathbb{R}_+^{p\times n}, \quad D \in \mathbb{R}_+^{p\times m}.$$

Definition 9.2 [8, 11] The positive discrete-time linear system (9.1) is called asymptotically stable if

$$\lim_{i\to\infty} x_i = 0 \quad \text{for any} \quad x_0 \in \mathbb{R}_+^n.$$

Theorem 9.2 *The positive discrete-time linear system (9.1) is asymptotically stable if and only if one of the following equivalent conditions is satisfied:*

1. *all coefficients of the polynomial*

$$p_n(z) = \det[\mathbb{I}_n(z+1) - A] = z^n + a_{n-1}z^{n-1} + \cdots + a_1 z + a_0$$

are positive, i.e. $a_i > 0$ *for* $i = 0, 1, \ldots, n-1$.

2. *all principal minors of the matrix* $\overline{A} = \mathbb{I}_n - A = [\overline{a}_{ij}]$ *are positive, i.e.*

$$M_1 = |\overline{a}_{11}| > 0, \quad M_2 = \begin{vmatrix} \overline{a}_{11} & \overline{a}_{12} \\ \overline{a}_{21} & \overline{a}_{22} \end{vmatrix} > 0, \; \ldots, \; M_n = \det\overline{A} > 0.$$

Proof The proof is given in [11].

Consider the continuous-time linear system

$$\dot{x} = Ax + Bu, \tag{9.2a}$$

$$y = Cx + Du, \tag{9.2b}$$

where $x = x(t) \in \mathbb{R}^n$, $u = u(t) \in \mathbb{R}^m$, $y = y(t) \in \mathbb{R}^p$ are the state, input and output vectors and $A \in \mathbb{R}^{n\times n}$, $B \in \mathbb{R}^{n\times m}$, $C \in \mathbb{R}^{p\times n}$, $D \in \mathbb{R}^{p\times m}$.

Definition 9.3 [8, 11] The continuous-time linear system (9.2) is called (internally) positive if $x \in \mathbb{R}_+^n$, $y \in \mathbb{R}_+^p$, $t \ge 0$ for any initial conditions $x_0 \in \mathbb{R}_+^n$ and all inputs $u \in \mathbb{R}_+^m$, $t \ge 0$.

Theorem 9.3 [8, 11] *The continuous-time linear system (9.2) is positive if and only if*

$$A \in \mathbb{M}_n, \quad B \in \mathbb{R}_+^{n\times m}, \quad C \in \mathbb{R}_+^{p\times n}, \quad D \in \mathbb{R}_+^{p\times m},$$

Definition 9.4 [8, 11] The positive continuous-time linear system (9.2) is called asymptotically stable if

$$\lim_{t\to\infty} x(t) = 0 \quad \text{for all} \quad x_0 \in \mathbb{R}_+^n.$$

Theorem 9.4 *The positive continuous-time linear system (9.2) is asymptotically stable if and only if one of the following equivalent conditions is satisfied:*

1. *all coefficients of the polynomial*

$$p_n(s) = \det[\mathbb{I}_n s - A] = s^n + \hat{a}_{n-1}s^{n-1} + \cdots + \hat{a}_1 s + \hat{a}_0$$

are positive, i.e. $\hat{a}_k > 0$ *for* $k = 0, 1, \ldots, n-1$.

2. *all principal minors of the matrix* $\hat{A} = -A = [\hat{a}_{ij}]$ *are positive, i.e.*

$$\hat{M}_1 = |\hat{a}_{11}| > 0, \quad \hat{M}_2 = \begin{vmatrix} \hat{a}_{11} & \hat{a}_{12} \\ \hat{a}_{21} & \hat{a}_{22} \end{vmatrix} > 0, \; \ldots, \; \hat{M}_n = \det\hat{A} > 0.$$

Proof The proof is given in [11].

Theorem 9.5 *The matrix $A \in \mathbb{M}_n$ satisfies the condition*

$$-A^{-1} \in \mathbb{R}_+^{n\times n}$$

if and only if the positive system (9.2) is asymptotically stable.

Proof The proof is given in [11].

9.3 Positivity of the Fractional Nonlinear Systems

Consider the fractional discrete-time nonlinear system

$$\Delta^\alpha x_i = Ax_i + f(x_{i-1}, u_i), \tag{9.3a}$$

$$y_i = g(x_i, u_i) \tag{9.3b}$$

and $0 < \alpha \le 1$, $i \in \mathbb{Z}_+ = \{0, 1, \ldots\}$, where

$$\Delta^\alpha x_i = \sum_{j=0}^{i} a_j^\alpha x_{i-j} \tag{9.4a}$$

with

$$a_j^\alpha = (-1)^j \binom{\alpha}{j}, \quad \binom{\alpha}{j} = \begin{cases} 1 & \text{for } k = 0 \\ \frac{\alpha(\alpha-1)\cdots(\alpha-j+1)}{j!} & \text{for } k = 1, 2, 3, \ldots \end{cases} \tag{9.4b}$$

is the α-order difference of x_i, $x_i \in \mathbb{R}^n$, $u_i \in \mathbb{R}^m$, $y_i \in \mathbb{R}^p$ are the state, input and output vectors, $A \in \mathbb{R}^{n\times n}$ and $f(x_{i-1}, u_i) \in \mathbb{R}^n$, $g(x_i, u_i) \in \mathbb{R}^p$ are vector functions continuous in x_i and u_i.

Note that the fractional difference (9.4a) is defined in the point "i" not as usually in the point "$i + 1$" [13, 22].

Substituting (9.4a) into (9.3a) we obtain

$$\sum_{j=0}^{i} a_j^\alpha x_{i-j} = Ax_i + f(x_{i-1}, u_i)$$

and

$$x_i = \sum_{j=1}^{i} A_1 c_j^\alpha x_{i-j} + f_1(x_{i-1}, u_i), \quad i \in \mathbb{Z}_+, \tag{9.5}$$

where

$$c_j^\alpha = -a_j^\alpha, \quad j = 1, \ldots, i; \quad A_1 = [\mathbb{I}_n - A]^{-1} \in \mathbb{R}^{n\times n},$$
$$f_1(x_{i-1}, u_i) = A_1 f(x_{i-1}, u_i).$$

Assuming $x_i = 0$, $i = 1, 2, \ldots$ from (9.5) for $i = 0$ we have

$$x_0 = f_1(0, u_0). \tag{9.6}$$

Therefore, the initial condition x_0 is related with u_0 by (9.6).

Lemma 9.1 *The matrix*

$$A_1 = [\mathbb{I}_n - A]^{-1} \in \mathbb{R}_+^{n\times n} \tag{9.7}$$

if and only if the positive linear system

$$x_{i+1} = Ax_i, \quad A \in \mathbb{R}_+^{n\times n} \tag{9.8}$$

is asymptotically stable.

Proof By Theorem 9.2 the positive discrete-time linear system (9.8) is asymptotically stable if and only if the matrix $A - \mathbb{I}_n \in \mathbb{M}_n$ is asymptotically stable (is Hurwitz) and by Theorem 9.5 the condition (9.7) is satisfied if the system (9.8) is asymptotically stable. □

Theorem 9.6 *The solution x_i of the Eq. (9.5) for given initial condition $x_0 \in \mathbb{R}^n$ and input $u_i \in \mathbb{R}^m$, $i \in \mathbb{Z}_+$ has the form*

$$x_i = \Phi_i x_0 + \sum_{j=1}^{i} \Phi_{i-j} f_1(x_{j-1}, u_j), \tag{9.9a}$$

where

$$\Phi_j = \sum_{k=1}^{j} c_k^\alpha A_1 \Phi_{j-k}, \quad j = 1, 2, \ldots, i; \quad \Phi_0 = \mathbb{I}_n. \tag{9.9b}$$

Proof The proof can be accomplished by induction or by checking that (9.9) satisfies the Eq. (9.5). □

In particular case for linear system

$$x_i = \sum_{j=1}^{i} A_1 c_j^\alpha x_{i-j} + B_1 u_i, \quad i \in \mathbb{Z}_+, \quad B_1 \in \mathbb{R}^{n\times m} \tag{9.10a}$$

the solution x_i has the form

$$x_i = \Phi_i x_0 + \sum_{j=1}^{i} \Phi_{i-j} B_1 u_j \tag{9.10b}$$

and the matrix Φ_j is given by (9.9b).

Remark 9.1 The solution x_i of the Eq. (9.5) can be computed using the formulae (9.9) iteratively for $i = 1, 2, \ldots$ and substituting x_{j-1} given by (9.9a) into the vector function $f_1(x_{j-1}, u_j)$ for $i = 1, 2, \ldots$.

Definition 9.5 The discrete-time nonlinear system (9.3) is called (internally) positive if $x_i \in \mathbb{R}_+^n$, $y_i \in \mathbb{R}_+^p$, $i \in \mathbb{Z}_+$ for any initial conditions $x_0 \in \mathbb{R}_+^n$ and all inputs $u_i \in \mathbb{R}_+^m$, $i \in \mathbb{Z}_+$.

Theorem 9.7 *The discrete-time nonlinear system (9.3) is positive if and only if* $0 < \alpha \leq 1$, *the matrix* $A \in \mathbb{R}_+^{n \times n}$ *is asymptotically stable and*

$$f(x_{i-1}, u_i) \in \mathbb{R}_+^n \quad \text{for} \quad x_i \in \mathbb{R}_+^n \quad \text{and} \quad u_i \in \mathbb{R}_+^m,\ i \in \mathbb{Z}_+, \tag{9.11}$$

$$g(x_i, u_i) \in \mathbb{R}_+^p \quad \text{for} \quad x_i \in \mathbb{R}_+^n \quad \text{and} \quad u_i \in \mathbb{R}_+^m,\ i \in \mathbb{Z}_+. \tag{9.12}$$

Proof Sufficiency. By Lemma 9.1 if $A \in \mathbb{R}_+^{n \times n}$ is asymptotically stable then $A_1 \in \mathbb{R}_+^{n \times n}$. It is well-known [13] that if $0 < \alpha \leq 1$ then $c_j^\alpha > 0$ for $j = 1, 2, \ldots$. Therefore, from (9.9b) we have $\Phi_j \in \mathbb{R}_+^{n \times n}$ for $j = 0, 1, 2, \ldots$ and from (9.9a) $x_i \in \mathbb{R}_+^n$ for $i = 1, 2, \ldots$, since by assumption (9.11) $f_1(x_{i-1}, u_i) = A_1 f(x_{i-1}, u_i) \in \mathbb{R}_+^n$ for $x_i \in \mathbb{R}_+^n$ and $u_i \in \mathbb{R}_+^m$, $i \in \mathbb{Z}_+$. If (9.12) holds then from (9.3b) we have $y_i \in \mathbb{R}_+^p$ for $i \in \mathbb{Z}_+$.

Necessity. If $f(x_{i-1}, u_i) = 0$ then $x_i \in \mathbb{R}_+^n$, $i \in \mathbb{Z}_+$ only if $A_1 \in \mathbb{R}_+^{n \times n}$ and by Lemma 9.1 implies the asymptotic stability of the matrix $A \in \mathbb{R}_+^{n \times n}$. Note that $x_i \in \mathbb{R}_+^n$ for $i \in \mathbb{Z}_+$ implies the condition (9.11). Similarly, $y_i \in \mathbb{R}_+^p$ for $i \in \mathbb{Z}_+$ implies the condition (9.12). □

9.4 Stability of the Positive Nonlinear Systems

Consider the fractional nonlinear system for zero inputs ($u_i = 0$ and $f(x_{i-1}, 0) = \bar{f}_2(x_{i-1})$ in the form

$$\Delta^\alpha x_i = A x_i + \bar{f}_2(x_{i-1}), \quad i \in \mathbb{Z}_+, \quad 0 < \alpha \leq 1 \tag{9.13}$$

or

$$x_i = \sum_{j=1}^{i} A_1 c_j^\alpha x_{i-j} + f_2(x_{i-1}), \quad i \in \mathbb{Z}_+, \quad 0 < \alpha \leq 1, \tag{9.14a}$$

where

$$f_2(x_{i-1}) = A_1 \bar{f}_2(x_{i-1}), \quad i \in \mathbb{Z}_+ \tag{9.14b}$$

and A_1 is defined by (9.7).

Definition 9.6 The fractional positive nonlinear system (9.13) is called asymptotically stable in the region $D \in \mathbb{R}_+^n$ if $x_i \in \mathbb{R}_+^n$, $i \in \mathbb{Z}_+$ and

$$\lim_{i \to \infty} x_i = 0 \quad \text{for} \quad x_0 \in D \in \mathbb{R}_+^n.$$

To test the asymptotic stability of the system the Lyapunov method will be used. As a candidate of the Lyapunov function we choose

$$V(x_i) = c^T x_i > 0 \quad \text{for} \quad x_i \in \mathbb{R}_+^n, \quad i \in \mathbb{Z}_+, \tag{9.15}$$

where $c \in \mathbb{R}_+^n$ is a vector with strictly positive components $c_i > 0$ for $i = 1, \ldots, n$.

Using (9.14) and (9.15) we obtain

$$\begin{aligned}
\Delta V(x_i) &= V(x_{i+1}) - V(x_i) = c^T x_{i+1} - c^T x_i \\
&= c^T \left[\sum_{j=1}^{i+1} A_1 c_j^\alpha x_{i-j+1} + f_2(x_i) - \left(\sum_{j=1}^{i} A_1 c_j^\alpha x_{i-j} + f_2(x_{i-1}) \right) \right] \\
&= c^T \left[\sum_{j=1}^{i} A_1 c_j^\alpha (x_{i-j+1} - x_{i-j}) + A_1 c_{i+1}^\alpha x_0 + f_2(x_i) - f_2(x_{i-1}) \right] < 0
\end{aligned}$$

and

$$\sum_{j=1}^{i} A_1 c_j^\alpha (x_{i-j+1} - x_{i-j}) + A_1 c_{i+1}^\alpha x_0 + f_2(x_i) - f_2(x_{i-1}) < 0, \quad x_i \in D \in \mathbb{R}_+^n \tag{9.16}$$

$i \in \mathbb{Z}_+$, since $c \in \mathbb{R}_+^n$ is strictly positive.

Therefore, the following theorem has been proved.

Theorem 9.8 *The positive discrete-time nonlinear system (9.13) is asymptotically stable in the region $D \in \mathbb{R}_+^n$ if the condition (9.16) is satisfied.*

Example 9.1 Consider the discrete-time nonlinear system (9.13) with

$$x_i = \begin{bmatrix} x_{1,i} \\ x_{2,i} \end{bmatrix}, \quad A = \begin{bmatrix} 0.3 & 0.1 \\ 0.2 & 0.4 \end{bmatrix}, \quad f_2(x_i) = \begin{bmatrix} x_{1,i} x_{2,i} \\ x_{2,i}^2 \end{bmatrix}.$$

In this case

$$A_1 = [\mathbb{I}_2 - A]^{-1} = \begin{bmatrix} 0.7 & -0.1 \\ -0.2 & 0.6 \end{bmatrix}^{-1} = \frac{1}{0.4} \begin{bmatrix} 0.6 & 0.1 \\ 0.2 & 0.7 \end{bmatrix} = \frac{1}{4} \begin{bmatrix} 6 & 1 \\ 2 & 7 \end{bmatrix} \in \mathbb{R}_+^{2 \times 2}.$$

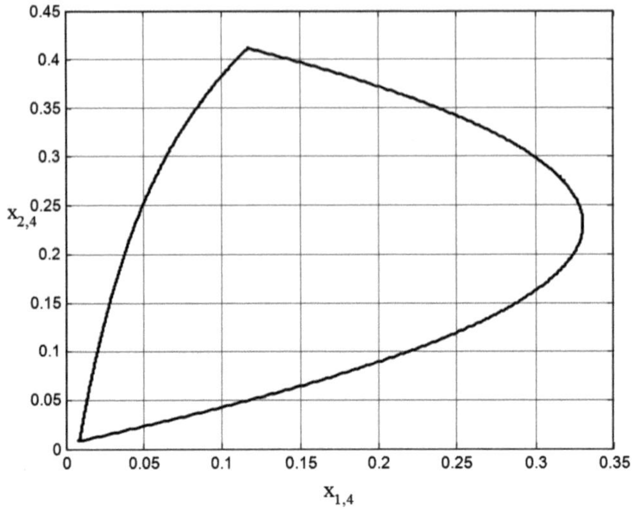

Fig. 9.1 Stability region (inside the *curved line*)

The nonlinear system is positive, since the matrix $A \in \mathbb{R}_+^{2\times 2}$ is asymptotically stable and $f_2(x_i) \in \mathbb{R}_+^2$ for all $x_i \in \mathbb{R}_+^2$, $i \in \mathbb{Z}_+$.

The region $D \in \mathbb{R}_+^2$ is defined by

$$D := \{x_{1,i}, x_{2,i}\} = \sum_{j=1}^{i} A_1 c_j^\alpha x_{i-j+1} + A_1 c_{i+1}^\alpha x_0 - x_i + f_2(x_i)$$
$$= \begin{bmatrix} 1.5\left(\sum_{j=1}^{i} c_j^\alpha x_{1,i-j+1} + c_{i+1}^\alpha x_{10}\right) + 0.25\left(\sum_{j=1}^{i} c_j^\alpha x_{2,i-j+1} + c_{i+1}^\alpha x_{20}\right) - x_{1,i} + x_{1,i}x_{2,i} \\ 0.5\left(\sum_{j=1}^{i} c_j^\alpha x_{1,i-j+1} + c_{i+1}^\alpha x_{10}\right) + 1.75\left(\sum_{j=1}^{i} c_j^\alpha x_{2,i-j+1} + c_{i+1}^\alpha x_{20}\right) - x_{2,i} + x_{2,i}^2 \end{bmatrix} \tag{9.17}$$

Let us assume

$$x_{10} = 0.1, \quad x_{20} = 0.2, \quad \alpha = 0.5, \quad i = 4. \tag{9.18}$$

The region defined by (9.17) with (9.18) is shown in Fig. 9.1.

9.5 Concluding Remarks

The positivity and asymptotic stability of the discrete-time nonlinear systems have been addressed. Necessary and sufficient conditions for the positivity of the discrete-time nonlinear systems have been established (Theorem 9.7). Using the Lyapunov direct method the sufficient conditions for asymptotic stability of the discrete-time nonlinear systems have been proposed (Theorem 9.8). The effectiveness of the conditions has been demonstrated on Example 9.1. The considerations can be extended to fractional continuous-time nonlinear systems.

An open problem is an extension of the conditions to the descriptor fractional discrete-time and continuous-time nonlinear systems.

Acknowledgements This work was supported by National Science Centre in Poland under work No. 2014/13/B/ST7/03467.

References

1. Czornik, A.: Perturbation Theory for Lyapunov Exponents of Discrete Linear Systems, vol. 17. AGH University of Science and Technology Press, Cracow (2012)
2. Czornik, A., Klamka, J., Niezabitowski, M.: On the set of Perron exponents of discrete linear systems. IFAC Proc. **47(4)**, 11740–11742 (2014)
3. Czornik, A., Newrat, A., Niezabitowski, M.: On the Lyapunov exponents of a class of the second-order discrete time linear systems with bounded perturbations. Int. J. Dyn. Syst. **4**(28), 473–483 (2013)
4. Czornik, A., Newrat, A., Niezabitowski, M., Szyda A.: On the Lyapunov and Bohl exponent of time-varying discrete linear system. In: Mediterranean Conference on Control and Automation (MED), pp. 194–197, Barcelona (2012)
5. Czornik, A., Niezabitowski, M.: Lyapunov exponents for system with unbounded coefficients. Int. J. Dyn. Syst. **2**(28), 140–153 (2013)
6. Czornik, A., Niezabitowski, M.: On the spectrum of discrete time-varying linear systems. Nonlinear Anal. Hybrid Syst. **9**, 27–41 (2013)
7. Czornik, A., Niezabitowski, M.: On the stability of Lyapunov exponents of discrete linear system. In: Proceedings of European Control Conference, pp. 2210–2213, Zurich (2013)
8. Farina, L., Rinaldi, S.: Positive Linear Systems: Theory and Applications. Wiley, New York (2000)
9. Kaczorek, T.: Positive singular discrete time linear systems. Bull. Pol. Acad. Sci. Tech. Sci. **45**(4), 619–631 (1997)
10. Kaczorek, T.: Positive descriptor discrete-time linear systems. Probl. Nonlinear Anal. Eng. Syst. **1**(7), 38–54 (1998)
11. Kaczorek, T.: Positive 1D and 2D Systems. Springer, London (2001)
12. Kaczorek, T.: Positive linear systems consisting of n subsystems with different fractional orders. IEEE Trans. Circ. Syst. **58**(6), 1203–1210 (2011)
13. Kaczorek, T.: Selected Problems of Fractional Systems Theory. Springer, Berlin (2012)
14. Kaczorek, T.: Descriptor positive discrete-time and continuous-time nonlinear systems. Proc. SPIE **9290** (2014). doi:10.1117/12.2074558
15. Kaczorek, T.: Minimum energy control of descriptor positive discrete-time linear systems. COMPEL **33**(3), 976–988 (2014)
16. Kaczorek, T.: Minimum energy control of fractional descriptor positive discrete-time linear systems. Int. J. Appl. Math. Comput. Sci. **24**(4), 735–743 (2014)
17. Kaczorek, T.: Necessary and sufficient conditions for minimum energy control of positive discrete-time linear systems with bounded inputs. Bull. Pol. Acad. Sci. Tech. Sci. **62**(1), 85–89 (2014)
18. Kaczorek, T.: Positivity and linearization of a class of nonlinear discrete-time systems by state feedbacks. Logistyka **6**, 5078–5083 (2014)

19. Kaczorek, T.: Analysis of the positivity and stability of discrete-time and continuous-time nonlinear systems. Comput. Prob. Electr. Eng. **5**(1), 11–16 (2015)
20. Kaczorek, T.: Descriptor standard and positive discrete-time nonlinear systems. Arch. Control Sci. **25**(2), 227–235 (2015)
21. Kaczorek, T.: Positivity and stability of discrete-time nonlinear systems. In: Proceedings of IEEE 2nd International Conference on Cybernetics (CYBERCONF) (2015)
22. Ostalczyk, P.: Discrete Fractional Calculus. Applications in Control and Image Processing. World Scientific (2015)
23. Zhang, J., Han, Z., Wu, H., Huang, J.: Robust stabilization of discrete-time positive switched systems with uncertainties and average dwell time switching. Circ. Syst. Signal Process. **33**(1), 71–95 (2014)
24. Zhang, H., Xie, D., Zhang, H., Wang, G.: Stability analysis for discrete-time switched systems with unstable subsystems by a mode-dependent average dwell time approach. ISA Trans. **53**(4), 1081–1086 (2014)

Chapter 10
Continuous-Time Compartmental Switched Systems

Maria Elena Valcher and Irene Zorzan

Abstract In this chapter we investigate state-feedback and output-feedback stabilization of compartmental switched systems, under the additional requirement that the resulting switched system is in turn compartmental. Necessary and sufficient conditions for the solvability of the two problems are given. Subsequently, affine compartmental switched systems are considered, and a characterization of all the switched equilibria that can be "reached" under some stabilizing switching law σ is provided.

Keywords Compartmental system · Linear/affine switched system · Stabilization · Switched equilibrium point

10.1 Introduction

Compartmental switched systems (CSSs) are positive switched systems whose subsystems are (linear) compartmental models. While the general class of positive switched systems has attracted a great deal of attention over the last 10 years [4, 9, 12, 15, 18], a systematic analysis of the class of CSSs was initiated only recently in [25, 26].

This class of systems provides an effective mathematical description of real phenomena/processes that are characterized by some distinguished features: firstly, they undergo different working conditions, each of them captured by a different linear state-space model; secondly, their describing variables are intrinsically nonnegative and obey some conservation law (e.g., mass, energy, fluid).

This is the case, for instance, when modeling a fluid network: different open/closed configurations of the pipes connecting the tanks correspond to different operating

M.E. Valcher (✉) · I. Zorzan
Dipartimento di Ingegneria dell'Informazione, Università di Padova,
via Gradenigo 6/B, 35131 Padova, Italy
e-mail: meme@dei.unipd.it

I. Zorzan
e-mail: irene.zorzan@studenti.unipd.it

F. Cacace et al. (eds.), *Positive Systems*, Lecture Notes in Control
and Information Sciences 471, DOI 10.1007/978-3-319-54211-9_10

conditions, and the state variables representing fluid levels in the tanks evolve in accordance with a fluid conservation law [4]. Analogously, a compartmental switched system may be adopted to describe a thermal system: state variables represent relative temperatures in the various rooms with respect to the external temperature, and each subsystem corresponds to a specific set of heat transmission coefficients that depend on the open/closed configurations of doors and windows between the rooms [4].

As a third example, consider a multicompartmental model describing the lung functioning: the behaviour of the lung, regarded as an agglomeration of subunits, differs significantly in the inspiration and in the expiration phases, and the time evolution of the state variables representing the volume of each subunit is governed by a mass conservation law [14, 16, 26]. Another practical example arises, for instance, when describing certain economical systems [17].

In [25, 26] stability under arbitrary switching or under dwell-time switching and stabilizability of CSSs with autonomous subsystems have been addressed. Specifically, it has been shown that, differently from the broader class of positive switched systems, for CSSs asymptotic stability is equivalent to the fact that all the subsystem matrices are Hurwitz. On the other hand, a CSS is stabilizable if and only if there exists a Hurwitz convex combination of the subsystem matrices.

In this chapter, we consider CSSs whose subsystems are compartmental and non-autonomous. Specifically, we assume that all the subsystems are single-input compartmental state-space models. For this class of CSSs we investigate state-feedback and output-feedback stabilization. In the final part of the chapter, we address affine compartmental switched systems (ACSS) [3, 5] and provide some results on the "reachability" of their switched equilibrium points, by means of switching control laws.

10.2 Notation

Given $k, n \in \mathbb{Z}$, with $k \leq n$, the symbol $[k, n]$ denotes the integer set $\{k, k+1, \ldots, n\}$. $\mathbb{R}_+$ is the semiring of nonnegative real numbers. In the sequel, the (i, j)th entry of a matrix A is denoted by $[A]_{ij}$, while the ith entry of a vector $\mathbf{v}$ by $[\mathbf{v}]_i$. A matrix A_+ with entries in $\mathbb{R}_+$ is a *nonnegative matrix* ($A_+ \geq 0$); if $A_+ \geq 0$ and at least one entry is positive, A_+ is a *positive matrix* ($A_+ > 0$), while if all its entries are positive it is a *strictly positive matrix* ($A_+ \gg 0$). The same notation is adopted for nonnegative, positive and strictly positive vectors. We let $\mathbf{e}_i$ denote the ith vector of the canonical basis in $\mathbb{R}^n$ (where n is always clear from the context), whose entries are all zero except for the ith one that is unitary. $\mathbf{1}_n$ is the n-dimensional vector with all entries equal to 1 (the dimension n will be omitted if it is clear from the context). $\mathscr{A}_M$ denotes the set of nonnegative vectors $\alpha \in \mathbb{R}_+^M$ such that $\mathbf{1}^\top \alpha = 1$. Given a matrix $A \in \mathbb{R}^{n \times m}$, its *nonzero pattern* $\overline{\mathrm{ZP}}(A)$ is the set $\{(i, j) \in [1, n] \times [1, m] : [A]_{ij} \neq 0\}$. For a vector $\mathbf{v} \in \mathbb{R}^n$, the nonzero pattern is defined as $\overline{\mathrm{ZP}}(\mathbf{v}) := \{i \in [1, n] \colon [\mathbf{v}]_i \neq 0\}$.

A real square matrix A is *Hurwitz* if all its eigenvalues lie in the open left complex halfplane, i.e., for every λ belonging to the spectrum $\sigma(A)$ of A we have $\mathrm{Re}(\lambda) < 0$. A *Metzler matrix* is a real square matrix, whose off-diagonal entries are nonnegative. If A is an $n \times n$ Metzler matrix, then [22] it exhibits a real dominant eigenvalue, known as *Frobenius eigenvalue* and denoted by $\lambda_F(A)$. This means that $\lambda_F(A) > \mathrm{Re}(\lambda)$, $\forall\, \lambda \in \sigma(A)$, $\lambda \neq \lambda_F(A)$, and there exists a positive eigenvector (*Frobenius eigenvector*) $\mathbf{v}_F$ corresponding to $\lambda_F(A)$. Moreover, for a Metzler matrix the following monotonicity property holds [22]: let $A, \bar{A} \in \mathbb{R}^{n\times n}$ be Metzler matrices such that $A \leq \bar{A}$, then $\lambda_{\max}(A) \leq \lambda_{\max}(\bar{A})$; if in addition $\bar{A}$ is irreducible, then $A < \bar{A}$ implies $\lambda_{\max}(A) < \lambda_{\max}(\bar{A})$.

An $n \times n$, $n \geq 2$, nonzero matrix A is *reducible* [10] if there exists a permutation matrix Π such that (s.t.)

$$\Pi^\top A \Pi = \begin{bmatrix} A_{11} & A_{12} \\ 0 & A_{22} \end{bmatrix},$$

where A_{11} and A_{22} are square (nonvacuous) matrices, otherwise it is *irreducible*. In general, given a Metzler matrix A, a permutation matrix Π can be found s.t.

$$\Pi^\top A \Pi = \begin{bmatrix} A_{11} & A_{12} & \dots & A_{1s} \\ 0 & A_{22} & \dots & A_{2s} \\ \vdots & & \ddots & \vdots \\ 0 & \dots & & A_{ss} \end{bmatrix}, \tag{10.1}$$

where each diagonal block A_{ii}, of size $n_i \times n_i$, is either scalar ($n_i = 1$) or irreducible. (10.1) is usually known as *Frobenius normal form* of A [11, 19].

A single-input (linear) *compartmental system* is a linear state-space model:

$$\dot{\mathbf{x}}(t) = A\mathbf{x}(t) + Bu(t), \tag{10.2}$$

where B is a nonnegative vector in $\mathbb{R}^n_+$, while the state matrix $A \in \mathbb{R}^{n\times n}$ is Metzler and the entries of each of its columns sum up to a nonpositive number, i.e., $\mathbf{1}_n^\top A \leq 0$. A square matrix endowed with these two properties is called *compartmental matrix* (see [13, 21]). For any such matrix the Frobenius eigenvalue $\lambda_F(A)$ is nonpositive, and if $\lambda_F(A) = 0$ then A is simply stable, by this meaning that it has the constant mode associated with $\lambda_F(A) = 0$, but no unstable modes. Moreover, a compartmental irreducible matrix A is non-Hurwitz if and only if $\mathbf{1}_n^\top A = 0$ [23].

Given a Metzler matrix $A \in \mathbb{R}^{n\times n}$, we associate with it [6, 7, 20, 24] a *digraph* $\mathscr{D}(A) = \{\mathscr{V}, \mathscr{E}\}$, where $\mathscr{V} = \{1, \dots, n\}$ is the set of vertices and $\mathscr{E}$ is the set of arcs (or edges). There is an arc $(j, \ell) \in \mathscr{E}$ from j to ℓ, with $j \neq \ell$, if and only if $[A]_{\ell j} > 0$. A sequence $j_1 \to j_2 \to \cdots \to j_k \to j_{k+1}$ is a *path* of length k from j_1 to j_{k+1} provided that $(j_1, j_2), \dots, (j_k, j_{k+1})$ are elements of $\mathscr{E}$.

We say that vertex ℓ is *accessible* from j, with $j \neq \ell$, if there exists a path in $\mathscr{D}(A)$ from j to ℓ (equivalently, $\exists\, k \in \mathbb{N}$ s.t. $[A^k]_{\ell j} > 0$). Two distinct vertices ℓ and j are said to *communicate* if each of them is accessible from the other. Each vertex is assumed to communicate with itself. The concept of communicating vertices allows to partition the set of vertices $\mathscr{V}$ into *communicating classes*, say $\mathscr{C}_1, \mathscr{C}_2, \ldots, \mathscr{C}_s$. Class $\mathscr{C}_j$ accesses class $\mathscr{C}_i$ if there is a path from some vertex $k \in \mathscr{C}_j$ to some vertex $h \in \mathscr{C}_i$. Each class $\mathscr{C}_i$ has clearly access to itself.

10.3 Linear Compartmental Switched Systems

In this chapter, by a (*continuous-time, single-input*) *linear compartmental switched system* (LCSS) we mean a system described by the following equation:

$$\dot{\mathbf{x}}(t) = A_{\sigma(t)}\mathbf{x}(t) + B_{\sigma(t)}u(t), \qquad t \in \mathbb{R}_+, \tag{10.3}$$

where $\mathbf{x}(t) \in \mathbb{R}^n_+$ and $u(t) \in \mathbb{R}_+$ denote the value at time t of the n-dimensional state variable and of the scalar input, respectively. $\sigma : \mathbb{R}_+ \to [1, M]$ is a switching function. For every $i \in [1, M]$, $A_i \in \mathbb{R}^{n\times n}$ is a compartmental matrix and $B_i \in \mathbb{R}^n_+$ is a positive vector, namely $B_i > 0$ for every $i \in [1, M]$.[1] Consequently, each ith subsystem (A_i, B_i) is a linear *non-autonomous* compartmental system. We assume that σ is right continuous and in every finite interval it has a finite number of discontinuities.

The problem we want to address is the following one: assume that all the subsystem matrices A_i, $i \in [1, M]$, are non-Hurwitz and that the switching function σ is arbitrary but known at every time instant $t \geq 0$. We want to determine, if possible, a feedback control law, depending at each time $t \geq 0$ on the value of the switching function σ at t, such that the resulting feedback switched system is still compartmental and its state trajectories converge to zero for every $\mathbf{x}(0) > 0$ and every σ, namely it is an asymptotically stable LCSS [25, 26].

We will investigate two kinds of feedback stabilization: state-feedback stabilization in Sect. 10.4 and output-feedback stabilization in Sect. 10.5.

10.4 State-Feedback Stabilization

State-feedback stabilization problem: determine, if possible, a state-feedback control law

$$u(t) = K_{\sigma(t)}\mathbf{x}(t), \tag{10.4}$$

with $K_i \in \mathbb{R}^{1\times n}$ for every $i \in [1, M]$, that makes the state trajectory converge to zero for every initial condition $\mathbf{x}(0) > 0$ and every switching function σ, while preserving the compartmental property of the resulting closed-loop switched system.

[1] As a matter of fact, this assumption is only for the sake of simplicity. Some of the B_i's could be zero, but in that case the corresponding matrices A_i's should be necessarily Hurwitz, in order for stabilization to be possible.

First of all, we observe that, under the state-feedback law (10.4), the resulting closed-loop switched system takes the following form

$$\dot{\mathbf{x}}(t) = (A_{\sigma(t)} + B_{\sigma(t)}K_{\sigma(t)})\mathbf{x}(t), \tag{10.5}$$

and hence the control input (10.4) solves the state-feedback stabilization problem if and only if (10.5) is an asymptotically stable LCSS. On the other hand, it has been proven in Proposition 1 of [25] that for LCSSs stability under arbitrary switching is equivalent to the fact that all the subsystems matrices are Hurwitz. Hence, solving the state-feedback stabilization problem means determining state-feedback matrices K_i, $i \in [1, M]$, such that for every $i \in [1, M]$ the matrix $A_i + B_iK_i$ is compartmental and Hurwitz. We first observe that if K_i is a positive matrix, then $A_i + B_iK_i > A_i$ and by the monotonicity of the Frobenius eigenvalue we can claim that $\lambda_F(A_i + B_iK_i) \geq \lambda_F(A_i)$. So, as A_i is not Hurwitz, then $A_i + B_iK_i$ is not Hurwitz for every choice of $K_i > 0$. On the other hand, if there exists a matrix $K_i \in \mathbb{R}^{1\times n}$, with both positive and negative entries, that makes $A_i + B_iK_i$ compartmental and Hurwitz, we can always introduce a permutation matrix Π such that

$$K_i\Pi = \begin{bmatrix} K_{i+} \ K_{i-} \end{bmatrix}, \qquad K_{i+} > 0, \text{ and } K_{i-} \ll 0.$$

It is clearly seen that if $A_i + B_iK_i$ (and hence $\Pi^\top A_i\Pi + \Pi^\top B_iK_i\Pi$) is compartmental and Hurwitz, then also $\Pi^\top A_i\Pi + \Pi^\top B_i\begin{bmatrix}0 \ K_{i-}\end{bmatrix}$ is compartmental and Hurwitz. So, we can always restrict our attention to matrices $K_i < 0$.

Now that we have focused our attention on negative state-feedback matrices, we can show that the solvability of the state-feedback stabilization problem only depends on the nonzero patterns of the pairs (A_i, B_i), $i \in [1, M]$.

Proposition 10.1 *For every $i \in [1, M]$, let Π_i be an $n \times n$ permutation matrix such that*

$$\Pi_i^\top A_i\Pi_i = \begin{bmatrix} A_{11}^{(i)} & A_{12}^{(i)} & \dots & A_{1s_i}^{(i)} \\ 0 & A_{22}^{(i)} & \dots & A_{2s_i}^{(i)} \\ \vdots & & \ddots & \vdots \\ 0 & \dots & \dots & A_{s_is_i}^{(i)} \end{bmatrix}, \quad \Pi_i^\top B_i = \begin{bmatrix} B_1^{(i)} \\ B_2^{(i)} \\ \vdots \\ B_{s_i}^{(i)} \end{bmatrix}, \tag{10.6}$$

where $A_{jj}^{(i)} \in \mathbb{R}^{n_j^{(i)}\times n_j^{(i)}}$, $j \in [1, s_i]$, are either scalar or irreducible matrices and $B_j^{(i)} \in \mathbb{R}_+^{n_j^{(i)}}$. For every $i \in [1, M]$, set $r_i := \max\{j \in [1, s_i] : B_j^{(i)} \neq 0\}$. The state-feedback stabilization problem is solvable if and only if for every $i \in [1, M]$ the following three conditions hold:

(a) $A_{jj}^{(i)}$ is (compartmental and) Hurwitz for every $j \neq r_i$;
(b) $B_j^{(i)} = 0$ for every $j \neq r_i$;
(c) there exists $\ell \in [1, n_{r_i}^{(i)}]$ such that $\overline{\mathrm{ZP}}(B_{r_i}^{(i)}) \setminus \{\ell\} \subseteq \overline{\mathrm{ZP}}(col_\ell(A_{r_ir_i}^{(i)})) \setminus \{\ell\}$.

Proof We will prove that the existence of $K_i < 0$ such that $A_i + B_iK_i$ is compartmental and Hurwitz is equivalent to the fact that conditions (a), (b) and (c) hold for the pair (A_i, B_i). Since the asymptotic stability of system (10.5) is equivalent to the fact that all matrices $A_i + B_iK_i$, $i \in [1, M]$, are compartmental and Hurwitz [25], the result will immediately follow.

Consider the pair (A_i, B_i) for a specific value of the index $i \in [1, M]$. For the sake of simplicity, in the following we will drop the dependence on the index i, and hence refer to the pair as (A, B). Also, we will assume that the pair (A, B) is already in the form (10.6) (namely $\Pi = I_n$). This does not affect the substance of the proof, only the notation.

(Necessity) Let $K \in \mathbb{R}^{1\times n}$, $K < 0$, be any state-feedback matrix such that $A + BK$ is compartmental and Hurwitz, and let us partition K in a way consistent with A and B, namely as

$$K = \begin{bmatrix} K_1 & K_2 & \dots & K_s \end{bmatrix},$$

with $K_j \in \mathbb{R}^{1\times n_j}$, $K_j \leq 0$. We first prove necessity. Necessity of condition (a) can be proven by following the same lines as those of Proposition 1 in [27]. Specifically, for every $K \in \mathbb{R}^{1\times n}$ the matrix $A + BK$ takes the block-triangular form given in (10.7).

$$A + BK = \left[\begin{array}{ccc|ccc} A_{11} + B_1K_1 & \dots & A_{1r} + B_1K_r & A_{1r+1} + B_1K_{r+1} & \dots & A_{1s} + B_1K_s \\ \vdots & \ddots & \vdots & \vdots & \ddots & \vdots \\ B_rK_1 & \dots & A_{rr} + B_rK_r & A_{rr+1} + B_rK_{r+1} & \dots & A_{rs} + B_rK_s \\ \hline & & & A_{r+1r+1} & \dots & A_{r+1s} \\ & & & \vdots & \ddots & \vdots \\ & & & 0 & \dots & A_{ss} \end{array}\right] \tag{10.7}$$

As $K < 0$, if $r > 1$ a necessary condition for this matrix to be compartmental is that $B_rK_j = 0$ for every $j \in [1, r-1]$, and since $B_r > 0$, this means that $K_j = 0$ for every $j \in [1, r-1]$ (if $r = 1$ the result is trivially true). Consequently, the matrix $A + BK$ takes the same block triangular form as A, with each diagonal block $A_{jj} + B_jK_j, j \neq r$, coinciding with the corresponding diagonal block A_{jj} in A. So, in order for the matrix $A + BK$ to be Hurwitz, all the diagonal blocks A_{jj}, $j \neq r$, must be (compartmental and) Hurwitz.

To prove necessity of condition (b) notice that since A_{rr} is compartmental, irreducible and non-Hurwitz, $\mathbf{1}^\top A_{rr} = 0$, and hence, by the compartmental property of A, it must be $A_{jr} = 0$ for every $j < r$. But then, since $K_r < 0$, for every $j < r$ the matrix $A_{jr} + B_jK_r = B_jK_r \geq 0$ if and only if $B_j = 0$ for every $j < r$, namely condition (b) holds.

To prove necessity of condition (c) notice that for every $\ell \in [1, n_r]$, if there exists $h \in \overline{\text{ZP}}(B_r) \setminus \{\ell\}$ such that $h \notin \overline{\text{ZP}}(\text{col}_\ell(A_{rr})) \setminus \{\ell\}$, then $[A_{rr} + B_rK_r]_{h\ell} \geq 0$ if and only if $[A_{rr} + B_rK_r]_{h\ell} = 0$ namely if and only if $[K_r]_\ell = 0$. Hence, if there does not exist $\ell \in [1, n_r]$ such that $\overline{\text{ZP}}(B_r) \setminus \{\ell\} \subseteq \overline{\text{ZP}}(\text{col}_\ell(A_{rr})) \setminus \{\ell\}$, then $K_r = 0$ and the matrix $A_{rr} + B_rK_r = A_{rr}$ cannot be Hurwitz.

(Sufficiency) We now prove that when conditions (*a*), (*b*) and (*c*) hold, a matrix $K < 0$ such that $A + BK$ is compartmental and Hurwitz exists. Let $\ell \in [1, n_r]$ be such that $\overline{\mathrm{ZP}}(B_r) \setminus \{\ell\} \subseteq \overline{\mathrm{ZP}}(\mathrm{col}_\ell(A_{rr})) \setminus \{\ell\}$ and set $k_\ell^* := \min_{\substack{j \in [1,n_r] \\ j \neq \ell}} \frac{[A_{rr}]_{j\ell}}{[B_r]_j}$. Then, for every k_ℓ with $-k_\ell^* \leq k_\ell < 0$, the matrix

$$\bar{K}_r = \begin{bmatrix} 0 \dots k_\ell \dots 0 \end{bmatrix} = k_\ell \mathbf{e}_\ell^\top$$

is such that $A_{rr} + B_r\bar{K}_r < A_{rr}$ is still compartmental. Moreover, recalling that A_{rr} is irreducible, by the monotonicity property of the spectral abscissa, $\lambda_F(A_{rr} + B_r\bar{K}_r) < \lambda_F(A_{rr}) = 0$. Finally, set $K = \begin{bmatrix} 0 \dots 0 \ \bar{K}_r \ 0 \dots 0 \end{bmatrix}$. Condition (*b*) ensures that $A + BK$ is still compartmental, while condition (*a*) ensures that $A + BK$ is also Hurwitz.

Remark 10.1 Proposition 10.1 extends the results about stabilization of continuous-time positive systems obtained in [8] to the class of compartmental switched systems. Indeed, in Theorem 1 of [8] necessary and sufficient conditions for the stabilization of a single pair (A_i, B_i), with A_i Metzler and irreducible, and $B_i > 0$, meanwhile preserving the Metzler property of the resulting matrix $A_i + B_iK_i$, have been provided, while Theorem 2 of [8] addresses the same problem without the irreducibility assumption on A_i. The compartmental assumption on both A_i and the closed loop matrix $A_i + B_iK_i$ had two consequences: on the one hand it allowed us to derive the previous characterization without introducing restrictive assumptions as in [8] (see, in particular, the hypothesis that there exists $\bar{\mathbf{x}} \gg 0$ such that $A_i\bar{\mathbf{x}} = 0$). On the other hand it led to a set of conditions that are slightly more restrictive than those derived in Theorem 2 of [8].

Remark 10.2 As an immediate consequence of the previous proof (see necessity of condition (*b*)), it follows that if A is a compartmental and reducible matrix in Frobenius normal form (10.1), A is non-Hurwitz if and only if there exists $i \in [1, s]$ such that A_{ii} is compartmental, irreducible (or scalar) and non-Hurwitz. For every such block A_{ii}, it must be $A_{ji} = 0$ for every $j \in [1, i-1]$. Hence, in the general case, a reducible compartmental matrix is non-Hurwitz if and only if a permutation matrix Π can be found such that $\Pi^\top A \Pi$ has the following structure

$$\Pi^\top A \Pi = \begin{bmatrix} A_{11} & 0 & \dots & 0 & A_{1q+1} & \dots & A_{1s} \\ 0 & A_{22} & \dots & 0 & A_{2q+1} & \dots & A_{2s} \\ \vdots & & \ddots & \vdots & \vdots & & \vdots \\ 0 & \dots & \dots & A_{qq} & A_{qq+1} & \dots & A_{qs} \\ 0 & \dots & \dots & 0 & A_{q+1q+1} & \dots & A_{q+1s} \\ \vdots & & & \vdots & \vdots & \ddots & \vdots \\ 0 & \dots & \dots & 0 & \dots & \dots & A_{ss} \end{bmatrix}, \quad \begin{array}{c} \mathbf{1}^\top A_{jj} = 0, \forall\, j \in [1, q], \\ A_{jj} \text{ Hurwitz}, \forall\, j \in [q+1, s]. \end{array} \tag{10.8}$$

10.5 Output-Feedback Stabilization

In this section we assume that for the LCSS (10.3) a scalar output measurement $y(t) = C_{\sigma(t)}\mathbf{x}(t)$ is available, namely we consider a single-input single-output LCSS taking the following form:

$$\dot{\mathbf{x}}(t) = A_{\sigma(t)}\mathbf{x}(t) + B_{\sigma(t)}u(t), \tag{10.9.1}$$

$$y(t) = C_{\sigma(t)}\mathbf{x}(t), \qquad t \in \mathbb{R}_+, \tag{10.9.2}$$

where for every $i \in [1, M]$ the matrices $A_i \in \mathbb{R}^{n\times n}$ are compartmental, and the vectors $B_i \in \mathbb{R}^n_+$ and $C_i \in \mathbb{R}^{1\times n}_+$ are positive. In this new set-up, we consider a problem similar to the one considered in the previous section, but the control input is now an output-feedback control input, i.e., $u(t) = k_{\sigma(t)}y(t) = k_{\sigma(t)}C_{\sigma(t)}\mathbf{x}(t)$, where we can constrain our attention (by the same reasoning we adopted in the previous section) to the case $k_i < 0$ for every $i \in [1, M]$.

Output-feedback stabilization problem: determine, if possible, an output-feedback control input

$$u(t) = k_{\sigma(t)}y(t), \tag{10.10}$$

with $k_i < 0$ for every $i \in [1, M]$, that makes the state trajectory converge to zero for every initial condition $\mathbf{x}(0) > 0$ and every switching function σ, while preserving the compartmental property of the resulting closed-loop switched system.

Notice that, when the control input (10.10) is applied to system (10.9), the resulting closed-loop switched system is given by

$$\dot{\mathbf{x}}(t) = (A_{\sigma(t)} + k_{\sigma(t)}B_{\sigma(t)}C_{\sigma(t)})\mathbf{x}(t). \tag{10.11}$$

By following the same reasoning as before, we can claim that the control input (10.10) solves the output-feedback stabilization problem if and only if the matrix $A_i + k_iB_iC_i$ is compartmental and Hurwitz for every $i \in [1, M]$. Of course, solving the output-feedback stabilization problem means solving the state-feedback stabilization problem with the additional constraint that every feedback matrix $K_i := k_iC_i$ is a scaled version of the output matrix C_i. It is then clear that conditions (a)–(c) of Proposition 10.1 are necessary conditions also for the solvability of the output-feedback stabilization problem (however, they are not sufficient). To investigate the solvability of the output-feedback stabilization problem, let us assume that, when the pair (A_i, B_i) is described as in (10.6), also the output matrix C_i is partitioned in a way consistent with A_i and B_i, namely as

$$C_i = \left[C_1^{(i)}\ C_2^{(i)}\ \dots\ C_{s_i}^{(i)}\right], \tag{10.12}$$

with $C_j^{(i)} \in \mathbb{R}_+^{1\times n_j^{(i)}}$. Again, it turns out that the solvability of the output-feedback stabilization problem only depends on the nonzero pattern of all triples (A_i, B_i, C_i), $i \in [1, M]$.

Proposition 10.2 *For every $i \in [1, M]$, let Π_i be an $n \times n$ permutation matrix such that the pair $(\Pi_i^\top A_i \Pi_i, \Pi_i^\top B_i)$ and the output matrix $C_i \Pi_i$ are described as in (10.6) and (10.12), respectively. Assume that for every $i \in [1, M]$ conditions (a) and (b) of Proposition 10.1 are satisfied, where r_i is the index of the unique nonzero block in $\Pi_i^\top B_i$. For every $i \in [1, M]$, set*

$$t_i := \min\{j \in [1, s_i] \colon C_j^{(i)} \neq 0\}.$$

The output-feedback stabilization problem is solvable if and only if for every $i \in [1, M]$ the following three conditions hold:

(c_1) *the first nonzero block in $C_i \Pi_i$ corresponds to the unique nonzero block in $\Pi_i^\top B_i$, namely $t_i = r_i$;*
(c_2) *for every $j \in [r_i + 1, s_i]$ such that $C_j^{(i)} \neq 0$ the following property holds:*

$$\overline{\text{ZP}}(B_{r_i}^{(i)}) \times \overline{\text{ZP}}(C_j^{(i)}) \subseteq \overline{\text{ZP}}(A_{r_i j}^{(i)}); \tag{10.13}$$

(c_3) $\left(\overline{\text{ZP}}(B_{r_i}^{(i)}) \times \overline{\text{ZP}}(C_{r_i}^{(i)})\right) \setminus \{(\ell, \ell) : \ell \in [1, n_{r_i}^{(i)}]\} \subseteq \overline{\text{ZP}}(A_{r_i r_i}^{(i)}) \setminus \{(\ell, \ell) : \ell \in [1, n_{r_i}^{(i)}]\}$.

Proof As in the proof of Proposition 10.1, we will show that there exists $k_i < 0$ such that $A_i + k_i B_i C_i$ is compartmental and Hurwitz if and only if conditions (c_1), (c_2) and (c_3) hold for the triple (A_i, B_i, C_i). Since asymptotic stability of system (10.11) is equivalent to the fact that all matrices $A_i + k_i B_i C_i$, $i \in [1, M]$, are compartmental and Hurwitz, the result follows.

Consider the triple (A_i, B_i, C_i) for a specific value of the index $i \in [1, M]$. For the sake of simplicity, similarly to what we did in the proof of the previous Proposition 10.1, we drop the dependence on the index i, and hence refer to the triple as (A, B, C), and we assume that the triple is already in the desired block form (i.e., $\Pi = I_n$).

As condition (b) of Proposition 10.1 holds, namely $B_j = 0$ for every $j \neq r$, then for every scalar k the matrix $A + kBC$ takes the following form

$$A + kBC = \left[\begin{array}{ccc|ccc} A_{11} & \dots & A_{1r} & A_{1r+1} & \dots & A_{1s} \\ \vdots & \ddots & \vdots & \vdots & \ddots & \vdots \\ kB_r C_1 & \dots & A_{rr} + kB_r C_r & A_{rr+1} + kB_r C_{r+1} & \dots & A_{rs} + kB_r C_s \\ \hline & & & A_{r+1r+1} & \dots & A_{r+1s} \\ & & & \vdots & \ddots & \vdots \\ & & & 0 & \dots & A_{ss} \end{array}\right].$$

Recalling that $B_r \neq 0$, a negative scalar k is such that $A + kBC$ is compartmental if and only if the following conditions hold:

(1) $kB_rC_j = 0$ for every $j \in [1, r-1]$, namely $C_j = 0$ for every $j \in [1, r-1]$ (i.e., condition (c_1) holds);
(2) $A_{rr} + kB_rC_r$ is compartmental;
(3) $A_{rj} + kB_rC_j$ is a nonnegative matrix for every $j \in [r+1, s]$.

It is easy to verify that there exists $k < 0$ satisfying condition (3) if and only if for every $j \in [r+1, s]$ such that $C_j \neq 0$ condition (10.13) holds. On the other hand, since by hypothesis condition (a) of Proposition 10.1 holds true, $A + kBC$ is compartmental and Hurwitz if and only if the matrix $A_{rr} + kB_rC_r$ is compartmental and Hurwitz. Recalling that A_{rr} is compartmental, irreducible and non-Hurwitz, by the monotonicity property of the spectral abscissa it follows that there exists $k < 0$ such that $A_{rr} + kB_rC_r$ is compartmental and Hurwitz if and only if also condition (c_3) holds true.

10.6 Affine Compartmental Switched Systems

In this section we consider (*continuous-time, single-input*) *affine compartmental switched systems* (ACSS), i.e., compartmental systems of type (10.3) for which the input function takes a constant value, i.e., $u(t) = \bar{u}, \forall t \in \mathbb{R}_+$. An ACSS is thus described by

$$\dot{\mathbf{x}}(t) = A_{\sigma(t)}\mathbf{x}(t) + b_{\sigma(t)}, \qquad (10.14)$$

where $b_{\sigma(t)} := B_{\sigma(t)}\bar{u}$. Clearly, for every $i \in [1, M]$, A_i is compartmental and b_i is positive. We assume that all pairs (A_i, b_i), $i \in [1, M]$, are stabilizable, in the standard sense of linear systems. This amounts to saying that when the matrix A_i is not Hurwitz, and hence $\lambda_F(A_i) = 0$, then the Hautus test matrix evaluated at zero, $[sI_n - A_i | b_i]|_{s=0}$, has rank n, namely b_i cannot be expressed as a linear combination of the columns of A_i. In particular, $b_i \neq 0$.

We say that a state $\bar{\mathbf{x}} > 0$ is a *switched equilibrium point* of (10.14) if the origin is included in the convex hull of the vectors $A_i\bar{\mathbf{x}} + b_i$, $i \in [1, M]$ (see [1] for details on discontinuous differential equations). Notice that, in general, $\bar{\mathbf{x}}$ is not an equilibrium point of any of the affine subsystems $\dot{\mathbf{x}}(t) = A_i\mathbf{x}(t) + b_i$, $i \in [1, M]$. However, if $\bar{\mathbf{x}} > 0$ is a switched equilibrium point, there exists $\alpha = \begin{bmatrix}\alpha_1 & \dots & \alpha_M\end{bmatrix}^\top \in \mathscr{A}_M$ such that

$$0 = \sum_{i=1}^{M} \alpha_i (A_i\bar{\mathbf{x}} + b_i) = A(\alpha)\bar{\mathbf{x}} + b(\alpha), \qquad (10.15)$$

where $A(\alpha) := \sum_{i=1}^{M} \alpha_i A_i$, and $b(\alpha) := \sum_{i=1}^{M} \alpha_i b_i$. By exploiting Theorem 2 in [25], we want to provide a characterization of all the switched equilibria that can be "reached" under some stabilizing switching law σ (see also [5]), by this meaning that for every $\varepsilon > 0$ we can ensure that there exists $\bar{t} > 0$ such that for every $t \geq \bar{t}$ the distance between the state trajectory and the switched equilibrium point is smaller than ε. To this end we need a preliminary lemma.

Lemma 10.1 *Let $A \in \mathbb{R}^{n\times n}$ be a reducible, non-Hurwitz, compartmental matrix in Frobenius normal form:*

$$A = \begin{bmatrix} A_{11} & A_{12} & \dots & A_{1s} \\ 0 & A_{22} & \dots & A_{2s} \\ \vdots & & \ddots & \vdots \\ 0 & \dots & & A_{ss} \end{bmatrix}, \tag{10.16}$$

where the diagonal blocks $A_{ii} \in \mathbb{R}^{n_i\times n_i}$, $i \in [1, s]$, are either scalar ($n_i = 1$) or irreducible matrices. Let $\mathscr{C}_i$ denote the communication class in $\mathscr{D}(A)$ associated with the block A_{ii}. The matrix A admits as left Frobenius eigenvector $\mathbf{v}_F$, corresponding to $\lambda_F(A) = 0$, a positive vector that can be partitioned according to the block-partition of A

$$\mathbf{v}_F^\top = \begin{bmatrix} \mathbf{v}_1^\top & \mathbf{v}_2^\top & \dots & \mathbf{v}_s^\top \end{bmatrix},$$

and whose blocks $\mathbf{v}_i \in \mathbb{R}_+^{n_i}$ satisfy the following conditions:

(1) if $\mathscr{C}_i$ is a conservative class, *namely it is associated with a non-Hurwitz block A_{ii}, then $\mathbf{v}_i = \mathbf{1}_{n_i}$;*
(2) if $\mathscr{C}_i$ is associated with a Hurwitz block A_{ii}, and $\mathscr{C}_i$ has not access to any conservative class, then $\mathbf{v}_i = 0$;
(3) if $\mathscr{C}_i$ is associated with a Hurwitz block A_{ii}, and $\mathscr{C}_i$ has access to some conservative class, then $\mathbf{v}_i \gg 0$.

Proof We first note that, by the assumptions on A (see Remark 10.2), a permutation matrix Π can be found such that $\Pi^\top A \Pi$ is described as in (10.8), where the first q (irreducible or scalar) diagonal blocks are singular and satisfy $\mathbf{1}_{n_i}^\top A_{ii} = 0$, while the remaining $s - q$ diagonal blocks are Hurwitz. It entails no loss of generality assuming that A has the structure given in (10.8) (namely $\Pi = I_n$), since this can be achieved by simply permuting the blocks of A and hence those of $\mathbf{v}_F$.

If we denote by $\mathscr{I}_1$, $\mathscr{I}_2$ and $\mathscr{I}_3$ the set of indices of the classes in (1), (2) and (3), respectively, then clearly $\mathscr{I}_1 = [1, q]$, while $\mathscr{I}_2 \cup \mathscr{I}_3 = [q + 1, s]$. Moreover, no class $\mathscr{C}_i$, $i \in \mathscr{I}_2$, has access to any class $\mathscr{C}_j, j \in \mathscr{I}_1 \cup \mathscr{I}_3$.

(1) The conservative classes are those corresponding to the first q diagonal blocks, and we have already pointed out that $\mathbf{1}_{n_i}^\top A_{ii} = 0$, $\forall\, i \in \mathscr{I}_1 = [1, q]$. So, the (essential) uniqueness of the left Frobenius eigenvector of an irreducible matrix, ensures that $\mathbf{v}_i = \mathbf{1}_{n_i}$, $\forall\, i \in \mathscr{I}_1$.
(2) We prove this result by induction. Let $i \in [q + 1, s]$ be the smallest index in $\mathscr{I}_2$. Then $\mathscr{C}_i$ is a communication class associated with a Hurwitz block and it has access to no other class, namely $A_{ji} = 0$ for every $j < i$. So, condition

$$\mathbf{v}_1^\top A_{1i} + \mathbf{v}_2^\top A_{2i} + \dots + \mathbf{v}_j^\top A_{ji} + \dots + \mathbf{v}_i^\top A_{ii} = 0, \tag{10.17}$$

becomes $\mathbf{v}_i^\top A_{ii} = 0$, and since A_{ii} is nonsingular, then $\mathbf{v}_i = 0$.
Suppose, now, that $i \in [q+2, s], i \in \mathscr{I}_2$, and we have shown that for every $j \in \mathscr{I}_2, j < i$, condition (2) holds. Then for every $j < i$, if $j \in \mathscr{I}_1 \cup \mathscr{I}_3$ then $A_{ji} = 0$, if $j \in \mathscr{I}_2$ then $\mathbf{v}_j = 0$. Consequently, (10.17) becomes, again, $\mathbf{v}_i^\top A_{ii} = 0$, and since A_{ii} is nonsingular, $\mathbf{v}_i = 0$.

(3) We prove also this fact by induction. Let $i \in [q+1, s]$ be the smallest index in $\mathscr{I}_3$. Then $\mathscr{C}_i$ is a communication class associated with a Hurwitz block and it has direct access to (distance 1 from) some conservative class $\mathscr{C}_j, j \in \mathscr{I}_1 = [1, q]$, by this meaning that there is an arc from some vertex in $\mathscr{C}_i$ to some vertex in $\mathscr{C}_j$. This amounts to saying that $A_{ji} > 0, \exists\, j \in \mathscr{I}_1$. On the other hand, for every $k < i, k \in \mathscr{I}_2$, (if any), we have already proved that $\mathbf{v}_k = 0$. So, condition (10.17) implies

$$\begin{aligned} \mathbf{v}_i^\top &= [\mathbf{v}_1^\top A_{1i} + \mathbf{v}_2^\top A_{2i} + \cdots + \mathbf{v}_j^\top A_{ji} + \cdots + \mathbf{v}_{i-1}^\top A_{i-1i}](-A_{ii})^{-1} \\ &\geq [\mathbf{v}_j^\top A_{ji}](-A_{ii})^{-1}, \end{aligned}$$

where we used the fact that A_{ii} is Hurwitz and irreducible (or scalar), and hence the matrix $(-A_{ii})^{-1}$ is strictly positive [2]. On the other hand, $\mathbf{v}_j = \mathbf{1}_{n_j} \gg 0$ and $A_{ji} > 0$. This ensures that $[\mathbf{v}_j^\top A_{ji}](-A_{ii})^{-1} \gg 0$, and hence $\mathbf{v}_i \gg 0$.

Suppose, now, that $i \in [q+2, s], i \in \mathscr{I}_3$, and we have shown that for every $j \in \mathscr{I}_3, j < i$, condition (3) holds. Then for every $j < i$, if $j \in \mathscr{I}_1 \cup \mathscr{I}_3$ then $\mathbf{v}_j \gg 0$ (and there exists $j \in \mathscr{I}_1 \cup \mathscr{I}_3$ such that $A_{ji} > 0$), if $j \in \mathscr{I}_2$ then $\mathbf{v}_j = 0$. Consequently, (10.17) leads, again, to $\mathbf{v}_i^\top \geq [\mathbf{v}_j^\top A_{ji}](-A_{ii})^{-1} \gg 0$.

This completes the proof.

We are now in a position to introduce the main result of this section, that adapts to the class of affine compartmental switched systems the characterization first given in [3].

Theorem 10.1 *Suppose that the switched compartmental system (10.14) is exponentially stabilizable [3, 25], by this meaning that when all the b_i's are set to zero in (10.14) the state trajectories can be driven to zero (by resorting to some switching control law). Also, assume that each pair $(A_i, b_i), i \in [1, M]$, in (10.14) is stabilizable. Then the set of all switched equilibrium points of system (10.14) that can be reached by resorting to some switching control law σ is given by*

$$\mathscr{E} = \{\bar{\mathbf{x}} > 0 : \bar{\mathbf{x}} = -A(\alpha)^{-1} b(\alpha), \exists\, \alpha \in \mathscr{A}_M^H\},$$

where $\mathscr{A}_M^H := \{\alpha \in \mathscr{A}_M : A(\alpha)$ is Hurwitz$\}$.

Proof We preliminary notice that the exponential stabilizability assumption on the switched compartmental system

$$\dot{\mathbf{x}}(t) = A_{\sigma(t)}\mathbf{x}(t), \tag{10.18}$$

ensures by Theorem 2 in [25] that the set $\mathscr{A}_M^H$ is not empty, and hence $\mathscr{E} \neq \emptyset$. Clearly, all elements of $\mathscr{E}$ are switched equilibrium points, since they satisfy Eq. (10.15). We now prove the converse, namely that all equilibria belong to $\mathscr{E}$. This amounts to saying that if $\bar{\mathbf{x}} > 0$ satisfies $A(\alpha)\bar{\mathbf{x}} + b(\alpha) = 0$ for some $\alpha \in \mathscr{A}_M$, then $A(\alpha)$ is Hurwitz.

Suppose, by contradiction, it is not. Then, being the convex combination of compartmental matrices, it will be compartmental with $\lambda_F(A(\alpha)) = 0$. If $A(\alpha)$ is irreducible, then $\mathbf{1}_n^\top A(\alpha) = 0$. Consequently, for every $i \in [1, M]$ such that $\alpha_i > 0$, one has $\mathbf{1}_n^\top A_i = 0$, thus implying that A_i is not Hurwitz. On the other hand, $\mathbf{1}_n^\top b(\alpha) = \mathbf{1}_n^\top (A(\alpha)\bar{\mathbf{x}} + b(\alpha)) = 0$, and this implies that for every $i \in [1, M]$ such that $\alpha_i > 0$, one has $b_i = 0$. This contradicts the stabilizability assumption on the pairs (A_i, b_i), $i \in [1, M]$, such that $\alpha_i > 0$.

Suppose, now, that $A(\alpha)$ is reducible. It entails no loss of generality assuming that $A(\alpha)$ is in Frobenius normal form (10.16) and $b(\alpha)$ is accordingly partitioned as

$$b(\alpha) = \begin{bmatrix} B_1(\alpha) \\ B_2(\alpha) \\ \vdots \\ B_s(\alpha) \end{bmatrix},$$

where $A_{ii}(\alpha) \in \mathbb{R}^{n_i \times n_i}$, $i \in [1, s]$, are either scalar ($n_i = 1$) or irreducible matrices, and $B_i(\alpha) \in \mathbb{R}_+^{n_i}$. This is a not restrictive assumption, since we can always reduce ourselves to this situation by resorting to a suitable permutation matrix Π, and hence moving from the pair $(A(\alpha), b(\alpha))$ to the pair $(\Pi^\top A(\alpha)\Pi, \Pi^\top b(\alpha))$. Now consider the left Frobenius eigenvector of $A(\alpha)$, $\mathbf{v}_F$, corresponding to $\lambda_F(A(\alpha)) = 0$ and given in Lemma 10.1, partitioned accordingly to the block-partition of $A(\alpha)$ and $b(\alpha)$ as

$$\mathbf{v}_F^\top = \begin{bmatrix} \mathbf{v}_1^\top \ \mathbf{v}_2^\top \ \dots \ \mathbf{v}_s^\top \end{bmatrix}, \qquad \text{with } \mathbf{v}_i \in \mathbb{R}_+^{n_i}.$$

By the previous lemma, we know that $\mathbf{v}_i \neq 0$ if and only if the class $\mathscr{C}_i$ is either conservative ($A_{ii}(\alpha)$ is singular) or it has access to some conservative class, and if $\mathbf{v}_i \neq 0$ then $\mathbf{v}_i \gg 0$. We denote by $\mathscr{I}$ the set of indices $i \in [1, s]$ such that $\mathbf{v}_i \gg 0$. According to the notation used within the proof of Lemma 10.1, $\mathscr{I} = \mathscr{I}_1 \cup \mathscr{I}_3$. So, condition

$$\mathbf{v}_F^\top (A(\alpha)\bar{\mathbf{x}} + b(\alpha)) = 0$$

implies $\mathbf{v}_F^\top b(\alpha) = 0$, and hence $B_i(\alpha) = 0$ for every $i \in \mathscr{I}$. This allows to say that a (new) permutation matrix $\tilde{\Pi}$ can be found such that

$$\tilde{\Pi}^\top A(\alpha)\tilde{\Pi} = \begin{bmatrix} D_{11}(\alpha) & 0 & D_{13}(\alpha) \\ 0 & D_{22}(\alpha) & D_{23}(\alpha) \\ 0 & 0 & D_{33}(\alpha) \end{bmatrix}, \qquad \tilde{\Pi}^\top B(\alpha) = \begin{bmatrix} 0 \\ E_2(\alpha) \\ 0 \end{bmatrix},$$

where $D_{11}(\alpha)$ is a block diagonal matrix that groups together all the diagonal blocks $A_{ii}(\alpha)$ in $A(\alpha)$ that are irreducible and conservative, $D_{22}(\alpha)$ is a block triangular matrix that groups together all the diagonal blocks $A_{ii}(\alpha)$ in $A(\alpha)$ that are irreducible, Hurwitz and correspond to classes that have no access to conservative classes, and finally $D_{33}(\alpha)$ is a block triangular matrix that groups together all the diagonal blocks $A_{ii}(\alpha)$ in $A(\alpha)$ that are irreducible, Hurwitz and correspond to classes that have access to some conservative class. Also, $D_{13}(\alpha) > 0, D_{23}(\alpha) \geq 0$ and $E_2(\alpha) > 0$. It is easily seen that for every $j \in [1, M]$ such that $\alpha_j > 0$ one has

$$\tilde{\Pi}^\top A_j \tilde{\Pi} = \begin{bmatrix} D_{11}^{(j)} & 0 & D_{13}^{(j)} \\ 0 & D_{22}^{(j)} & D_{23}^{(j)} \\ 0 & 0 & D_{33}^{(j)} \end{bmatrix}, \qquad \tilde{\Pi}^\top b_j = \begin{bmatrix} 0 \\ E_2^{(j)} \\ 0 \end{bmatrix},$$

and that $D_{11}^{(j)}$ is a block diagonal matrix whose diagonal blocks are conservative and hence singular. It is also clear that for every $K_j \tilde{\Pi} = \begin{bmatrix} K_1^{(j)} & K_2^{(j)} & K_3^{(j)} \end{bmatrix}$ one has

$$\begin{aligned} \tilde{\Pi}^\top (A_j + b_j K_j) \tilde{\Pi} &= \tilde{\Pi}^\top A_j \tilde{\Pi} + \tilde{\Pi}^\top b_j K_j \tilde{\Pi} \\ &= \begin{bmatrix} D_{11}^{(j)} & 0 & D_{13}^{(j)} \\ E_2^{(j)} K_1^{(j)} & D_{22}^{(j)} + E_2^{(j)} K_2^{(j)} & D_{23}^{(j)} + E_2^{(j)} K_3^{(j)} \\ 0 & 0 & D_{33}^{(j)} \end{bmatrix}, \end{aligned}$$

and hence $0 \in \sigma(A_j + b_j K_j)$ for every K_j, thus contradicting the stabilizability assumption on the pair (A_j, b_j). Therefore $A(\alpha)$ must be Hurwitz and hence $\bar{\mathbf{x}}$ belongs to $\mathscr{E}$.

The second part of the proof proceeds like the one in [4] and we report it here only for the sake of completeness. We now want to prove that all points in $\mathscr{E}$ are equilibria achievable by means of some stabilizing switching control law. Let $A(\alpha), \alpha \in \mathscr{A}_M^H$, be a Hurwitz matrix and let $P = P^\top$ be a positive definite matrix such that $A^\top(\alpha)P + PA(\alpha)$ is negative definite. Let $\bar{\mathbf{x}}$ be the element of $\mathscr{E}$ corresponding to $A(\alpha)$, and consider the control Lyapunov function $V(\mathbf{x} - \bar{\mathbf{x}}) := (\mathbf{x} - \bar{\mathbf{x}})^\top P(\mathbf{x} - \bar{\mathbf{x}})$ and the control strategy $\sigma(t) = u(\mathbf{x}(t))$ where

$$\begin{aligned} u(\mathbf{x}) &= \arg\min_i (A_i\mathbf{x} + b_i)^\top P(\mathbf{x} - \bar{\mathbf{x}}) + (\mathbf{x} - \bar{\mathbf{x}})^\top P(A_i\mathbf{x} + b_i) \\ &= \arg\min_i 2(\mathbf{x} - \bar{\mathbf{x}})^\top P(A_i\mathbf{x} + b_i). \end{aligned} \tag{10.19}$$

Keeping in mind that $A(\alpha)\bar{\mathbf{x}} = -b(\alpha)$, we have for $\mathbf{x} \neq \bar{\mathbf{x}}$

$$\begin{aligned} (\mathbf{x} - \bar{\mathbf{x}})^\top P(A_i\,\mathbf{x} + b_i) &= \underbrace{(\mathbf{x} - \bar{\mathbf{x}})^\top PA(\alpha)(\mathbf{x} - \bar{\mathbf{x}})}_{<0} \\ &+ (\mathbf{x} - \bar{\mathbf{x}})^\top P[(A_i\mathbf{x} + b_i) - (A(\alpha)\mathbf{x} + b(\alpha))]. \end{aligned}$$

The first term on the right hand side is negative. On the other hand, by construction, the vector $A(\alpha)\mathbf{x} + b(\alpha)$ belongs to the convex hull of the vectors $A_i\mathbf{x} + b_i$, and hence

$$\min_i (\mathbf{x}-\bar{\mathbf{x}})^\top P[(A_i\mathbf{x}+b_i)-(A(\alpha)\mathbf{x}+b(\alpha))]\leq 0.$$

This ensures that $\min_i \dot{V}(\mathbf{x}-\bar{\mathbf{x}}) = \min_i 2(\mathbf{x}-\bar{\mathbf{x}})^\top P(A_i\mathbf{x}+b_i)$ is negative, and hence we have a stabilizing switching law that leads the system evolution to $\bar{\mathbf{x}}$.

References

1. Aubin, J.P., Cellina, A.: Differential Inclusions. Set Valued Maps and Viability Theory. Springer (1984)
2. Berman, A., Plemmons, R.J.: Nonnegative Matrices in the Mathematical Sciences. Academic Press, New York (1979)
3. Blanchini, F., Colaneri, P., Valcher, M.E.: Co-positive Lyapunov functions for the stabilization of positive switched systems. IEEE Trans. Autom. Control **57**(12), 3038–3050 (2012)
4. Blanchini, F., Colaneri, P., Valcher, M.E.: Switched linear positive systems. Found. Trends Syst. Control **2**(2), 101–273 (2015)
5. Bolzern, P., Colaneri, P., De Nicolao, G.: On almost sure stability of discrete-time Markov jump linear systems. In: Proceedings of 43rd IEEE Conference on Decision and Control, Atlantis, Paradise Island, The Bahamas, pp. 3204–3208 (2004)
6. Bru, R., Romero, S., Sánchez, E.: Canonical forms for positive discrete-time linear control systems. Linear Algebra Appl. **310**, 49–71 (2000)
7. Brualdi, R.A., Ryser, H.J.: Combinatorial Matrix Theory. Cambridge University Press (1991)
8. de Leenheer, P., Aeyels, D.: Stabilization of positive linear systems. Syst. Control Lett. **44**, 259–271 (2001)
9. Fornasini, E., Valcher, M.E.: Linear copositive Lyapunov functions for continuous-time positive switched systems. IEEE Trans. Autom. Control **55**(8), 1933–1937 (2010)
10. Frobenius, G.F.: Über Matrizen aus nicht Negativen Elementen. Sitzungsber. Kon. Preuss. Akad. Wiss. Berlin, 1912, pp. 456–477, Ges. Abh., Springer, vol. 3:546–567 (1968)
11. Gantmacher, F.R.: The Theory of Matrices. Chelsea Pub. Co. (1960)
12. Gurvits, L., Shorten, R., Mason, O.: On the stability of switched positive linear systems. IEEE Trans. Autom. Control **52**(6), 1099–1103 (2007)
13. Haddad, W.M., Chellaboina, V., Hui, Q.: Nonnegative and Compartmental Dynamical Systems. Princeton University Press (2010)
14. Hou, S.P., Meskin, N., Haddad, W.M.: A general multicompartment lung mechanics model with nonlinear resistance and compliance respiratory parameters. In: Proceedings of 2014 American Control Conference, Portland, OR, pp. 566–571 (2014)
15. Knorn, F., Mason, O., Shorten, R.N.: On linear co-positive Lyapunov functions for sets of linear positive systems. Automatica **45**(8), 1943–1947 (2009)
16. Li, H., Haddad, W.M.: Optimal determination of respiratory airflow patterns using a non-linear multicompartment model for a lung mechanics system. Comput. Math. Meth. Med. **2012**(165946) (2012)
17. Lin, L.: Stabilization analysis for economic compartmental switched systems based on quadratic Lyapunov function. Nonlinear Anal. Hybrid Syst. **2**, 1187–1197 (2008)
18. Mason, O., Shorten, R.N.: On linear copositive Lyapunov functions and the stability of switched positive linear systems. IEEE Trans. Autom. Control **52**(7), 1346–1349 (2007)
19. Minc, H.: Nonnegative Matrices. Wiley, New York (1988)
20. Schneider, H.: The influence of the marked reduced graph of a nonnegative matrix on the Jordan form and on related properties. Linear Algebra Appl. **84**, 161–189 (1986)
21. Smith, H.L.: Linear Compartmental Systems—The Basics (2006)
22. Son, N.K., Hinrichsen, D.: Robust stability of positive continuous time systems. Numer. Funct. Anal. Opt. **17**(5 & 6), 649–659 (1996)

23. Taussky, O.: A recurring theorem on determinants. Am. Math. Mon. **56**(10), 672–676 (1949)
24. Valcher, M.E.: Controllability and reachability criteria for discrete time positive systems. Int. J. Control **65**, 511–536 (1996)
25. Valcher, M.E., Zorzan, I.: Stability and stabilizability of continuous-time compartmental switched systems. IEEE Trans. Autom. Control. **61**(12), 3885–3897 (2016). doi:10.1109/TAC.2016.2525016
26. Valcher, M.E., Zorzan, I.: On the stabilizability of continuous-time compartmental switched systems. In: Proceedings of the 54th IEEE Conf. on Decision and Control, pp. 4246–4251, Osaka, Japan (2015)
27. Valcher, M.E., Zorzan, I.: New results on the solution of the positive consensus problem. In: Proceedings of the 55th IEEE Conf. on Decision and Control, pp. 5251–5256, Las Vegas, Nevada (2016)

Chapter 11
Improved Controller Design for Positive Systems and Its Application to Positive Switched Systems

Junfeng Zhang, Linli Ma, Qian Wang, Yun Chen and Shaosheng Zhou

Abstract This chapter will address a new controller design approach for positive systems. First, we decompose the feedback gain matrix $K_{m\times n}$ into $m \times n$ nonnegative components and $m \times n$ non-positive components. For the nonnegative components, each component contains only one positive element and the other ones are zero. Similarly, each non-positive component contains only one negative element and the other ones are zero. Then, a simple but effective controller design of positive systems is proposed by incorporating the decomposed feedback gain matrix into the resulting closed-loop systems. The present approach is thus applied to positive switched systems. It is shown that the designed controller for positive switched systems is less conservative than those ones in the literature.

Keywords Positive systems · Controller design · Linear programming · Positive switched systems.

J. Zhang (✉) · L. Ma · Q. Wang · Y. Chen · S. Zhou
Key Lab for IOT and Information Fusion Technology of Zhejiang,
Hangzhou Dianzi University, Hangzhou 310018, China
e-mail: jfz5678@126.com

L. Ma
e-mail: malinlity@163.com

Q. Wang
e-mail: wq@hdu.edu.cn

Y. Chen
e-mail: cloudscy@hdu.edu.cn

S. Zhou
e-mail: sszhou@hdu.edu.cn

J. Zhang
Institute of Information and Control, School of Automation,
Hangzhou Dianzi University, Hangzhou 310018, China

F. Cacace et al. (eds.), *Positive Systems*, Lecture Notes in Control
and Information Sciences 471, DOI 10.1007/978-3-319-54211-9_11

11.1 Introduction

Positive systems are a special class of control systems [1]. Over past two decades, positive systems have gained increasing interests due to their extensive applications in practice and theoretical complexes in control theory [2–9]. Compared with general systems, positive systems do not receive much attention until this century. This leads to that many issues of positive systems are open.

As general systems, stabilization is also a fundamental issue of positive systems. There have been some significant results on the stabilization of positive systems. A linear programming approach to controller design of positive systems was proposed in [10, 11]. The output-feedback controller of positive systems [12] was proposed by using the approach in [10, 11]. The problem of ℓ_1-induced state-feedback controller design for positive systems was investigated by using a linear copositive Lyapunov function in [13]. In [14], a static output-feedback controller design was presented, where an iterative linear matrix inequality algorithm was provided to compute the feedback gain matrix. In [15], the output-feedback controller was designed by virtue of an iterative convex optimization algorithm. More results on positive systems can refer to [16–23].

As far as the stabilization of positive systems is concerned, it is clear that there is still much room for improvements in the above mentioned works. This motivates us to carry out the present work. This chapter will further provide a new controller design approach to remove some restrictions in the heavy computational burden, the controller gain matrix, and the unreliability algorithms in the literature. By decomposing the feedback gain matrix into parts, the new approach removes those restrictions in the literature. Our developed design approach is very efficient in solving the control synthesis problems of positive systems. An application to positive switched systems is also given to show the efficiency of the proposed approach. The rest of the chapter is organized as follows: Sect. 11.2 provides the problem statements; Sect. 11.3 gives main results; Sect. 11.4 concludes the chapter.

Notations Let $\Re$, $\Re^n$, $\Re^{n\times n}$ be the sets of real numbers, n-dimensional vectors and $n \times n$ matrices, respectively. Denote by $\mathbb{N}$, $\mathbb{N}^+$ the sets of nonnegative and positive integers. For a vector $x = (x_1, \dots, x_n)^T$, $x \succeq 0$ ($\succ 0$) means that $x_i \geq 0$ ($x_i > 0$) $\forall i = 1, \dots, n$. Similarly, $x \preceq 0$ ($\prec 0$) means that $x_i \leq 0$ ($x_i < 0$) $\forall i = 1, \dots, n$. For a matrix $A = [a_{ij}] \in \Re^{n\times n}$, $A \succeq 0$ ($\succ 0$) means that $a_{ij} \geq 0$ ($a_{ij} > 0$) $\forall i, j = 1, \dots, n$. Similarly, $A \preceq 0$ ($\prec 0$) means that $a_{ij} \leq 0$ ($a_{ij} < 0$) $\forall i, j = 1, \dots, n$. A matrix A is called as Metzler if all its non-diagonal elements are nonnegative. I is the identical matrix with proper dimension. $\Re^n_+ \triangleq \{x | x \in \Re^n, x \succeq 0\}$. Let $\mathbf{1}_n = (\underbrace{1, \dots, 1}_{n})^T$ and $\mathbf{1}_n^{(i)} = (\underbrace{0, \dots, 0}_{i-1}, 1, \underbrace{0, \dots, 0}_{n-i})^T$. Throughout the chapter, the dimensions of vectors and matrices are assumed to be compatible if not stated.

11.2 Problem Formulation

Consider the following system:

$$\begin{aligned} \dot{x}(t) &= Ax(t) + Bu(t), \\ y(t) &= Cx(t), \end{aligned} \tag{11.1}$$

where $x(t) \in \Re^n$, $u(t) \in \Re^m$ and $y(t) \in \Re^r$ are system state, control input, and output, respectively. Assume that $A \in \Re^{n\times n}$ is a Metzler matrix, $B \succeq 0$ with $B \in \Re^{n\times m}$, and $C \succeq 0$ with $C \in \Re^{r\times n}$.

The following preliminaries are first introduced for later use.

Definition 11.1 [3, 6] System (11.1) is positive if its state and output are nonnegative for all time t whenever the initial condition $x(t_0)$ and control input $u(t)$ are nonnegative.

Lemma 11.1 [3, 6] *System (11.1) is positive if and only if A is a Metzler matrix, $B \succeq 0$ and $C \succeq 0$.*

Noting the assumptions for system (11.1), it follows that system (11.1) is positive by Lemma 11.1.

Lemma 11.2 *A matrix M is Metzler if and only if there exists a positive constant ς such that $M + \varsigma I \succeq 0$.*

11.3 Main Results

In this section, we will address the stabilization of positive systems and positive switched systems (PSSs). The objective of the stabilization is to design a controller such that the resulting closed-loop system is positive and stable.

11.3.1 Stabilization of Positive Systems

We first consider the stabilization of system (11.1).

Theorem 11.1 *If there exist constants $\varsigma > 0, k_{ij}^+ > 0, k^+ > 0, k_{ij}^- < 0, k^- < 0$ and vectors $v \succ 0$ with $v \in \Re^n$ such that*

$$A^T v + \zeta^+ + \zeta^- \prec 0, \tag{11.2a}$$

$$\begin{aligned} A\mathbf{1}_m^T B^T v + B \textstyle\sum_{i=1}^m \sum_{j=1}^r \mathbf{1}_m^{(i)} (\zeta_{ij}^+ \\ + \zeta_{ij}^-)^T + \varsigma I \succeq 0, \end{aligned} \tag{11.2b}$$

$$k_{ij}^{+} < k^{+}, \tag{11.2c}$$

$$k_{ij}^{-} < k^{-}, \tag{11.2d}$$

hold for $i = 1, \ldots, m,\ j = 1, \ldots, n,$ *where* $\zeta_{ij}^{\pm} = (\underbrace{0, \ldots, 0}_{j-1}, k_{ij}^{\pm}, \underbrace{0, \ldots, 0}_{n-j})^T \in \Re^n$ *and* $\zeta^{\pm} = (k^{\pm}, \ldots, k^{\pm})^T \in \Re^n$*, then under the state-feedback control law*

$$\begin{aligned} u(t) &= Kx(t) \\ &= \frac{\Sigma_{i=1}^{m} \Sigma_{j=1}^{n} \mathbf{1}_m^{(i)} (\zeta_{ij}^{+} + \zeta_{ij}^{-})^T}{\mathbf{1}_m^T B^T v} x(t) \end{aligned} \tag{11.3}$$

the resulting closed-loop system (11.1) is positive and asymptotically stable.

Proof By $\mathbf{1}_m \succ 0$ with $\mathbf{1}_m \in \Re^m$, $B \succeq 0$ with $B \in \Re^{n \times m}$, and $v \succ 0$ with $v \in \Re^n$, we have $\mathbf{1}_r^T B^T v > 0$. This together with (11.2b) gives that

$$\begin{aligned} A + B &\frac{\Sigma_{i=1}^{m} \Sigma_{j=1}^{n} \mathbf{1}_m^{(i)} (\zeta_{ij}^{+} + \zeta_{ij}^{-})^T}{\mathbf{1}_m^T B^T v} \\ &+ \frac{\varsigma}{\mathbf{1}_m^T B^T v} I \succeq 0. \end{aligned} \tag{11.4}$$

Using (11.3), it follows that

$$A + BK + \frac{\varsigma}{\mathbf{1}_m^T B^T v} I \succeq 0 \tag{11.5}$$

By Lemma 11.2, $A + BK$ is a Metzler matrix. Then, the closed-loop system (11.1) is positive by Lemma 11.1, that is, $x(t) \succeq 0\ \forall t \geq 0$.

Choose a linear copositive Lyapunov function candidate $V(x(t)) = x(t)^T v$. Then

$$\dot{V}(x(t)) = x(t)^T (A^T v + K^T B^T v). \tag{11.6}$$

By (11.2c) and (11.2d), we get

$$\begin{aligned} &\textstyle\sum_{i=1}^{m} \sum_{j=1}^{n} \mathbf{1}_m^{(i)} (\zeta_{ij}^{+} + \zeta_{ij}^{-})^T \\ &= \textstyle\sum_{i=1}^{m} \mathbf{1}_m^{(i)} \sum_{j=1}^{n} (\zeta_{ij}^{+} + \zeta_{ij}^{-})^T \\ &\preceq \textstyle\sum_{i=1}^{m} \mathbf{1}_m^{(i)} (\zeta^{+} + \zeta^{-})^T \\ &= \mathbf{1}_m (\zeta^{+} + \zeta^{-})^T. \end{aligned} \tag{11.7}$$

Furthermore,

$$\begin{aligned}&K^TB^Tv\\&=\frac{\Sigma_{i=1}^m\Sigma_{j=1}^n(\zeta_{ij}^+ +\zeta_{ij}^-)\mathbf{1}_m^{(i)T}B^Tv}{\mathbf{1}_m^TB^Tv}\\&\preceq\frac{(\zeta^+ +\zeta^-)\mathbf{1}_m^TB^Tv}{\mathbf{1}_m^TB^Tv}\\&=\zeta^+ +\zeta^-.\end{aligned}\tag{11.8}$$

With the fact $x(t)\succeq 0$ in mind, one can obtain from (11.6) that

$$\dot{V}(x(t))\le x(t)^T(A^Tv+\zeta^+ +\zeta^-).\tag{11.9}$$

By (11.2a), we have $\dot{V}(x(t))<0$. This completes the proof. □

Remark 11.1 In Theorem 11.1, the gain matrix K is decomposed into

$$\begin{aligned}K&=\frac{1}{\mathbf{1}_m^TB^Tv}\begin{pmatrix}k_{11}^+ & k_{12}^+ & \cdots & k_{1n}^+\\ k_{21}^+ & k_{22}^+ & \cdots & k_{2n}^+\\ \vdots & \vdots & \ddots & \vdots\\ k_{m1}^+ & k_{m2}^+ & \cdots & k_{mn}^+\end{pmatrix}\\&\quad+\frac{1}{\mathbf{1}_m^TB^Tv}\begin{pmatrix}k_{11}^- & k_{12}^- & \cdots & k_{1n}^-\\ k_{21}^- & k_{22}^- & \cdots & k_{2n}^-\\ \vdots & \vdots & \ddots & \vdots\\ k_{m1}^- & k_{m2}^- & \cdots & k_{mn}^-\end{pmatrix}\\&=\frac{\sum_{i=1}^m\sum_{j=1}^n\mathbf{1}_m^{(i)}(\zeta_{ij}^+ +\zeta_{ij}^-)^T}{\mathbf{1}_m^TB^Tv}.\end{aligned}\tag{11.10}$$

Thus, the term K^TB^Tv is transformed into the linear programming form. It should be pointed out that the rank of the gain matrix K is general without any restrictions. The condition (2) is solvable by using the linear programming technique.

Remark 11.2 (i) Theorem 11.1 gives the sufficient condition for the existence of feedback controller of positive systems whereas in the literature [10–15] some necessary and sufficient conditions were established. Then, Theorem 11.1 is more conservative than those results in the literature.

(ii) In [13–15], some iterative algorithms were addressed to compute the controller gain matrix. These algorithms contain some complexities and unreliability such as the introduction of some additional parameters and an initial controller gain. The design in [10–12] is nice if one only considers the stabilization of positive systems. In our opinion, the design in [10–12] seems to be restricted if applying it to hybrid positive systems.

(iii) Aiming to these restrictions in those literature, Theorem 11.1 is presented. The advantages of Theorem 11.1 lie in: (a) the implemental algorithm is easy, (b) the restriction in the gain matrix is removed, and (c) it can be easily applied to other control issues of hybrid positive systems.

The following corollary gives the output-feedback controller design of positive systems and its proof is omitted.

Corollary 11.1 *If there exist constants* $\varsigma > 0, k_{ij}^{+} > 0, k^{+} > 0, k_{ij}^{-} < 0, k^{-} < 0$ *and vectors* $\upsilon \succ 0$ *with* $\upsilon \in \Re^n$ *such that*

$$\begin{aligned}
&A^T\upsilon + C^T\zeta^{+} + C^T\zeta^{-} \prec 0,\\
&A\mathbf{1}_m^T B^T\upsilon + B\sum_{i=1}^{m}\sum_{j=1}^{r}\mathbf{1}_m^{(i)}(\zeta_{ij}^{+}\\
&\qquad + \zeta_{ij}^{-})^T C + \varsigma I \succeq 0,\\
&k_{ij}^{+} < k^{+},\\
&k_{ij}^{-} < k^{-},
\end{aligned} \tag{11.11}$$

hold for $i = 1, \ldots, m,\ \ j = 1, \ldots, r,$ *where* $\zeta_{ij}^{\pm} = (\underbrace{0, \ldots, 0}_{j-1}, k_{ij}^{\pm}, \underbrace{0, \ldots, 0}_{r-j})^T \in \Re^r$ *and* $\zeta^{\pm} = (k^{\pm}, \ldots, k^{\pm})^T \in \Re^r$*, then under the output-feedback control law*

$$\begin{aligned}
u(t) &= Ky(t)\\
&= \frac{\Sigma_{i=1}^{m}\Sigma_{j=1}^{r}\mathbf{1}_m^{(i)}(\zeta_{ij}^{+} + \zeta_{ij}^{-})^T}{\mathbf{1}_m^T B^T\upsilon}y(t)
\end{aligned} \tag{11.12}$$

the resulting closed-loop system (11.1) is positive and asymptotically stable.

11.3.2 Stabilization of PSSs

In this subsection, we propose the feedback controller design of PSSs by applying the present approach in Theorem 11.1. Consider the switched system:

$$\begin{aligned}
\dot{x}(t) &= A_{\sigma(t)}x(t) + B_{\sigma(t)}u(t),\\
y(t) &= C_{\sigma(t)}x(t),
\end{aligned} \tag{11.13}$$

where $x(t) \in \Re^n$, $u(t) \in \Re^m$, and $y(t) \in \Re^r$ are system state, control input, and output, respectively. The function $\sigma(t)$ represents the switching law, which is right continuous takes values in a finite set $S = \{1, 2, \ldots, J\}$, $J \in \mathbb{N}^{+}$. The $\sigma(t_i)$th subsystem is active for $t \in [t_i, t_{i+1})$, $i \in \mathbb{N}$, where t_i and t_{i+1} are the switching time instants. The states of system (11.1) are continuous and do not jump in the switching time instants. For system (11.1), assume that $A_p \in \Re^{n\times n}$ is a Metzler matrix and $B_p \succeq 0$ with $B_p \in \Re^{n\times m}$, $C_p \succeq 0$ with $C_p \in \Re^{r\times n}$ for each $p \in S$.

Theorem 11.2 *If there exist constants* $\varsigma_p > 0, k^+_{pij} > 0, k^+_p > 0, k^-_{pij} < 0, k^-_p < 0$ *and vectors* $v_p \succ 0$ *with* $v_p \in \Re^n$ *such that*

$$A^T_p v_p + \zeta^+_p + \zeta^-_p + \mu v_p \prec 0, \tag{11.14a}$$

$$\begin{aligned} A_p \mathbf{1}^T_m B^T_p v_p + B_p \textstyle\sum_{i=1}^m \sum_{j=1}^n \mathbf{1}^{(i)}_m (\zeta^+_{pij} \\ + \zeta^-_{pij})^T + \varsigma_p I \succeq 0, \end{aligned} \tag{11.14b}$$

$$k^+_{pij} < k^+_p, \tag{11.14c}$$

$$k^-_{pij} < k^-_p, \tag{11.14d}$$

$$v_p \prec \lambda v_q, \tag{11.14e}$$

hold for $i = 1, \ldots, m,\ j = 1, \ldots, n$, *where* $\zeta^{\pm}_{pij} = (\underbrace{0, \ldots, 0}_{j-1}, k^{\pm}_{pij}, \underbrace{0, \ldots, 0}_{n-j})^T \in R^n$ *and* $\zeta^{\pm}_p = (k^{\pm}_p, \ldots, k^{\pm}_p)^T \in R^n$, *then under the state-feedback control law*

$$\begin{aligned} u(t) &= K_p x(t) \\ &= \frac{\Sigma^m_{i=1} \Sigma^n_{j=1} \mathbf{1}^{(i)}_m (\zeta^+_{pij} + \zeta^-_{pij})^T}{\mathbf{1}^T_m B^T_p v_p} x(t) \end{aligned} \tag{11.15}$$

the resulting closed-loop system (11.13) is positive and asymptotically stable with the average dwell time satisfying

$$\tau > \tfrac{\ln \lambda}{\mu}. \tag{11.16}$$

Sketch of Proof From the proof of Theorem 11.1, we can get that, for each $p \in S$, the subsystem is positive and asymptotically stable under the state-feedback control law (11.15). Choose multiple linear copositive Lyapunov functions $V(x(t)) = x(t)^T v_{\sigma(t)}$, then

$$\begin{aligned} \dot{V}(x(t)) = x(t)^T \big(A^T_{\sigma(t_\iota)} v_{(\sigma(t_\iota))} \\ + K^T_{\sigma(t_\iota)} B^T_{\sigma(t_\iota)} v_{(\sigma(t_\iota))}\big) \end{aligned} \tag{11.17}$$

for $t \in [t_\iota, t_{\iota+1})$. From (11.14c), (11.14d), and (11.15), we can have

$$K^T_{\sigma(t_\iota)} B^T_{\sigma(t_\iota)} v_{(\sigma(t_\iota))} \preceq \zeta^+_{\sigma(t_\iota)} + \zeta^-_{\sigma(t_\iota)}. \tag{11.18}$$

With $x(t) \succeq 0$ in mind, substituting (11.18) into (11.17) gives

$$\begin{aligned} \dot{V}(x(t)) \le x(t)^T \big(A^T_{\sigma(t_\iota)} v_{(\sigma(t_\iota))} \\ + \zeta^+_{\sigma(t_\iota)} + \zeta^-_{\sigma(t_\iota)}\big). \end{aligned} \tag{11.19}$$

This together with (11.14a) yields

$$\dot{V}(x(t)) \leq -\mu V(x(t)) \tag{11.20}$$

for $t \in [t_\iota, t_{\iota+1})$. Then,

$$V(x(t)) \leq e^{-\mu(t-t_\iota)} V(x(t_\iota)) \tag{11.21}$$

for $t \in [t_\iota, t_{\iota+1})$. By (11.14e), it follows that

$$V(x(t)) \leq \lambda e^{-\mu(t-t_\iota)} V(x(t_\iota^-)). \tag{11.22}$$

By recursive deduction, we get

$$\begin{aligned} V(x(t)) &\leq \lambda^2 e^{-\mu(t-t_{\iota-1})} V(x(t_{\iota-2})) \\ &\leq \cdots \\ &\leq \lambda^{N_{\sigma(t_0,t)}} e^{-\mu(t-t_0)} V(x(t_0)), \end{aligned} \tag{11.23}$$

where $N_{\sigma(t_0,t)}$ is the number of the switching in $[t_0, t]$. Noting $\lambda > 1$, (11.23) is transformed into

$$\begin{aligned} V(x(t)) &\leq \lambda^{N_0+\frac{t-t_0}{\tau}} e^{-\mu(t-t_0)} V(x(t_0)) \\ &= \lambda^{N_0} e^{(\frac{\ln\lambda}{\tau}-\mu)(t-t_0)} V(x(t_0)), \end{aligned} \tag{11.24}$$

where N_0 is the chatter bound. Then

$$||x(t)||_1 \leq \frac{\varrho_2 \lambda^{N_0}}{\varrho_1} e^{(\frac{\ln\lambda}{\tau}-\mu)(t-t_0)} ||x(t_0)||_1, \tag{11.25}$$

where ϱ_1 and ϱ_2 are the minimal and maximal elements of υ_p $\forall p \in S$. By (11.15), $\frac{\ln\lambda}{\tau} - \mu < 0$. In addition, $\frac{\varrho_2 \lambda^{N_0}}{\varrho_1} > 0$ is obvious. So, the resulting closed-loop system (11.13) is positive and exponentially stable. □

Remark 11.3 In [24, 25], the state-feedback controllers of PSSs and nonlinear PSSs were proposed. In should be pointed out that the controller gain matrices contain the restriction on the rank. In [26], we remove the restriction in [24, 25]. However, the method in [26] contain a new restriction on average dwell time. Theorem 11.2 has removed the restrictions in [24–26].

Remark 11.4 It is also worthy noting that the approach in Theorem 11.2 can be applied to positive time-delay systems [27] and thus the restriction in [27] can be removed. Up to now, there have been many interesting results on hybrid positive systems referring to positive Markovian jump systems and positive T-S fuzzy systems. We notice that, when considering the issues of hybrid positive systems, a common restriction is just the one stated in Remark 11.3. Therefore, Theorem 11.2 can be further extended for those issues.

Corollary 11.2 *If there exist constants $\varsigma_p > 0, k_{pij}^+ > 0, k_p^+ > 0, k_{pij}^- < 0, k_p^- < 0$ and vectors $v_p \succ 0$ with $v_p \in \Re^n$ such that*

$$\begin{aligned}
&A_p^T v_p + C_p^T \zeta_p^+ + C_p^T \zeta_p^- + \mu v_p \prec 0,\\
&A_p \mathbf{1}_m^T B_p^T v_p + B_p \sum_{i=1}^m \sum_{j=1}^r \mathbf{1}_m^{(i)} (\zeta_{pij}^+ \\
&\qquad\qquad + \zeta_{pij}^-)^T C_p + \varsigma_p I \succeq 0,\\
&k_{pij}^+ < k_p^+,\\
&k_{pij}^- < k_p^-,\\
&v_p \prec \lambda v_q,
\end{aligned} \tag{11.26}$$

hold for $i = 1, \ldots, m,\ j = 1, \ldots, r$, where $\zeta_{pij}^{\pm} = (\underbrace{0, \ldots, 0}_{j-1}, k_{pij}^{\pm}, \underbrace{0, \ldots, 0}_{n-j})^T \in \Re^r$ and $\zeta_p^{\pm} = (k_p^{\pm}, \ldots, k_p^{\pm})^T \in \Re^r$, then under the output-feedback control law

$$\begin{aligned}
u(t) &= K_p y(t)\\
&= \frac{\Sigma_{i=1}^m \Sigma_{j=1}^r \mathbf{1}_m^{(i)} (\zeta_{pij}^+ + \zeta_{pij}^-)^T}{\mathbf{1}_m^T B_p^T v_p} y(t)
\end{aligned} \tag{11.27}$$

the resulting closed-loop system (11.13) is positive and asymptotically stable with the average dwell time satisfying (11.15).

11.4 Conclusions and Future Work

This chapter has addressed a new approach to control synthesis of positive systems. Sufficient conditions for the feedback controller of positive systems are established by using a linear copositive Lyapunov function associated with linear programming technique. Then, the approach is applied to the controller design of PSSs. It is shown that the restrictions in the literature are removed.

Further work refers to two aspects. On one hand, some extension of the approach in the chapter can be proceeded. On the other hand, necessary and sufficient conditions for the approach are expected.

Acknowledgements This work was supported in part by the National Nature Science Foundation of China (61503107, 61503105, 61473107, U1509203, U1509205), the Zhejiang Provincial Natural Science Foundation of China (LY16F030005, LR16F030003).

References

1. Luenberger, D.: Introduction to Dynamic Systems: Theory, Models, and Applications. Springer, Berlin (1979)
2. Bru, R., Romero, S., Sánchez, E.: Canonical forms for positive discrete-time linear control systems. Linear Algebra Appl. **310**(1), 49–71 (2000)

3. Farina, L., Rinaldi, S.: Positive Linear Systems: Theory and Applications. Wiley, New York (2000)
4. Fornasini, E., Valcher, M.: Controllability and reachability of 2-D positive systems: a graph theoretic approach. IEEE Trans. Circ. Syst. I Regul. Pap. **52**(3), 576–585 (2005)
5. H$\ddot{a}$rdin, H., van Schuppen, J.: System Reduction of Nonlinear Positive Systems by Linearization and Truncation. Springer, London (2006)
6. Kaczorek, T.: Positive 1D and 2D Systems. Springer, London (2002)
7. Kaczorek, T., Buslowicz, M.: Minimal realization for positive multivariable linear systems with delay. Int. J. Appl. Math. Comput. Sci. **14**(2), 181–188 (2004)
8. Shorten, R., Wirth, F., Leith, D.: A positive systems model of tcp-like congestion control: asymptotic results. IEEE/ACM Trans. Netw. **14**(3), 616–629 (2006)
9. Xie, G., Wang, L.: Reachability and Controllability of Positive Linear Discrete-Time Systems with Time-delays. Springer, Berlin (2003)
10. Rami, A.M., Tadeo, F.: Controller synthesis for positive linear systems with bounded controls. IEEE Trans. Circ. Syst. II Express Briefs **54**(2), 151–155 (2007)
11. Rami, M.A., Tadeo, F., Benzaouia, A.: Control of constrained positive discrete systems. In: Proceedings of 2007 American Control Conference, pp. 5851–5856 (2007)
12. Rami, A.M.: Solvability of static output-feedback stabilization for lti positive systems. Syst. Control Lett. **60**(9), 704–708 (2011)
13. Chen, X., Lam, J., Li, P., Shu, Z.: ℓ_1-induced norm and controller synthesis of positive systems. Automatica **49**(5), 1377–1385 (2013)
14. Wang, C., Huang, T.: Static output feedback control for positive linear continuous-time systems. Int. J. Robust Nonlinear Control **23**(14), 1537–1544 (2013)
15. Shen, J., Lam, J.: On static output-feedback stabilization for multi-input multi-output positive systems. Int. J. Robust Nonlinear Control **25**(16), 3154–3162 (2015)
16. Arneson, H., Langbort, C.: A linear programming approach to routing control in networks of constrained linear positive systems. Automatica **48**(5), 800–807 (2012)
17. Busłowicz, M., Kaczorek, T.: Robust stability of positive discrete-time interval systems with time-delays. Bull. Pol. Acad. Sci. Tech. Sci. **52**(2), 99–102 (2004)
18. Commault, C., Marchand, N.: Positive systems. In: Proceedings of the Second Multidisciplinary International Symposium on Positive Systems: Theory and Applications (POSTA 06) (2006)
19. Fornasini, E., Valcher, M.: Linear copositive lyapunov functions for continuous-time positive switched systems. IEEE Trans. Autom. Control **55**(8), 1933–1937 (2010)
20. Fornasini, E., Valcher, M.: Stability and stabilizability criteria for discrete-time positive switched systems. IEEE Trans. Autom. Control **57**(5), 1208–1221 (2012)
21. Liu, X.: Constrained control of positive systems with delays. IEEE Trans. Autom. Control **54**(7), 1596–1600 (2009)
22. Charalambous, T., Feyzmahdavian, R.H., Johansson, M.: Asymptotic stability and decay rates of homogeneous positive systems with bounded and unbounded delays. SIAM J. Control Optim. **52**(4), 2623–2650 (2014)
23. Zhao, X., Zhang, L., Shi, P., Liu, M.: Stability of switched positive linear systems with average dwell time switching. Automatica **48**(6), 1132–1137 (2012)
24. Zhang, J., Han, Z., Zhu, F., Huang, J.: Feedback control for switched positive linear systems. IET Control Theory Appl. **7**(3), 464–469 (2013)
25. Zhang, J., Han, Z., Zhu, F., Zhao, X.: Absolute exponential stability and stabilization of switched nonlinear systems. Systems Control Lett. **66**, 51–57 (2014)
26. Zhang, J., Huang, J., Zhao, X.: Further results on stability and stabilisation of switched positive systems. IET Control Theory Appl. **9**(14), 2132–2139 (2014)
27. Xiang, M., Xiang, Z.: Stability, l_1-gain and control synthesis for positive switched systems with time-varying delay. Nonlinear Anal. Hybrid Syst. **9**, 9–17 (2013)

Part IV
Positive Distributed Parameters and Positive Multidimensional Systems

Chapter 12
Polyhedral Invariance for Convolution Systems over the Callier-Desoer Class

Jean Jacques Loiseau

Abstract BIBO stability is a central concept for convolution systems, introduced in control theory by Callier, Desoer and Vidyasagar, in the seventies. It means that a bounded input leads to a bounded output, and is characterized by the fact that the kernel of the system is integrable. We generalize this result in this chapter, giving conditions for the output of a convolution system to evolve in a given polyhedron, for any input evolving in another given convex polyhedron. The conditions are formulated in terms of integrals deduced from the kernel of the considered system and the given polyhedra. The condition is exact. It permits to construct exact inner and outer polyhedral approximations of the reachable set of a linear system. The result is compared to various known results, and illustrated on the example of a system with two delays.

Keywords Convolution systems · Callier-Desoer class · Invariance · Reachable set · Polyhedra · Approximations

12.1 Introduction

The evaluation of the reachable space of a dynamical system is important for the verification of properties [4], planification of trajectories and design of control laws to achieve closed-loop specifications [7]. Exact formulae can not always be determined, so that various methods have been developed to compute approximations of the reachable set. The case of linear finite dimensional systems has been deeply investigated [19, 24]. The basic approach consists in reformulating the problem in terms of optimal control, which can be extended to the case of nonlinear systems [11] and hybrid systems [4, 9]. The effect of uncertainties or disturbances can also be handled using similar ideas and interval analysis [15].

The case of distributed systems has also been addressed. Systems with state delays are considered in [8], where a bounding ellipsoid of the reachable state is derived

J.J. Loiseau (✉)
Université Bretagne Loire, École Centrale de Nantes, LS2N CNRS UMR 6004,
1 rue de la Noë, 44321 Nantes cedex 03, France
e-mail: Jean-Jacques.Loiseau@irccyn.ec-nantes.fr

F. Cacace et al. (eds.), *Positive Systems*, Lecture Notes in Control and Information Sciences 471, DOI 10.1007/978-3-319-54211-9_12

using Linear Matrix Inequalities. This idea gave rise to many generalizations, to distributed delays and variable delays, see e.g. [2] and the references therein. The question is generalized in [18] to that of the determination of invariant sets, for a class of discrete systems with delays.

A different approach was recently introduced. The question is formulated in [16] in an input-output setting. This is the basis of the present work. It concerns a large class of convolution systems, that includes localized or distributed time delay systems, ordinary or neutral time-delay systems, fractional systems and many other distributed systems. The basic idea is to observe that the input-output gain of a convolution system is bounded by the L_1 norm of its kernel. This can be reinterpreted in terms of reachability: the output of a system with input in the unit ball is included into the ball which radius is the L_1 norm of the kernel. When the underlying topology is the infinite norm, this observation comes down to a polytopic bound of the reachable set of a constrained system. The aim of this communication is to develop this idea, and to provide basic tools for the determination of polytopic approximations of the reachable set for a large class of convolution systems. For a multivariable convolution system, which input is constrained in a given polyhedron, we formulate conditions for the output of the system to evolve in another given polyhedron. The conditions are formulated in terms of integrals deduced from the kernel of the considered system and the given polyhedra. The conditions are necessary and sufficient, which shows that the bounds are in some sense exact.

The article is organized as follows. In Sect. 12.2, we recall the basic concepts that are used, in particular the definition of the Wiener algebra, and of a polytope. We identify bounds for the output of a given constrained system over the Wiener algebra in Sect. 12.3. These bounds are used to design overapproximations and underapproximations of the reachable set of the system at a given time horizon. In Sect. 12.4, the result is discussed, and illustrated on examples. Section 12.5 is a short conclusion.

12.2 Background Concepts

12.2.1 Convolution Kernels

An input-output linear system given in the form of a convolution,

$$y = h \star u \ , \tag{12.1}$$

is BIBO-stable if its kernel h belongs to the class $\mathscr{A}$ of generalized functions of the form

$$h(t) = h_a(t) + \sum_{i \in \mathbb{N}} h_i \delta(t - t_i) \ , \tag{12.2}$$

where h_a is in L_1, $h_i \in \mathbb{R}$, $t_i \in \mathbb{R}_+$, $t_i < t_{i+1}$ for $i \geq 0$, and $\sum_{i \in \mathbb{N}} |h_i| < \infty$. The set $\mathscr{A}$ endowed with the convolution product forms a Banach commutative algebra for the norm

$$||h||_{\mathscr{A}} = \int_0^{+\infty} |h_a(t)|dt + \sum_{i \in \mathbb{N}} |h_i| . \tag{12.3}$$

This norm was shown to be the induced norm when h is seen as an operator over L_∞. We indeed have

$$\sup_{u \neq 0} \frac{||h \star u||_\infty}{||u||_\infty} = ||h||_{\mathscr{A}} , \tag{12.4}$$

for every h in $\mathscr{A}$. Here, as usually, $|| \cdot ||_\infty$ denotes the sup-norm on L_∞, say $||u||_\infty = \text{ess}\sup_{t \geq 0} |u(t)|$, $||y||_\infty = \text{ess}\sup_{t \geq 0} |y(t)|$. This shows that every bounded input leads to a bounded output, and that $||h||_{\mathscr{A}}$ gives an exact bound on the output $y(t)$.

The set $\mathscr{A}$ is sometimes called Wiener algebra (see, e.g. [20]). Many properties of the set $\mathscr{A}$ are exposed in [10], and its use in control theory was gradually introduced by various authors, among them Desoer [1, 5, 6], Callier [1, 5] and Vidyasagar [6]. The set of fractions of elements of $\mathscr{A}(\sigma) = \mathrm{e}^{-\sigma t} \mathscr{A}$ is called the Callier-Desoer class and is a key concept to describe robust stabilization methods for a large class of distributed systems. The matter continues to generate interesting results, see for instance Quadrat [20], or Lakkonen [13] for a recent survey.

The transfer of a system of the form (12.1) is the Laplace transform $\hat{h}(s)$ of the kernel $h(t)$. For instance, the class $\mathscr{A}$ includes:

- the class of linear finite dimensional systems with rational transfer, e.g.

$$\hat{h}(s) = (sI - A)^{-1} \ , \ h(t) = \mathrm{e}^{At} \ ,$$

- the class of time-delay systems, e.g.

$$\hat{h}(s) = \frac{\mathrm{e}^{-\theta s}}{1 + sT} \ , \ h(t) = \begin{cases} 0 & , \text{ for } t < \theta \ , \\ \mathrm{e}^{t - \theta} & , \text{ for } t \geq \theta \ , \end{cases} \ ,$$

  that are important models in many applications,

- the class of systems with distributed delays, e.g.

$$\hat{h}(s) = \frac{1 - \mathrm{e}^{\theta a} \mathrm{e}^{-\theta s}}{s - a} \ , \ h(t) = \begin{cases} \mathrm{e}^{at} & , \text{ for } t \leq \theta \ , \\ 0 & , \text{ for } t > \theta \ , \end{cases} \ ,$$

  that are important for the stabilization of time-delay systems,

- BIBO stable diffusive systems, e.g.

$$\hat{h}(s) = \frac{1 - \mathrm{e}^{-\alpha\sqrt{s}}}{\sqrt{s}} \ , \ h(t) = 1 - \mathrm{erfc}\left(\frac{\alpha}{2\sqrt{t}}\right) \ .$$

This short list is not exhaustive. The class also includes many other linear distributed systems, and covers many application fields [3, 22].

The system (12.2) is said to be regular if $h(t) = h_a(t)$, or equivalently if the singular part is absent, say $h_i = 0$ for $i \in \mathbb{N}$. Notice that the class of regular systems is also very large, for instance the four examples of transfer functions mentioned above belong to this family.

Finally notice that in the present work, we basically consider systems with kernels of the form (12.2) that are well defined, in the sense that the kernel $h(t)$ is integrable over every finite interval $[0, t]$. This includes the Callier-Desoer class, which justifies the use of this expression in the title of the chapter. In Sect. 12.3.3, we shall assume that the kernel of the system is defined over $\mathscr{A}$.

12.2.2 Reachable Sets

We now consider a multivariable convolution system, defined by a kernel H, say

$$y = H \star u \ , \tag{12.5}$$

where $u(t) \in \mathscr{U} \subset \mathbb{R}^m$, for $t \geq 0$. Recall that the convolution product $\star$ is defined as

$$y_i(t) = \int_0^t \sum_j H_{ij}(t - \tau)u_j(\tau)d\tau \ . \tag{12.6}$$

We consider a system with entries of the form (12.2), so that $H_{ij}(t) = h_{a_{ij}}(t) + \sum_{k\in\mathbb{N}} h_{kij}(t - t_k)$, for $i = 1$ to p and $j = 1$ to m. We hence have, for $i = 1$ to p:

$$y_i(t) = \sum_j \left(\int_0^t h_{a_{ij}}(t - \tau)u_j(\tau)d\tau + \sum_{k|t_k\leq t} h_{kij}u_j(t - t_k) \right) \ .$$

We are interested into the characterization of the range of system (12.5). The basic concept is that of reachable set.

Definition 12.1 System (12.5) and a subset $\mathscr{U}$ of $\mathbb{R}^m$ being given, we say that an input function u is admissible, if $u(t) \in \mathscr{U}$, for $t \geq 0$. The reachable set $\mathscr{R}(\mathscr{U})$ is then defined as the set of vectors $x \in \mathbb{R}^p$ for which there exists an admissible control u such that the output $y(t)$ defined by (12.5) satisfies $y(t) = x$ for some $t \geq 0$. We also define the set $\mathscr{R}(\mathscr{U}, t)$ of vectors x that are reachable at t, so that $x = y(t)$ for some admissible input u, and the set $\mathscr{R}_t(\mathscr{U})$ of the vectors x reachable within t, so that $x = y(\tau)$, for some instant τ satisfying $0 \leq \tau \leq t$.

These definitions are taken from [24], a seminal paper on the computation of reachable sets for systems without memory. We remark that

$$\mathscr{R}_t(\mathscr{U}) = \bigcup_{\tau\in[0,t]} \mathscr{R}(\mathscr{U}, \tau)\,,$$

and

$$\mathscr{R}(\mathscr{U}) = \bigcup_{t>0} \mathscr{R}(\mathscr{U}, t) = \bigcup_{t>0} \mathscr{R}_t(\mathscr{U})\,.$$

One can see that $\mathscr{R}(\mathscr{U}, t)$ is convex, if $\mathscr{U}$ is convex. In Sect. 12.3, we shall in particular study the case where $\mathscr{U}$ is given in the form of a polytope $\mathscr{C}(M)$. The sets $\mathscr{R}_t(\mathscr{U})$ and $\mathscr{R}(\mathscr{U})$ are not convex, in general. Let us discuss these aspects.

The sets $\mathscr{R}_t(\mathscr{U})$ and $\mathscr{R}(\mathscr{U})$ are not connected, in general. This is due to the singular part of the kernels of the form (12.2), that may cause discontinuity of the solution $y(t)$. Consider for instance the kernel $h(t) = \delta(t-\theta)$, where θ is any positive number, and $\mathscr{U} = \{1\}$. We have in this example $\mathscr{R}(\mathscr{U}) = \{0, 1\}$, that is not connected. One can find conditions under which the sets are connected, or convex.

Proposition 12.1 *System (12.5) being given, together with a subset $\mathscr{U}$ of $\mathbb{R}^m$, and a real number $t \geq 0$, the following claims are true.*

- *(i) The set $\mathscr{R}(\mathscr{U}, t)$ is convex if $\mathscr{U}$ is convex.*
- *(ii) The sets $\mathscr{R}_t(\mathscr{U})$ and $\mathscr{R}(\mathscr{U})$ are connected if $\mathscr{U}$ is convex and the kernel of the system is regular.*
- *(iii) The sets $\mathscr{R}_t(\mathscr{U})$ are growing with t if $0 \in \mathscr{U}$.*
- *(iv) The sets $\mathscr{R}_t(\mathscr{U})$ and $\mathscr{R}(\mathscr{U})$ are convex if $\mathscr{U}$ is convex, and $0 \in \mathscr{U}$.*

Proof Notice first that if $\mathscr{U}$ is convex, and y and y' are reached using the admissible input trajectories $u(t)$ and $u'(t)$, respectively, then $\alpha u(t) + (1-\alpha)u'(t)$ is admissible too, and permits to reach $\alpha y + (1-\alpha)y'$. This shows that $\mathscr{R}(\mathscr{U}, t)$ is convex if $\mathscr{U}$ is convex. Further, the trajectories $y(t)$ of the system are continuous when the kernel is regular. Consider now two points y and y' in $\mathscr{R}(U)$. There exist admissible inputs u and u', and two instants $t, t' \geq 0$ such that $y = (H \star u)(t)$ and $y' = (H \star u')(t')$. We can assume, without any limitation, that $t' < t$. Defining $y'' = (H \star u)(t')$, one can see that there is a path from y' to y'' in $\mathscr{R}(\mathscr{U}, t')$, since this set is convex. There is also a path from y'' to y in $\mathscr{R}_t(\mathscr{U})$, since $y(\tau)$ is continuous, and takes its values into $\mathscr{R}_t(\mathscr{U})$, by definition of this set. Therefore, since $\mathscr{R}(\mathscr{U}, t')$ is a subset of $\mathscr{R}_t(\mathscr{U})$, one deduces that there exists in the latter set a path from y' to y, which shows the second assertion of the proposition. The third assertion is obtained remarking that if $y \in \mathscr{R}(\mathscr{U}, t)$ and $0 \in \mathscr{U}$, then there exists an admissible function u, and an instant t, such that $y = (H \star u)(t)$. One can see that $y = (H \star u')(t')$, taking $u'(\tau) = 0$, for $\tau \in [0, t'-t[$, and $u'(\tau) = u(t+\tau-t')$, for $\tau \geq t'-t$. This shows that $y \in \mathscr{R}(\mathscr{U}, t')$, for every t' greater than t, and establishes the third assertion. The last assertion is a consequence of (i) and (iii). □

As a consequence to this remark, the hypotheses that $\mathscr{U}$ is convex and $0 \in \mathscr{U}$ are often formulated in the literature, even in the case of localized systems. Of course, these assumptions are limitative. The identification of more accurate conditions might be useful in certain applications with discontinuous behaviors.

12.2.3 Elements of Convex Analysis

We now recall the definition of a polytope and some basic concepts of convex analysis. These concepts are taken from [21] (see in particular Sects. 6 and 13), and will be useful to analyse the reachability of constrained convolution systems.

A convex set $\mathscr{C} \subset \mathbb{R}^n$ is such that, for every $x, y \in \mathscr{C}$, and every $\lambda \in [0, 1]$, the vector $z = \lambda x + (1-\lambda)y$ lies in $\mathscr{C}$. The support function of $\mathscr{C}$ is $f_{\mathscr{C}} : \mathbb{R}^n \to \mathbb{R}^n$, defined by

$$f_{\mathscr{C}}(v) = \sup_{x \in \mathscr{C}} v^T x \ ,$$

for $v \in \mathbb{R}^n$. Notice that $f_{\mathscr{C}}(v)$ takes only finite values if $\mathscr{C}$ is bounded. The ball of radius ε centered on $x \in \mathbb{R}^n$ is denoted $B(x, \varepsilon)$, as usually. A convex set $\mathbb{C}$ is open if there exists $\varepsilon > 0$ such that the ball $B(x, \varepsilon)$ is included into $\mathbb{C}$. It is closed if its complement is open. The least closed set containing $\mathbb{C}$ is called its closure, denoted $\overline{\mathscr{C}}$. The greatest open set included into $\mathscr{C}$ is called the interior of $\mathscr{C}$.

The concept of relative interior, that we now recall, is specific to the convex sets. The affine hull of a convex set $\mathscr{C}$ is denoted aff $\mathscr{C}$ and is defined as the set

$$\text{aff}\, \mathscr{C} = \{z \in \mathbb{R}^n \mid \exists x, y \in \mathscr{C}, \alpha \in \mathbb{R}, z = x + \alpha(y - x)\} \ .$$

An affine set can also be written as aff $\mathscr{C} = x + \text{lin}\, \mathscr{C}$, for any element $x \in \mathscr{C}$, where lin$\mathscr{C}$ is the vector space generated by the differences $y - x$, with $y \in \mathscr{C}$. The relative interior of $\mathscr{C}$, denoted ri $\mathscr{C}$, is the interior of $\mathscr{C}$ when it is considered as a subset of aff $\mathscr{C}$, say

$$\text{ri}\mathscr{C} = \{x \in \mathbb{R}^n \mid \exists \varepsilon > 0,\ B(x, \varepsilon) \cap \text{aff}\, \mathscr{C} \subset \mathscr{C}\} \ .$$

One says that $\mathscr{C}$ is relatively open if it equals its relative interior. If $\mathscr{C}$ is reduced to a unique point, then lin $\mathscr{C} = 0$ and ri$\mathscr{C} = \overline{\mathscr{C}} = \mathscr{C}$. In general, the three sets $\mathscr{C}$, ri$\mathscr{C}$, and $\overline{\mathscr{C}}$ are different, and we have the following.

Theorem 12.1 *Two convex sets $\mathscr{C}_1$ and $\mathscr{C}_2$ being given, the following claims are equivalent.*

(i) $f_{\mathscr{C}_1}(v) = f_{\mathscr{C}_2}(v)$, *for every vector* $v \in \mathbb{R}^n$,
(ii) $\overline{\mathscr{C}_1} = \overline{\mathscr{C}_2}$,
(iii) ri $\mathscr{C}_1 =$ ri $\mathscr{C}_2$.

For a convex set $\mathscr{C}$, and a vector $v \in \mathbb{R}^n$, we have the inclusion $\{v^T x \mid x \in \mathscr{C}\} \subset [-f_{\mathscr{C}}(-v), f_{\mathscr{C}}(v)]$, because $\inf_{x \in \mathscr{C}}\{v^T x\} = -\sup_{x \in \mathscr{C}}\{-v^T x\}$. The bounds are exact, and they belong or not to the set, depending on its topological property. The following theorem precises this aspect (this is Theorem 13.1 of [21]).

Theorem 12.2 *A nonempty convex set $\mathscr{C}$ being given, the following claims hold true.*

(i) $\mathscr{C}$ *is closed if and only if* $\forall v \in \mathbb{R}^n$, $\{v^T x \mid x \in \mathscr{C}\} = [-f_{\mathscr{C}}(-v), f_{\mathscr{C}}(v)]$.
(ii) $\mathscr{C}$ *is open if and only if* $\forall v \in \mathbb{R}^n$, $\{v^T x \mid x \in \mathscr{C}\} =]-f_{\mathscr{C}}(-v), f_{\mathscr{C}}(v)[$.
(iii) $\mathscr{C}$ *is relatively open if and only if* $\{v^T x \mid x \in \mathscr{C}\} =]-f_{\mathscr{C}}(-v), f_{\mathscr{C}}(v)[$, $\forall v \in \mathbb{R}^n$ *such that* $-f_{\mathscr{C}}(-v) < f_{\mathscr{C}}(v)$.

We finally recall basic facts and definitions concerning polytopes.

Definition 12.2 A matrix $M \in \mathbb{R}^{m \times n}$ being given, the convex polytope of $\mathbb{R}^m$ generated by the columns of M is the set denoted $\mathscr{C}(M)$, and defined by

$$\mathscr{C}(M) = \left\{ x \in \mathbb{R}^m \mid \exists v \in \mathbb{R}^n, v \geq 0, \sum_{i=1}^{n} v_i = 1, x = Mv \right\} .$$

The relatively open polytope generated by M is defined by

$$\mathscr{C}_{ro}(M) = \left\{ x \in \mathbb{R}^m \mid \exists v \in \mathbb{R}^n, v_i > 0, \sum_{i=1}^{n} v_i = 1, x = Mv \right\} .$$

In other words, introducing the notation $\Gamma = \{v \in \mathbb{R}^n, v \geq 0, \sum_{i=1}^{n} v_i = 1\}$, we have $\mathscr{C}(M) = M\Gamma$, and $\mathscr{C}_{ro}(M) = M \operatorname{ri} \Gamma = \operatorname{ri} M\Gamma$. The following result is then clear (see Theorems 6.6 and 6.9 from [21]).

Proposition 12.2 *For every matrix M, we have the equality*

$$\operatorname{ri} \mathscr{C}(M) = \mathscr{C}_{ro}(M) .$$

Definition 12.3 A matrix $P \in \mathbb{R}^{q \times p}$ and a vector $\pi \in \mathbb{R}^q$ being given, the polyhedron denoted $\mathscr{P}(P, \pi)$ is the set defined as

$$\mathscr{P}(P, \pi) = \left\{ z \in \mathbb{R}^p \mid Pz \leq \pi \right\} .$$

The relatively open polyhedron $\mathscr{P}_{ro}(P, \pi)$ is defined by

$$\mathscr{P}_{ro}(P, \pi) = \left\{ z \in \mathscr{P}(P, \pi) \mid \sum_{j=1}^{p} P_{ij} z_j < \pi_i, \text{ for } i \in J(P, \pi) \right\} ,$$

with $J(P, \pi) = \{i \mid \exists z \in \mathscr{P}(P, \pi), \sum_{j=1}^{p} P_{ij} z_j < \pi_i\}$.

In other words, some of the constraints corresponding to the rows of the matrix P and the vector π actually define the affine hull of $\mathscr{P}(P, \pi)$. The other ones, corresponding to the set $J(P, \pi)$, define a subset of aff $\mathscr{P}(P, \pi)$, with a nonempty interior that equals ri $\mathscr{P}(P, \pi)$. We hence have the following result.

Proposition 12.3 *A matrix $P \in \mathbb{R}^{p\times m}$ and a vector $\pi \in \mathbb{R}^p$ being given, the following equality holds true*

$$\text{ri}\ \mathscr{P}(P, \pi) = \mathscr{P}_{ro}(P, \pi)\ .$$

We are now ready to study the reachable set of convolution systems.

12.3 Polyhedral Bounds of the Reachable Set

12.3.1 Elementary Bounds

A basic question consists in determining the range of the output $y(t)$ of system (12.5). We are precisely interested in verifying whether or not the output $y(t)$ belongs to a given polyhedron, provided that the input $u(t)$ evolves in another given polyhedron. The following elementary remark will be useful in the sequel.

Lemma 12.1 *Let be given a vector $x \in \mathbb{R}^n$ and a vector v in the convex set $\Gamma = \{v \in \mathbb{R}^n, v \geq 0, \sum_{i=1}^n v_i = 1\}$ defined as in Definition 12.2. Then, we have*

$$\max_{v\in\Gamma} \left\{ \sum_{j=1}^n x_j v_j \right\} = \max_j\ x_j\ .$$

Proof Since $x_j \leq \max_j\ x_j$, it is clear that $\sum_{j=1}^n x_j v_j \leq (\max_j\ x_j)(\sum_{j=1}^n v_j)$. By definition of Γ, it appears that $\sum_{j=1}^n x_j v_j \leq \max_j\ x_j$, so that $\max_{v\in\Gamma} \left\{\sum_{j=1}^n x_j v_j\right\} \leq \max_j\ x_j$. This is an exact bound, which follows considering the vector v defined by $v_k = 1$ and $v_j = 0$, for $j \neq k$, with $k = \arg\max_j x_j$. This ends the proof. □

We are now able to formulate the basic result on polyhedral bounds of system (12.5).

Theorem 12.3 *System (12.5) being given, together with matrices $M \in \mathbb{R}^{m\times n}$, $P \in \mathbb{R}^{q\times p}$, a vector $\pi \in \mathbb{R}^q$, and $t > 0$, then $y(t)$ belongs to $\mathscr{P}(p, \pi)$ for every input satisfying $u(\tau) \in \mathscr{C}(M)$, for $\tau \geq 0$, if and only if the following condition holds true for $i = 1$ to q:*

$$\int_0^t \max_j \left\{(PH(\tau)M)_{ij}\right\} d\tau \leq \pi_i\ .$$

Proof To begin with the proof, we proceed by equivalences:

$$\begin{array}{rll} y(t) \in \mathscr{P}(P,\pi) & \Longleftrightarrow Py(t) \leq \pi & \text{, by definition of } \mathscr{P}(P,\pi), \\ & \Longleftrightarrow \int_0^t PH(\tau)u(t-\tau)d\tau \leq \pi & \text{, by definition of the system,} \\ & \Longleftrightarrow \int_0^t PH(\tau)Mv(t-\tau)d\tau \leq \pi & \text{, by definition of } \mathscr{C}(M), \\ & \Longleftrightarrow \int_0^t \max_j \left\{(PH(\tau)M)_{ij}\right\} d\tau \leq \pi_i & \text{, by Lemma 12.1.} \end{array}$$

Assuming that H is a matrix with inputs that are integrable over $[0, t]$, we observe that the integrals in these equivalences are well defined. They are indeed bounded by the product $p \cdot m \cdot \max_k\{|P_{ik}|\} \cdot B \cdot \max_l |\{M_{lj}|\}$, if B is a bound of the integrals of the entries of H, for instance $B = \max_{k,l} ||H_{kl}||_{\mathscr{A}}$, if H is a matrix over $\mathscr{A}$. In these statements, the vector $v(t-\tau)$ lies in Γ, by hypothesis, which permits to apply Lemma 12.1. The fact that this lemma gives exact bounds is essential to obtain the last equivalence, from which the theorem is deduced. □

A preliminary version of this result was obtained in [16]. We first remark that upper bounds and lower bounds of the behavior of the given system can be derived from Theorem 12.3. For this purpose, one defines

$$\lambda_i(t) = \int_0^t \min_j \left\{(H(\tau)M)_{ij}\right\} d\tau \ , \tag{12.7}$$

and

$$\mu_i(t) = \int_0^t \max_j \left\{(H(\tau)M)_{ij}\right\} d\tau \ . \tag{12.8}$$

Corollary 12.1 *The matrix M and the system (12.5) being given as in Theorem 12.3, and $\lambda_i(t)$, $\mu_i(t)$ being defined as in (12.7), (12.8), we have*

$$\lambda_i(t) \leq y_i(t) \leq \mu_i(t) \ ,$$

for $i = 1$ to p. In addition, the bounds are reached, so that the range of $y_i(t)$, when the input satisfies $u(\tau) \in \mathscr{C}(M)$, for $\tau \geq 0$, is exactly the closed interval $[\lambda_i(t), \mu_i(t)]$.

Proof The upper bound of Corollary 12.1 is obtained taking $P = I_p$ in Theorem 12.3. The lower bound is obtained with $P = -I_p$, since $\min_j\{x_j\} = -\max_j\{-x_j\}$, and $-\max_{v\in\Gamma}\{-x_j v_j\} = \min_{v\in\Gamma} x_j v_j$, with Γ defined as in Lemma 12.1.

To complete the proof, we remark that the upper bound $\mu_i(t)$ is indeed reached using the control defined by $u_k(\tau) = M_{kj(t-\tau)}$, for $k = 1$ to m and $\tau \in [0, t]$, with

$$j(\tau) = \arg\max_j \left\{(H(\tau)M)_{ij}\right\} \ .$$

Similarly, the lower bound is reached using the control that maximizes $-y_i(t)$, that is defined in terms of an argument of $\max_j \left\{-(H(\tau)M)_{ij}\right\}$. □

We can finally remark the following fact, that will be useful in Sect. 12.3.3.

Corollary 12.2 *Under the conditions of Corollary 12.1, the range of* $y_i(t)$ *when* $u(\tau) \in \mathscr{C}_{ro}(M)$ *equals the open interval* $]\lambda_i(t), \mu_i(t)[$, *if* $\lambda_i(t) < \mu_i(t)$, *and is reduced to* $\{\mu_i(t)\}$, *if* $\lambda_i(t) = \mu_i(t)$.

Proof If the equality $\lambda_i(t) = \mu_i(t)$ is satisfied, one can see that $\min_j \left\{(H(\tau)M)_{ij}\right\} = \max_j \left\{(H(\tau)M)_{ij}\right\}$ almost everywhere in the interval $[0, t]$, and therefore the kernels $(H(\tau)M)_{ij}$, for $j = 1$ to n, are equal almost everywhere in this interval. In this case, $y(t)$ takes a unique value, say $\int_0^t (H(\tau)M)_{i1} d\tau$. If $\lambda_i(t) < \mu_i(t)$, then the different kernels $(H(\tau)M)_{ij}$, for $j = 1$ to n, are not equal on a subset of $[0, t]$ having a nonzero measure. Taking an instant t from this set, we observe that $\min_j \left\{(H(\tau)M)_{ij}\right\} < H(\tau)u(t-\tau) < \max_j \left\{(H(\tau)M)_{ij}\right\}$ holds true, for every input $u(t-\tau) \in \mathscr{C}_{ro}(M)$, from which one deduces that $\lambda_i(t) < y_i(t) < \mu_i(t)$. The conclusion is obtained remarking that the bounds can be approached with an arbitrary precision. To this aim, define $K = \int_0^t \sum_k \left(\max_j \left\{(H(\tau)M)_{ij}\right\} - (H(\tau)M)_{ik}\right) d\tau$. We can see that K is positive, and has a finite value if the kernel is integrable over $[0, t]$. Taking $u(t-\tau) = Mv(t-\tau)$, with $v_j(t-\tau) = 1 - (n-1)\varepsilon/K$, and $v_k = \varepsilon/K$, for $k \neq j(t-\tau)$, we obtain $y_i(t) = \mu_i(t) - \varepsilon$. The lower bound $\lambda_i(t)$ is approached in the same way, using an argument $j(\tau)$ of $\max_j \left\{-(H(\tau)M)_{ij}\right\}$ and defining now $K = \int_0^t \sum_k \left((H(\tau)M)_{ik} - \min_j \left\{(H(\tau)M)_{ij}\right\}\right)$. One checks that the input defined by $u(t-\tau) = Mv(t-\tau)$, with $v_j(t-\tau) = 1 - (n-1)\varepsilon/K$, and $v_k = \varepsilon/K$, for $k \neq j(t-\tau)$ leads to an output verifying $y_i(t) = \lambda_i(t) + \varepsilon$, which ends the proof. □

12.3.2 Polyhedral Approximations of the Reachable Set

The previous results can be interpreted in terms of reachability.

Remark that the difference between the left and right members of the condition of Theorem 12.3 is the distance between the reachable set and the plan $\{y \in \mathbb{R}^p \mid \sum_j P_{ij} y_j = \pi_i\}$. The left member of the condition, say

$$\rho_i(t) = \int_0^t \max_j \left\{(PH(\tau)M)_{ij}\right\} d\tau \ , \tag{12.9}$$

is therefore so that the plan $\{y \in \mathbb{R}^p \mid \sum_j P_{ij} y_j = \rho_i(t)\}$ is tangent to the reachable space at t, say $\mathscr{R}(\mathscr{C}(M), t)$, of the system constrained by $\mathscr{U} = \mathscr{C}(M)$. If the matrix P is given, the polyhedron $\mathscr{P}(P, \rho(t))$ is the least polyhedron whose faces are oriented according to P, and that contains the reachable set. One can also compute a point of the intersection between the face and the reachable set. We first define the integers

$$j_k(\tau) = \arg\max_j \left\{(PH(\tau)M)_{kj}\right\} \ ,$$

for $k = 1$ to q, and the output vectors

$$\nu_i(k,t) = \int_0^t (H(\tau)M)_{ij_k(\tau)} d\tau \ ,$$

for $k = 1$ to q, and $i = 1$ to p. Then, N is defined as the matrix which columns are the vectors $\nu(k,t)$, say

$$N_{ij} = \nu_i(j,t) \ ,$$

for $i = 1$ to p, $j = 1$ to q. The following definitions are inspired by [24].

Definition 12.4 A compact convex set $\mathscr{R}$ being given, we say that a polyhedron is an exact outer approximation of $\mathscr{R}$ if its faces are tangent to $\mathscr{R}$, and that it is an exact inner approximation of $\mathscr{R}$, if its vertices are on the boundary of $\mathscr{R}$.

Theorem 12.4 *The system (12.1) being given, together with an integer q and two matrices $P \in \mathbb{R}^{q\times p}$ and $M \in \mathbb{R}^{m\times q}$, and taking N and ρ defined as above, the convex polytope $\mathscr{C}(N)$ is an exact inner approximation, and the polyhedron $\mathscr{P}(P,\rho(t))$ is an exact outer approximation, of $\mathscr{R}(\mathscr{C}(M),t)$.*

Proof For $k = 1$ to q, the control defined by $u^{(k)}(t-\tau) = Mv^{(k)}(\tau)$, with $v_j^{(k)}(\tau) = 1$, if $j = j_k(\tau)$, and $v_j^{(k)}(\tau) = 0$, if $j \neq j_k(\tau)$, satisfies $(Py)_k(t) = \rho_k(t)$. This shows that the faces of $\mathscr{P}(P,\rho(t))$ are tangent to $\mathscr{R}(\mathscr{C}(M),t)$, and the vertices of $\mathscr{C}(N)$ are on the boundary of $\mathscr{R}(\mathscr{C}(M),t)$, which ends the proof. □

In other words, we have the chain of inclusions $\mathscr{C}(N) \subset \mathscr{R}(\mathscr{C}(M),t) \subset \mathscr{P}(P,\rho(t))$, and the distance between the three sets is null:

$$\inf\{d(x,y) \mid x \in \mathscr{C}(N),\, y \in \mathscr{R}(\mathscr{C}(M),t)\} = 0 \ ,$$

and

$$\inf\{d(y,z) \mid y \in \mathscr{R}(\mathscr{C}(M),t),\, z \in \mathscr{P}(P,\rho(t))\} = 0.$$

The precision of the approximation can be defined as the Hausdorff distance between the upper and lower approximations, defined, since $\mathscr{C}(N)) \subset \mathscr{P}(P,\rho(t))$, as:

$$\max\{d(z,\mathscr{C}(N)) \mid z \in \mathscr{P}(P,\rho(t))\} \ .$$

This distance is decreasing when rows are added to the matrix P. This permits to reach an arbitrary precision choosing a matrix P that corresponds to plans in many different directions. In practice, the number of rows is rapidly growing with the dimension of the system. For this reason, one may prefer rough approximations in high dimension. Anyway, this formulation is well fitted for numerical computations. The integrals can be easily approximated using Matlab or Scilab, for instance, provided that the kernel $H(t)$ is explicitly known, or can be numerically computed. We shall give a simple example in Sect. 12.4.

We complete this study with remarks concerning the topological structure and the approximation of $\mathscr{R}_t(\mathscr{U})$ and $\mathscr{R}(\mathscr{U})$.

12.3.3 Additional Comments on the Structure of the Reachable Set

We first complete the previous results in terms of the reachable set at a given instant.

Proposition 12.4 *The matrix $M \in \mathbb{R}^{m\times n}$, an instant $t \geq 0$ and the system (12.5) being given, then the following claims are true.*

(i) The set $\mathscr{R}(\mathscr{C}(M), t)$ is closed.

(ii) The set $\mathscr{R}(\mathscr{C}_{ro}(M), t)$ is relatively open.

(iii) We have the equalities $\mathscr{R}(\mathscr{C}_{ro}(M), t) = \mathrm{ri}\,\mathscr{R}(\mathscr{C}(M), t)$.

Proof The proof uses Theorem 12.2 (that is Theorem 13.1 of [21]), and a variant of Theorem 12.3 and Corollary 12.2. According to claim (i) of Proposition 12.1, the set $\mathscr{R}(\mathscr{C}(M), t)$ is convex. We then remark that the support function of $\mathscr{R}(\mathscr{C}(M), t)$ is defined, in any direction $v \in \mathbb{R}^p$, by $f_{\mathscr{R}(\mathscr{C}(M),t)}(v) = \int_0^t \max_j\{(v^T H(\tau)M)_j\}d\tau$. Applying Corollary 12.2, one obtains that $v^T\mathscr{R}(\mathscr{C}(M), t)$ is either reduced to a single element, if $f_{\mathscr{R}(\mathscr{C}(M),t)}(v) = -f_{\mathscr{R}(\mathscr{C}(M),t)}(-v)$, or equal to the open interval $]-f_{\mathscr{R}(\mathscr{C}(M),t)}(-v), f_{\mathscr{R}(\mathscr{C}(M),t)}(v)[$, if $-f_{\mathscr{R}(\mathscr{C}(M),t)}(-v) < f_{\mathscr{R}(\mathscr{C}(M),t)}(v)$. The second claim is therefore deduced from claim (iii) of Theorem 12.2. In a similar way, one can see that for every $v \in \mathbb{R}^p$, the set $v^T\mathscr{R}(\mathscr{C}(M), t)$ is a closed interval. The claim (i) is then deduced from claim (i) of Theorem 12.2. From Corollaries 12.1 and 12.2, we conclude that actually $\mathscr{R}(\mathscr{C}(M), t)$ is the closure of $\mathscr{R}(\mathscr{C}_{ro}(M), t)$. We further obtain from Theorem 12.2 that the open interior of both sets are equal, and the conclusion follows since $\mathscr{R}(\mathscr{C}_{ro}(M), t)$ is equal to its relative interior, from claim (i). □

In other words, the set $\mathscr{R}(\mathscr{C}(M), t)$ is closed if the kernel $H(t)$ is integrable over $[0, t]$, because the limits are reached in the inequalities presented in Sect. 12.3.1, and its relative interior coincides with the set of points that are reachable using inputs in the relative interior of the polyhedron $\mathscr{U} = \mathscr{C}(M)$. When t tends to the infinity, the upper bound found for $y(t)$ when the system is subject to a bounded input $u(t)$ converges to a bounded limit (assuming that system (12.5) is over $\mathscr{A}$), but this limit may be reachable, or not, depending on $H(t)$, and the choosen direction v. As a consequence, $\mathscr{R}(\mathscr{C}(M))$ is not closed, in general. In the same way, when the kernels include delayed diracs, the function $\mu_i(t)$ may be discontinuous, so that the set of points that are reachable within a finite time, $\mathscr{R}_t(\mathscr{C}(M))$, is not always closed.

Consider for instance the system $y = h * u$, with $h(\tau) = f_a(\tau) - \delta(1-\tau)$, with $f_a(\tau) = 1$, for $\tau \in [0, 1]$, and $f_a(\tau) = 0$, for $\tau > 0$. We have

$$y(t) = \begin{cases} \int_0^t f_a(\tau)u(t-\tau)d\tau & , \text{ for } t < 1, \\ \int_0^t f_a(\tau)u(t-\tau)d\tau - u(t-1) & , \text{ for } t \geq 1. \end{cases} \tag{12.10}$$

One can verify that taking $u(\tau) = 1$ on this example, we obtain $y(t) = t$, for $t \in [0, 1[$, and $y(t) = 0$, for $t \geq 1$. The point $y = 1$ is not reachable within $t = 1$, if $\mathscr{U} = \{1\}$. We have in this case $M = (1)$, $\mathscr{C}(M) = \{1\}$, and $\mathscr{R}_t(\mathscr{C}(M)) = [0, t]$, for $t \in [0, 1[$, and $\mathscr{R}_t(\mathscr{C}(M)) = [0, 1[$, for $t \geq 1$.

A singular kernel may also cause that $\mathscr{R}_t(\mathscr{C}(M))$ and $\mathscr{R}(\mathscr{C}(M))$ are not connected set. The consequences of these remarks are different in terms of outer or inner approximations.

Remark 12.1 We can adapt Theorem 12.3 and Corollary 12.1 to have the constraint $y(\tau) \in \mathscr{P}(P, \pi)$ satisfied within a finite time interval, say $[0, t]$, or respectively for $t \geq 0$. For this purpose, one now defines

$$\rho_i(t) = \sup_{0 \leq \theta \leq t} \int_0^\theta \max_j \left\{(PH(\tau)M)_{ij}\right\} d\tau \ , \tag{12.11}$$

or, respectively,

$$\rho_i = \sup_{t \geq 0} \int_0^t \max_j \left\{(PH(\tau)M)_{ij}\right\} d\tau \ . \tag{12.12}$$

We then obtain the following bounds within t:

$$y_i(\theta) \leq \rho_i(t) \ ,$$

for $\theta \in [0, t]$, or, respectively

$$y_i(t) \leq \rho_i$$

for $t \geq 0$.

Going on in this direction, we remark that the polyhedra $\mathscr{P}(P, \rho(t))$, or $\mathscr{P}(P, \rho)$, respectively, are outer approximations of $\mathscr{R}_t(\mathscr{C}(M))$ and $\mathscr{R}(\mathscr{C}(M))$, respectively. As introduced in Proposition 12.1, additional hypotheses can be introduced to be able to calculate inner approximations of the reachable sets.

Remark 12.2 The integral that appears in (12.12) is an increasing function of the time t, when its integrand is non-negative. This is always the case when 0 lies in $\mathscr{C}(M)$, or when the kernel $H(t)$ and the matrix M are non-negative. In this case, the bound (12.12) is equal to

$$\rho_i = \int_0^\infty \max_j \left\{(PH(\tau)M)_{ij}\right\} d\tau \ ,$$

that is well-defined if $H(t)$ is defined over $\mathscr{A}$.

Under the same hypothesis, that $0 \in \mathscr{C}(M)$, we observe that $\rho_i(t)$ is actually given by (12.9), and $\mathscr{R}_t(\mathscr{C}(M)) = \mathscr{R}(\mathscr{C}(M), t)$. In this case, the procedure presented in

Sect. 12.3.2 can be used to calculate a matrix N that corresponds to a lower approximation of the reachable set $\mathscr{R}_t(\mathscr{C}(M))$. We can also adapt this procedure to the case of an indefinite integral. For $k = 1$ to q, we define N as in Sect. 12.3.2, with $t = \infty$. According to Definition 12.4, we have obtained an exact approximation of the closure of the reachable set. We may remark that $\mathscr{C}(N)$ is not included into $\mathscr{R}(\mathscr{C}(M))$, in general, but we have the inclusion $\mathscr{C}_{ro}(N) \subset \mathscr{R}(\mathscr{C}_{ro}(M)) \subset \mathscr{P}_{ro}(P, \rho)$. In this sense, the matrices P, N, and the vector ρ also define exact approximations of the relatively open reachable set.

Remark 12.3 In many applications, one wants to compute approximations of the tube $(\mathscr{R}(\mathscr{C}(M), t), t) \subset \mathbb{R}^p \times \mathbb{R}_+$. As suggested in claim (ii) of Proposition 12.1, this tube is well defined if the kernel of system (12.4) is regular. The tube is then approximated using polyhedral approximations of $\mathscr{R}(\mathscr{C}(M), t_i)$ at successive instants t_i.

12.4 Remarks and Examples

12.4.1 Positive Kernels

The classical characterization of the positivity of a system in terms of the positivity of its kernel can also be seen as a consequence of Theorem 12.3.

Definition 12.5 The system (12.1) is said to be non-negative if every non-negative input $u(t)$ leads to a non-negative output $y(t)$. The multivariable system (12.5) is non-negative if its entries are all non-negative.

Corollary 12.3 *The system (12.1) is non-negative if and only if its kernel (12.2) is non-negative almost everywhere. The system (12.5) is non-negative if and only if all the entries of its kernel $H(t)$ are non-negative almost everywhere.*

Proof By definition, the system (12.1) is non-negative if $\mathscr{R}(\mathscr{C}(M)) \subset \mathscr{P}(P, \pi)$, with $M = (0, 1)$, $P = (-1)$, and $\pi = (0)$. Applying Theorem 12.3, we conclude that $\int_0^t \max\{0, -h(\tau)\}d\tau \leq 0$, for $t \geq 0$, from which we deduce that $h(\tau)$ takes non-negative values almost everywhere. □

If the system (12.2) is positive, and $u(t)$ lies in $[\alpha, \beta]$, we have the following inequalities, for $t \geq 0$

$$\alpha \int_0^t h(\tau)d\tau \leq y(t) \leq \beta \int_0^t h(\tau)d\tau \ .$$

In addition, these bounds are exact, in the sense that they are reached. If in addition the kernel $h(t)$ is an element of $\mathscr{A}$, then we have $y(t) \in [\alpha\|h\|_{\mathscr{A}}, \beta\|h\|_{\mathscr{A}}]$. The limits of this interval may be reached or not, but they are exact in the sense of the discussion of Sect. 12.3.3, for instance we have $\mathscr{R}(\mathscr{C}_{ro}(M)) =]\alpha\|h\|_{\mathscr{A}}, \beta\|h\|_{\mathscr{A}}[$.

This first result can be generalized to kernels that are not necessarily positive. Every measure h in $\mathscr{A}$ can be uniquely decomposed into a difference $h = h^+ - h^-$, where h^+ and h^- are two positive measures in $\mathscr{A}$ with disjoint supports. Then if $\alpha \le u(t) \le \beta$, the range of $y(t)$ is given by

$$\alpha \int_0^t h^+(\tau)d\tau - \beta \int_0^t h^-(\tau)d\tau \le y(t) \le \beta \int_0^t h^+(\tau)d\tau - \alpha \int_0^t h^-(\tau)d\tau \ ,$$

for any positive t, that can be rewritten as

$$\int_0^t \min\{\alpha h(\tau), \beta h(\tau)\}\, d\tau \le y(t) \le \int_0^t \max\{\alpha h(\tau), \beta h(\tau)\}\, d\tau \ ,$$

that in turns appears to be a consequence of Theorem 12.3. The latter formulation is well fitted for numerical computations, since it avoids the computation of h^+ and h^-. Indeed the infinite integral can be easily approximated using Matlab or Scilab, provided that $h(t)$ is explicitly known, or can be numerically computed. We also remark that this formula gives the way to calculate a control law $u^{\max}$ that maximizes the output. This control law is given by

$$u^{\max}(t-\tau) = \begin{cases} \alpha \ , & \text{if } \max\{\alpha h(\tau), \beta h(\tau)\} = \alpha h(\tau) \ , \\ \beta \ , & \text{else} \ , \end{cases} \tag{12.13}$$

for any positive t. In the same way, the control given by

$$u^{\min}(t-\tau) = \begin{cases} \alpha \ , & \text{if } \min\{\alpha h(\tau), \beta h(\tau)\} = \alpha h(\tau) \ , \\ \beta \ , & \text{else} \ , \end{cases}$$

permits to reach the lower value of the output. When t goes to infinity, we obtain the results that follow. They are well-known and often used (or rediscovered) in the literature.

(i) If $u(t) \in]-u_{\max}, +u_{\max}[$, then $y(t) \in]-y_{\max}, +y_{\max}[$, with $y_{\max} = \|h\| u_{\max}$.

(ii) If $h(t)$ is positive, and $u(t) \in [0, +u_{\max}[$, then $y(t) \in [0, y_{\max}[$.

(iii) If $h(t) = h_+(t) - h_-(t)$ with h_+ and h_- positive, and $u(t) \in]-u_{\min}, +u_{\max}[$, then $y(t) \in]-y_{\min}, +y_{\max}[$, with $y_{\min} = \|h_+\| u_{\min} + \|h_-\| u_{\max}$, and $y_{\max} = \|h_+\| u_{\max} + \|h_-\| u_{\min}$.

12.4.2 Constrained Control and $\mathscr{D}$-Invariance

We give here a simple example of the explicit computation of the bounds of input-output systems. It illustrates that these techniques may be useful to design control laws for constrained systems.

We consider the following model, which was introduced by Simon some years ago [23]. The inventory level $y(t)$ of a simple logistic system follows the law

$$\dot{y}(t) = u(t-\theta) - d(t) ,$$

where $u(t)$ is the production rate order and $d(t)$is the instantaneous demand. We assume that for $t < \theta$, we have $\dot{y}(t) = \phi(t) - d(t)$, where $\phi(t)$ corresponds to some initial condition. We choose the control law in the form

$$u(t) = K(y_c - z(t)) ,$$

with

$$z(t) = \begin{cases} y(t) + \int_{t-\theta}^{t} u(\tau)d\tau & \text{for } t \geq \theta , \\ y(t) + \int_{t}^{\theta} \phi(\tau)d\tau + \int_{0}^{t} u(\tau)d\tau & \text{for } t < \theta . \end{cases}$$

One can show that the solution is written

$$\hat{y}(s) = \frac{1 + K\frac{1-\mathrm{e}^{-s\theta}}{s}}{s+K}\left(y_0 + \hat{\phi}(s) - \hat{d}(s)\right) + \frac{K\mathrm{e}^{-s\theta}}{s+K}\cdot\left(\frac{y_c}{s} + \hat{\phi}\right) .$$

We therefore introduce the notations

$$\hat{h}_1(s) = \frac{\left(1 + K\frac{1-\mathrm{e}^{-s\theta}}{s}\right)}{s+K} \quad \hat{h}_2(s) = \frac{K\mathrm{e}^{-s\theta}}{s+K}$$

that are the Laplace transform of the kernels

$$h_1(t) = \begin{cases} 1 & , \text{ for } t \in [0, \theta[, \\ \mathrm{e}^{-K(t-\theta)} & , \text{ for } t \geq \theta , \end{cases} \quad h_2(t) = \begin{cases} 0 & , \text{ for } t \in [0, \theta[, \\ \mathrm{e}^{-K(t-\theta)} & , \text{ for } t \geq \theta , \end{cases}$$

and we notice that $\|h_1\|_{\mathscr{A}} = \theta + 1/K$, and $\|h_2\|_{\mathscr{A}} = 1$. Assuming that the range of the external demand $d(t)$ is $[0, d_{\max}]$, we deduce the bounds

$$-d_{\max}\|h_1\|_{\mathscr{A}} + y_c\|h_2\|_{\mathscr{A}} \leq y(t) \leq y_c\|h_2\|_{\mathscr{A}}$$

that lead to explicit bounds on $y(t)$

$$y_c - d_{\max}\left(\theta + \frac{1}{K}\right) \leq y(t) \leq y_c ,$$

for $t \geq \theta$, and on the admissible initial conditions

$$y_0 + wip_0 - \theta \leq y(t) \leq y_0 + wip_0 .$$

over the initial period $t \in [0, \theta[$, with $wip_0 = \int_0^\theta \phi(\tau)d\tau$. From these bounds, one can easily deduce conditions to meet the constraints on the production and inventory capacity, that are given as $u(t) \in [0, u_{\max}]$ and $y(t) \in [0, y_{\max}]$, for every demand in the range $d(t) \in [0, d_{\max}]$. The admissible values of the control parameters are:

$$y_c \in [\theta d_{\max}, y_{\max}] \ , \ K \geq \frac{d_{\max}}{y_c} - \theta \ ,$$

and the admissible values of the sizing parameters are:

$$u_{\max} \geq d_{\max} \ , \ \theta d_{\max} < y_{\max} \ .$$

These results were obtained using other methods in [17]. The same model can also be used to study the congestion control in communication networks, and similar results have been expounded in [12].

12.4.3 Example of Approximation of the Reachable Set

Let us consider the following time delay system

$$\dot{x}(t) = \begin{bmatrix} -2 & 0 \\ 0 & -1 \end{bmatrix} x(t) + \begin{bmatrix} 1 & 0 \\ -1 & 1 \end{bmatrix} x(t-1) + \begin{bmatrix} 0 & 1 \\ 0.5 & 0 \end{bmatrix} x(t-\pi) + \begin{bmatrix} -0.5 \\ 1 \end{bmatrix} u(t) \, ,$$

where the initial state of the system is taken as $x(t) = 0$ for $t \in [-\pi, 0]$, and $u(t)$ verifies $u(t) \in \mathscr{U} = \{0 \leq u(t) \leq 1\}$, for $t \geq 0$. Formally, this system can be rewritten in the form

$$x(t) = (H \star u)(t) \ ,$$

where $H \in \mathscr{A}^{2\times 1}$. The first step of the design is to numerically compute the kernels $H_{11}(t)$ and $H_{21}(t)$ using the solver dde23 of MATLAB. The result is plotted in Fig. 12.1. The second step of the design consists in the computation of the outer and inner approximations of the reachable set of the system. For this purpose, we consider the matrix P that is obtained by the concatenation of row vectors of the form $(\cos \frac{2k\pi}{K}, \sin \frac{2k\pi}{K})$, for $k = 1$ to K, and apply the procedure indicated in Sect. 12.3.2 to compute the vector ν and the matrix N, so that the outer and inner approximations of the reachable set are respectively $\mathscr{P}(P, \nu)$ and $\mathscr{C}(N)$. The polyhedra obtained for $K = 5$ are shown in Fig. 12.2. We also represent on the figure the reachable set, that was finely approximated using $K = 360$.

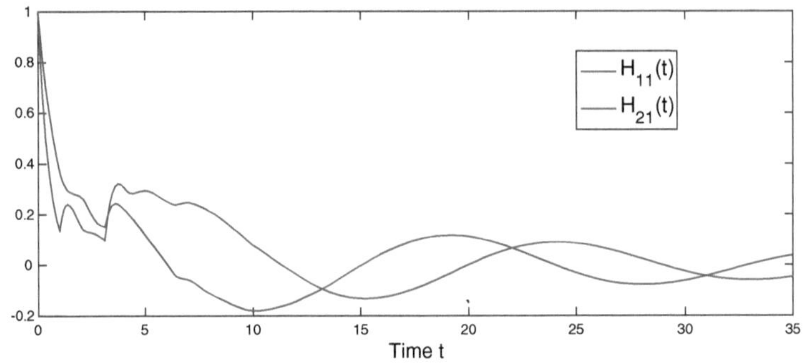

Fig. 12.1 Graphs of the kernels $H_{11}(t)$ and $H_{21}(t)$

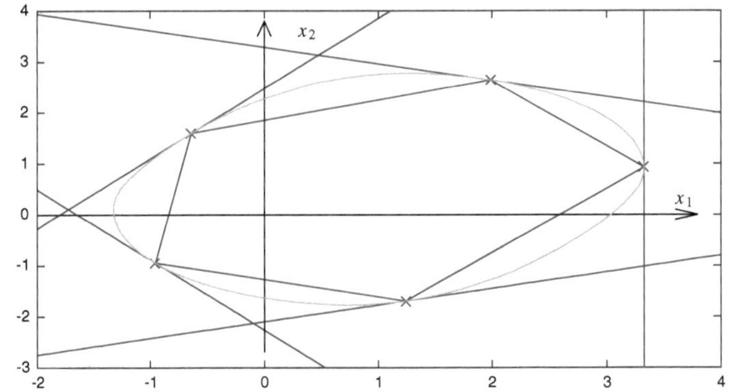

Fig. 12.2 Approximation of the reachable set

12.5 Conclusions

We have characterized bounds for a class of input-output systems defined by a convolution. They are derived from the concept of BIBO stability, and are given in terms of integrals that are easy to compute numerically. A method for the approximation of reachable sets of convolution systems was obtained from these bounds. We shortly commented the topological structure of the reachable set and the case of positive systems. The method was illustrated on a simple regulation problem of inventory level in a logistic system, and on an academic example of system with two non-commensurable delays.

Acknowledgements The author thanks very much Filippo Cacace and Joseph Winkin for their warm encouragements, which were crucial to produce this report.

References

1. Callier, F.M., Desoer, C.A.: An algebra of transfer functions for distributed linear time-invariant systems. IEEE Trans. Circuits Syst. **25**, 651–662 (1978)
2. Chen, H., Cheng, J., Zhong, S., Yang, J., Kang, W.: Improved results on reachable set bounding for linear systems with discrete and distributed delays. Adv. Differ. Equ. **145** (2015)
3. Chiasson, J., Loiseau, J.J. (eds.): Applications of Time Delay Systems. Springer, Berlin (2007)
4. Cousot, P., Halbwachs, N.: Automatic discovery of linear restraints among variables of a program. In: Conference Record of the Fifth Annual Symposium on Principles of Programming Languages. ACM Press, New York (1978)
5. Desoer, C.A., Callier, F.M.: Convolution feedback systems. SIAM J. Control **10**, 737–746 (1972)
6. Desoer, C.A., Vidyasagar, M.: Feedback Systems: Input-Output Properties. Academic Press, New York (1975)
7. Falcone, P., Ali, M., Sjoberg, J.: Predictive Threat assessment via reachability analysis and set invariance theory. IEEE Trans. Intell. Transp. Syst. **12**, 1352–1361 (2011)
8. Fridman, E., Shaked, U.: On reachable sets for linear systems with delay and bounded peak inputs. Automatica **39**, 2005–2010 (2003)
9. Guéguen, H., Lefebvre, M.-A., Zaytoon, J., Nasri, O.: Safety verification and reachability analysis for hybrid systems. Annu. Rev. Control **33**, 25–36 (2009)
10. Hille, E., Phillips, R.S.: Functional Analysis and Semi-Groups. American Mathematical Society, Providence (1957)
11. Hwang, I., Stipanović, D.M., Tomlin, C.J.: Polytopic approximations of reachable sets applied to linear dynamic games and to a class of nonlinear systems. In: Advances in Control, Communication Networks, and Transportation Systems, in Honor of Pravin Varaiya, pp. 3–19. Birkhäuser, Boston (2005)
12. Ignaciuk, P., Bartoszevicz, A.: Congestion Control in Data Transmission Networks. Sliding Modes and Other Designs. Springer, New York (2013)
13. Lakkonen, P.: Robust regulation for infinite-dimensional systems and signals in the frequency domain. Ph.D. Thesis, Tampere University of Technology, Finland (2013)
14. Lygeros, J., Tomlin, C.J., Sastry, S.: Controllers for reachability specifications for hybrid systems. Automatica **35**, 349–370 (1999)
15. Meslem, N., Ramdani, N., Candau, Y.: Approximation garantie de l'espace d'état atteignable des systèmes dynamiques continus incertains. JESA J. Européen des Systèmes Automatisés **43**, 241–266 (2009)
16. Moussaoui, C., Abbou, R., Loiseau, J.J.: On bounds of input-output systems. Reachability set determination and polyhedral constraints verification. In: Boje, S.O., Xia, X. (eds.) Proceedings of 19th IFAC World Congress, pp. 11012–11017. International Federation of Automatic Control, Cape Town (2014)
17. Moussaoui, C., Abbou, R., Loiseau, J.J.: Controller design for a class of delayed and constrained systems: application to supply chains. In: Seuret, A., Özbay, I., Bonnet, C., Mounier, H. (eds.) Low-Complexity Controllers for Time-Delay Systems, pp. 61–75. Springer, Berlin (2014)
18. Olaru, S., Stanković, N., Bitsoris, G., Niculescu, S.-I.: Low complexity invariant sets for time-delay systems: a set factorization approach. In: Seuret, A., Özbay, H., Bonnet, C., Mounier, H. (eds.) Low-Complexity Controllers for Time-Delay Systems, pp. 127–139. Springer, New York (2014)
19. Pecsvaradi, T., Narendra, K.S.: Reachable sets for linear dynamical systems. Inf. Control **19**, 319–344 (1971)
20. Quadrat, A.: A lattice approach to analysis and synthesis problems. Math. Control Signals Syst. **18**, 147–186 (2006)
21. Rockafellar, R.T.: Convex Analysis. Princeton University Press, Chichester (1970)
22. Sabatier, J., Agrawal, O.P., Tenreiro Machado, J.A. (eds.): Advances in Fractional Calculus. Springer, Berlin (2007)

23. Simon, H.A.: On the application of servomechanism theory in the study of production control. Econometrica **20**, 247–268 (1952)
24. Varaya, P.: Reach set computation using optimal control. In: Inan, M.K., Kurshan, R.P. (eds.) Verification of Digital and Hybrid Systems, pp. 323–331. Springer, Berlin (2000)

Chapter 13
On the Connection Between the Stability of Multidimensional Positive Systems and the Stability of Switched Positive Systems

Hugo Alonso and Paula Rocha

Abstract In this work, we study the connection of the stability of multidimensional positive systems with the stability of switched positive systems. In a previous work, we showed that the stability of a multidimensional positive system implies the stability of a related switched positive system. Here, we investigate the reciprocal implication.

Keywords Stability · Switched positive systems · Multidimensional positive systems

13.1 Introduction

The study of stability conditions for switched positive systems has attracted the attention of several researchers (see, for instance, [4, 5, 8]). By relating a switched positive system with a multidimensional positive system, in [1] we provided a simple sufficient condition, that could be stated in terms of the spectral radius of a single matrix. However, it turns out that this sufficient condition is not necessary. In order to understand how far sufficiency is from necessity, here we search for additional conditions under which the stability of a switched positive system implies the stability of the related multidimensional positive system.

The remainder of this chapter is organized as follows. In the next section, we make a brief introduction to multidimensional positive systems and their stability. The connection between the stability of these systems and the stability of switched

H. Alonso (✉)
CIDMA, Universidade de Aveiro, Campus Universitário de Santiago,
3810-193 Aveiro, Portugal
e-mail: hugo.alonso@ua.pt

H. Alonso
Universidade Lusófona do Porto, Rua Augusto Rosa, 24, 4000-098 Porto, Portugal

P. Rocha
SYSTEC, Faculty of Engineering, University of Porto, Rua Dr. Roberto Frias, s/n,
4200-465 Porto, Portugal
e-mail: mprocha@fe.up.pt

F. Cacace et al. (eds.), *Positive Systems*, Lecture Notes in Control
and Information Sciences 471, DOI 10.1007/978-3-319-54211-9_13

positive systems is studied in Sect. 13.3. Finally, the chapter ends with the conclusions in Sect. 13.4.

13.2 Multidimensional Positive Systems and Their Stability

The k-dimensional (kD) positive linear discrete systems of order n considered here are of the form

$$\Sigma^{kD}_{A_1,\ldots,A_k}: \quad \omega(i) = \sum_{j=1}^{k} A_j \omega(i - e_j), \tag{13.1}$$

where $\omega(i) \in \mathbb{R}^n$ represents the non-negative local state at $i = (i_1, \ldots, i_k) \in \mathbb{Z}^k$, $A_1, \ldots, A_k \in \mathbb{R}^{n\times n}$ are non-negative matrices, $e_j \in \mathbb{Z}^k$ is the j-th unit vector and so $i - e_j = (i_1, \ldots, i_{j-1}, i_j - 1, i_{j+1}, \ldots, i_k)$. Furthermore, letting $\bar{i} = \sum_{j=1}^{k} i_k$, the global state of $\Sigma^{kD}_{A_1,\ldots,A_k}$ at time $\ell \in \mathbb{Z}_0^+$ is defined as the set of local states $\Omega_\ell = \{\omega(i) : \bar{i} = \ell\}$. Note that the notions of local and global state only coincide in the particular case of $k = 1$, when (13.1) describes a 1D system Σ_A such that $\omega(\ell) = A\omega(\ell - 1)$. Now, it is obvious that, given a non-negative initial state Ω_0, a sequence $\Omega_1, \Omega_2, \ldots$ is uniquely determined by (13.1). The behavior of the global state sequences determines the stability properties of the system. In particular, $\Sigma^{kD}_{A_1,\ldots,A_k}$ is said to be asymptotically stable if for every non-negative Ω_0 such that $||\Omega_0|| < \infty$, one has $\lim_{\ell\to+\infty} ||\Omega_\ell|| = 0$, where $||\Omega_\ell|| = \sup\{||\omega(i)||_2 : \bar{i} = \ell\}$ and $||\cdot||_2$ denotes the usual Euclidean norm. In the area of multidimensional systems, it is well known that the following condition (which does not explore the fact that the system is positive) is necessary and sufficient for the asymptotic stability of $\Sigma^{kD}_{A_1,\ldots,A_k}$ [2]:

$$\det(I_n - \sum_{j=1}^{k} z_j A_j) \neq 0 \ \forall (z_1, \ldots, z_k) \in \mathbb{D}^k,$$

where $\mathbb{D}^k = \{(z_1, \ldots, z_k) \in \mathbb{C}^k : |z_j| \leq 1,\ j = 1, \ldots, k\}$ is the closed unit polydisc in $\mathbb{C}^k$. This condition is unpractical and is not in general easy to check. However, if we use the fact that the kD system is positive, then we get a simpler condition stated in the proposition below. The result was presented for $k = 2$ in [10]. We presented it for $k \geq 2$ in [1], but without a proof. We now prove it.

Proposition 13.1 *The kD positive system $\Sigma^{kD}_{A_1,\ldots,A_k}$ is asymptotically stable if and only if the 1D positive system Σ_A with $A = A_1 + \cdots + A_k$ is asymptotically stable.*

Proof Let us assume that the kD positive system $\Sigma^{kD}_{A_1,\ldots,A_k}$ is asymptotically stable. Suppose that the local states in Ω_0 are all equal to a non-negative $\omega_0 \in \mathbb{R}^n$, arbitrarily chosen. Then, it can be seen that the local states in Ω_ℓ are all equal to $(A_1 + \cdots + A_k)^\ell \omega_0$ and hence that $||\Omega_\ell|| = ||(A_1 + \cdots + A_k)^\ell \omega_0||_2$ for all $\ell \in \mathbb{Z}_0^+$.

The asymptotic stability of the kD positive system implies that $\lim_{\ell\to+\infty} ||\Omega_\ell|| = 0$ and, therefore, $\lim_{\ell\to+\infty} ||(A_1 + \cdots + A_k)^\ell \omega_0||_2 = 0$. Given that ω_0 is arbitrary, it follows that the 1D positive system Σ_A with $A = A_1 + \cdots + A_k$ is asymptotically stable.

Now, let us assume that the 1D positive system Σ_A with $A = A_1 + \cdots + A_k$ is asymptotically stable. Suppose that the global state Ω_0 of the kD positive system $\Sigma^{kD}_{A_1,\ldots,A_k}$ is non-negative and such that $||\Omega_0|| < \infty$. Then, there exists $L \in \mathbb{R}^+$ such that, if $\omega(i)$ with $\bar{i} = 0$ is a local state in Ω_0, then $0_n \leq \omega(i) \leq L_n$, where 0_n and L_n are vectors of length n with all components equal to 0 and L, respectively, and where the inequalities should be understood component-wise. Now, let $\Psi : (\mathbb{Z}_0^+)^k \mapsto \mathbb{R}^{n\times n}$ be the map whose value $\Psi(i) = \Psi(i_1, \ldots, i_k)$ corresponds to the matrix resulting from the sum of all products in $\{A_1, \ldots, A_k\}$ where A_j appears i_j times for $j = 1, \ldots, k$, usually known as the Hurwitz product of $A_1, \ldots, A_k$ associated with i. For instance, if $k = 2$, then $\Psi(0, 0) = I_n$, $\Psi(i_1, 0) = A_1^{i_1}$ when $i_1 > 0$, $\Psi(0, i_2) = A_2^{i_2}$ when $i_2 > 0$ and $\Psi(i_1, i_2) = A_1\Psi(i_1 - 1, i_2) + A_2\Psi(i_1, i_2 - 1)$ when $i_1, i_2 > 0$ [3]. With this notation, if $\omega(i)$ with $\bar{i} = \ell$ is a local state in Ω_ℓ, we have

$$\begin{aligned}
||\omega(i)||_2 &= ||\sum\nolimits_{\bar{j}=\ell} \Psi(j)\omega(i - j)||_2 \\
&\leq ||\sum\nolimits_{\bar{j}=\ell} \Psi(j)L_n||_2 \\
&= ||(\sum\nolimits_{\bar{j}=\ell} \Psi(j))L_n||_2 \\
&= ||(A_1 + \cdots + A_k)^\ell L_n||_2
\end{aligned}$$

and so $||\Omega_\ell|| \leq ||(A_1 + \cdots + A_k)^\ell L_n||_2$ for all $\ell \in \mathbb{Z}_0^+$. The asymptotic stability of the 1D positive system Σ_A with $A = A_1 + \cdots + A_k$ implies that $\lim_{\ell\to+\infty} ||(A_1 + \cdots + A_k)^\ell L_n||_2 = 0$ and, therefore, $\lim_{\ell\to+\infty} ||\Omega_\ell|| = 0$. Finally, minding that Ω_0 is arbitrary, it follows that the kD positive system $\Sigma^{kD}_{A_1,\ldots,A_k}$ is asymptotically stable. □

Remark 13.1 According to the proposition, checking the asymptotic stability of the kD positive system $\Sigma^{kD}_{A_1,\ldots,A_k}$ amounts to check the asymptotic stability of the 1D positive system Σ_A with $A = A_1 + \cdots + A_k$, but this is very easy, because Σ_A is asymptotically stable if and only if the spectral radius of A is less than one, that is, $\rho(A) < 1$.

13.3 On the Connection Between the Stability of Multidimensional Positive Systems and the Stability of Switched Positive Systems

A switched positive linear discrete-time system of order n composed of k subsystems can be described by

$$\Sigma_{A_1,\ldots,A_k}: \quad x(\ell) = A_{\sigma(\ell-1)}x(\ell-1), \quad A_{\sigma(\ell-1)} \in \{A_1,\ldots,A_k\}, \tag{13.2}$$

where $x(\ell) \in \mathbb{R}^n$ represents the non-negative state vector at time $\ell \in \mathbb{Z}_0^+$, $A_1,\ldots,A_k \in \mathbb{R}^{n\times n}$ are non-negative matrices associated with the k subsystems and $\sigma : \mathbb{Z}_0^+ \mapsto \{1,\ldots,k\}$ is the switching signal. It is clear that, given a non-negative initial state

$$x(0) = x_0 \tag{13.3}$$

and a switching signal σ, a sequence $x(1), x(2), \ldots$ is uniquely determined by (13.2). The behavior of the state sequences determines the stability properties of the system. In particular, $\Sigma_{A_1,\ldots,A_k}$ is said to be uniformly asymptotically stable if it is uniformly stable (u.s.) and globally uniformly attractive (g.u.a.), *i.e.*:

- $\forall \varepsilon > 0, \exists \delta > 0$: $||x(0)||_2 < \delta \Rightarrow ||x(\ell)||_2 < \varepsilon \ \forall \ell \in \mathbb{Z}_0^+, \sigma$ (u.s.);
- $\forall r, \varepsilon > 0, \exists \ell^\star \in \mathbb{Z}^+$: $||x(0)||_2 < r \Rightarrow ||x(\ell)||_2 < \varepsilon \ \forall \ell \geq \ell^\star, \sigma$ (g.u.a.).

As is known, $\Sigma_{A_1,\ldots,A_k}$ is uniformly asymptotically stable if there exists a common quadratic Lyapunov function (CQLF) $V(x) = x^T Px$ such that

$$P \succ 0 \quad \wedge \quad P - A_j^T PA_j \succ 0 \quad j = 1,\ldots,k, \tag{13.4}$$

where T denotes transposition and $P \succ 0$ means that P is positive definite [9].

Now, consider the kD positive system $\Sigma^{kD}_{A_1,\ldots,A_k}$ described by (13.1) and whose global state $\Omega_0 = \{\omega(i) : \bar{i} = 0\}$ is determined by

$$\omega(0) = x_0, \quad \omega(i) = 0 \quad \bar{i} = 0 \wedge i \neq 0. \tag{13.5}$$

Note that, in $\Sigma_{A_1,\ldots,A_k}$, the state is updated in each step in a single direction, corresponding to the variable ℓ. Moreover, $\Sigma_{A_1,\ldots,A_k}$ has k operation modes, and when the j-th mode is active, the state update is made according to $x(\ell) = A_j x(\ell-1)$. On the other hand, in $\Sigma^{kD}_{A_1,\ldots,A_k}$, the local state is updated in each step in k directions, corresponding to the variables $i_1,\ldots,i_k$ in i. In addition, the contribution of the j-th update direction to the overall update, given by

$$\begin{aligned}\omega(i_1,\ldots,i_j,\ldots,i_k) = {} & A_1\omega(i_1-1,\ldots,i_j,\ldots,i_k) + \cdots + \\ & A_j\omega(i_1,\ldots,i_j-1,\ldots,i_k) + \cdots + \\ & A_k\omega(i_1,\ldots,i_j,\ldots,i_k-1),\end{aligned}$$

is represented by $A_j\omega(i_1,\ldots,i_j-1,\ldots,i_k)$. Therefore, we can think of an update direction in $\Sigma^{kD}_{A_1,\ldots,A_k}$ as being associated with an operation mode in $\Sigma_{A_1,\ldots,A_k}$. Furthermore, it is easy to see that the local state $\omega(i) = \omega(i_1,\ldots,i_k)$ of $\Sigma^{kD}_{A_1,\ldots,A_k}$ equals the sum of all possibilities for the state $x(\ell)$ of the switching system $\Sigma_{A_1,\ldots,A_k}$ after $\ell = \bar{i}$ steps where the value of the switching signal is j for i_j times with $j = 1,\ldots,k$. Hence, the two systems have state evolutions that are closely related. This is illustrated in Fig. 13.1 for $k = 2$. Note for instance that the value of $\omega(i) = \omega(i_1, i_2)$ along the i_j-

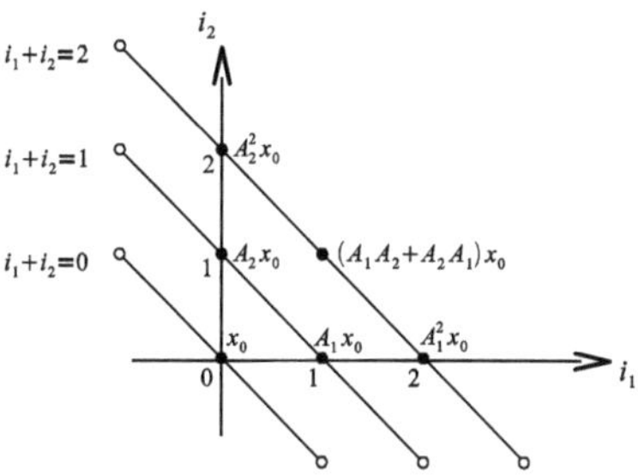

Fig. 13.1 State evolution of the 2D system $\Sigma^{2D}_{A_1,A_2}$ associated with the switching system Σ_{A_1,A_2}

axis evolves in the same manner as the value of $x(\ell)$ when the switching signal is such that $\sigma(\ell) = j$ for all ℓ. Also remark that the value of $\omega(1, 1) = (A_1A_2 + A_2A_1)x_0$ results from the sum of the possible values for $x(2)$ after two steps where the value of the switching signal is 1 in one step and 2 in the other. Given the close relation between the state evolutions of both systems, it is not surprising that their stability properties are also related. This is clarified in the next proposition.

Proposition 13.2 *The switched positive system $\Sigma_{A_1,\dots,A_k}$ described by (13.2), (13.3) is uniformly asymptotically stable if the associated kD positive system $\Sigma^{kD}_{A_1,\dots,A_k}$ described by (13.1), (13.5) is asymptotically stable.*

We presented this result in [1]. In the following, we study the reciprocal implication and identify conditions under which the uniform asymptotic stability of the switched positive system $\Sigma_{A_1,\dots,A_k}$ implies the asymptotic stability of the associated kD positive system $\Sigma^{kD}_{A_1,\dots,A_k}$.

Start by noting that, as explained in Remark 13.1, a kD positive system $\Sigma^{kD}_{A_1,\dots,A_k}$ is asymptotically stable if and only if $\rho(A_1 + \dots + A_k) < 1$. In [1], we showed that, if $\rho(A_1 + \dots + A_k) < 1$, then it is possible to find a CQLF for the switched positive system $\Sigma_{A_1,\dots,A_k}$. Unfortunately, the converse is not true, as shown in the next example.

Example 13.1 Consider the switched positive system Σ_{A_1,A_2} described by (13.2), (13.3) with $k = 2$ and

$$A_1 = \begin{pmatrix} 0.7 & 0 \\ 0 & 0.1 \end{pmatrix} \qquad A_2 = \begin{pmatrix} 0.4 & 0 \\ 0 & 0.1 \end{pmatrix}.$$

It is obvious that A_1 and A_2 are such that $\rho(A_1), \rho(A_2) < 1$ and commute. Therefore, it is possible to find a CQLF for Σ_{A_1,A_2} [7]. Moreover, it can be seen that $\rho(A_1 + A_2) = 1.1 \nless 1$.

At this point, a natural question arises: is there a relation between the existence of a CQLF for a switched positive system $\Sigma_{A_1,\dots,A_k}$ and the value of $\rho(A_1 + \dots + A_k)$? If the CQLF has no special form, then the answer is given by the following:

Proposition 13.3 *If the switched positive system $\Sigma_{A_1,\dots,A_k}$ described by (13.2), (13.3) has a CQLF, then $\rho(A_1 + \dots + A_k) < k$.*

Proof Let us assume that $V(x) = x^T P x$ is a CQLF for $\Sigma_{A_1,\ldots,A_k}$ such that $P \succ 0$ and

$$\begin{aligned} P - A_1^T P A_1 &\succ 0 \\ &\vdots \\ P - A_k^T P A_k &\succ 0. \end{aligned}$$

Then,

$$\begin{aligned} \left(P - A_1^T P A_1\right) + \cdots + \left(P - A_k^T P A_k\right) \succ 0 \Leftrightarrow \\ kP - \sum_{j=1}^{k} A_j^T P A_j \succ 0 \Leftrightarrow \\ k^2 \left(kP^{-1}\right)^{-1} - \sum_{j=1}^{k} A_j^T P A_j \succ 0 \Leftrightarrow \\ \left(kP^{-1}\right)^{-1} - \sum_{j=1}^{k} \left(\frac{1}{k} A_j\right)^T P \left(\frac{1}{k} A_j\right) \succ 0. \end{aligned}$$

According to [6], the latter condition implies that the kD positive system $\Sigma^{kD}_{\frac{1}{k}A_1,\ldots,\frac{1}{k}A_k}$ is asymptotically stable. This in turn implies that $\rho(\frac{1}{k}A_1 + \cdots + \frac{1}{k}A_k) < 1$ and so $\rho(A_1 + \cdots + A_k) < k$. □

In the proposition just presented, no special form was assumed for the CQLF. However, if the CQLF for the switched positive system $\Sigma_{A_1,\ldots,A_k}$ is of a certain type, then the bound on $\rho(A_1 + \cdots + A_k)$ can be tightened. This is clarified in the next result, which is the main contribution of this chapter. It identifies conditions under which the uniform asymptotic stability of the switched positive system $\Sigma_{A_1,\ldots,A_k}$ implies the asymptotic stability of the associated kD positive system $\Sigma^{kD}_{A_1,\ldots,A_k}$. The proof is omitted because it is based on arguments similar to those previously used.

Proposition 13.4 *If the switched positive system $\Sigma_{A_1,\ldots,A_k}$ described by (13.2), (13.3) is uniformly asymptotically stable and has a CQLF $V(x) = x^T P x$ such that $P \succ 0$ and*

$$\begin{aligned} \frac{1}{k^2} P - A_1^T P A_1 &\succ 0 \\ &\vdots \\ \frac{1}{k^2} P - A_k^T P A_k &\succ 0, \end{aligned}$$

then $\rho(A_1 + \cdots + A_k) < 1$ and the associated kD positive system $\Sigma^{kD}_{A_1,\ldots,A_k}$ described by (13.1), (13.5) is asymptotically stable.

Remark 13.2 It is easy to see that a matrix P in the conditions above also satisfies $P - A_j^T P A_j \succ 0$ for $j = 1, \ldots, k$. This means that in the previous proposition we are indeed asking for the existence of a CQLF for $\Sigma_{A_1,\ldots,A_k}$ of a special form.

The next example illustrates the application of Proposition 13.4.

Example 13.2 Consider the switched positive system $\Sigma_{A_1,\ldots,A_k}$ described by (13.2), (13.3) with non-negative diagonal matrices

$$A_j = \operatorname{diag}\left(\alpha_{j1}, \ldots, \alpha_{jn}\right) \qquad j = 1, \ldots, k.$$

Assume that $\Sigma_{A_1,\ldots,A_k}$ is uniformly asymptotically stable. Given that $\rho(A_1), \ldots, \rho(A_k) < 1$, since the system is stable only if each subsystem is stable, and $A_1, \ldots, A_k$ commute, $\Sigma_{A_1,\ldots,A_k}$ has a CQLF $V(x) = x^T P x$ with P of diagonal form [7]:

$$P = \operatorname{diag}\left(p_1, \ldots, p_n\right),$$

where $p_1, \ldots, p_n > 0$. Assume that $\frac{1}{k^2}P - A_j^T P A_j \succ 0$ for $j = 1, \ldots, k$, that is, that the CQLF is in the conditions of the previous proposition. Then,

$$\begin{aligned} \frac{1}{k^2}P - A_j^T P A_j \succ 0 &\Leftrightarrow \\ \operatorname{diag}\left(p_1\left(\frac{1}{k^2} - \alpha_{j1}^2\right), \ldots, p_n\left(\frac{1}{k^2} - \alpha_{jn}^2\right)\right) \succ 0 &\Leftrightarrow \\ 0 \le \alpha_{j1}, \ldots, \alpha_{jn} < \frac{1}{k} & \end{aligned}$$

for $j = 1, \ldots, k$. It is now simple to check that $\rho(A_1 + \cdots + A_k) < 1$ and hence the associated kD positive system $\Sigma^{kD}_{A_1,\ldots,A_k}$ described by (13.1), (13.5) is asymptotically stable.

13.4 Conclusions

In this chapter we studied the relation between the stability of multidimensional positive systems and the stability of switched positive systems. Motivated by the fact that the stability of the former implies the stability of the latter [1], but not vice-versa, we searched for additional conditions under which the stability of a switched positive system implies the stability of a related multidimensional positive system. As a preliminary result, we showed that if the switched positive system has a common quadratic Lyapunov function of a certain type, then the associated multidimensional positive system is stable. In our opinion, this might be a step forward to obtain necessary and sufficient conditions for the stability of a new class of switched positive systems.

Acknowledgements This work was supported by *FEDER* funds through *COMPETE* – Operational Programme Factors of Competitiveness ("Programa Operacional Factores de Competitividade") and by Portuguese funds through the Portuguese Foundation for Science and Technology ("FCT–Fundação para a Ciência e a Tecnologia"), within the project UID/MAT/04106/2013 associated with the *Center for Research and Development in Mathematics and Applications* (University of Aveiro) and the project POCI-01-0145-FEDER-006933 - SYSTEC - Research Center for Systems and Technologies.

References

1. Alonso, H., Rocha, P.: A general stability test for switched positive systems based on a multidimensional system analysis. IEEE Trans. Autom. Control **55**(11), 2660–2664 (2010)
2. Bose, N.K.: Multidimensional Systems Theory and Applications, 2nd edn. Kluwer Academic Publishers, The Netherlands (2003)
3. Fornasini, E., Valcher, M.E.: Matrix pairs in two-dimensional systems: an approach based on trace series and Hankel matrices. SIAM J. Control Optim. **33**(4), 1127–1150 (1995)
4. Fornasini, E., Valcher, M.E.: Stability properties of a class of positive switched systems with rank one difference. Syst. Control Lett. **64**, 12–19 (2014)
5. Gurvits, L., Shorten, R., Mason, O.: On the stability of switched positive linear systems. IEEE Trans. Autom. Control **52**(6), 1099–1103 (2007)
6. Ooba, T.: Stabilization of multidimensional systems using local state estimators. Multidimens. Syst. Signal Process. **12**, 49–61 (2001)
7. Ooba, T., Funahashi, Y.: On the simultaneous diagonal stability of linear discrete-time systems. Syst. Control Lett. **36**, 175–180 (1999)
8. Sun, Y.: Stability analysis of positive switched systems via joint linear copositive Lyapunov functions. Nonlinear Anal.: Hybrid Syst. **19**, 146–152 (2016)
9. Sun, Z., Ge., S.S.: Switched Linear Systems: Control and Design. Springer, UK (2005)
10. Valcher, M.E.: On the internal stability and asymptotic behavior of 2-D positive systems. IEEE Trans. Circuits Syst. I: Fundam. Theory Appl. **44**(7), 602–613 (1997)

Chapter 14
Positive Stabilization of a Diffusion System by Nonnegative Boundary Control

Jonathan N. Dehaye and Joseph J. Winkin

Abstract This chapter deals with the issue of considering nonnegative inputs in the positive stabilization problem. It is shown in two different ways why one cannot expect to positively stabilize a positive system by use of a nonnegative input, first by a classical approach with a formal proof, then by working on an extended system for which the new input corresponds to the time derivative of the nominal one, thus circumventing the sign restriction. However, it is shown via a classical example of positive system—the pure diffusion system—that positively stabilizing a positive system with a nonnegative input is in some way possible: using a boundary control, the input sign depends on whether the boundary control appears in the boundary conditions or in the dynamics. The chapter then provides a parameterization of all positively stabilizing feedbacks for a discretized model of the pure diffusion system, some numerical simulations and a convergence discussion which allows to extend the results to the infinite-dimensional case, where the system is described again by a parabolic partial differential equation and the input acts either in the dynamics or in the boundary conditions.

Keywords Positive systems · Nonnegative input · Diffusion equation · Positive stabilization · Feedback parameterization · Partial differential equations

14.1 Introduction

Positive linear systems are linear systems whose state variables are nonnegative at all time whenever so are the initial state and the input. Studying this kind of systems is of great importance as the nonnegativity property can be found frequently in numerous

J.N. Dehaye (✉) · J.J. Winkin
Department of Mathematics and naXys, University of Namur,
Rempart de la Vierge 8, 5000 Namur, Belgium
e-mail: jonathan.dehaye@unamur.be

J.J. Winkin
e-mail: joseph.winkin@unamur.be

F. Cacace et al. (eds.), *Positive Systems*, Lecture Notes in Control
and Information Sciences 471, DOI 10.1007/978-3-319-54211-9_14

fields like biology, chemistry, physics, ecology, economy or sociology (see e.g. [1, 4, 10, 11, 19] for particular examples).

It is known that positively stabilizing an unstable (lumped parameter) positive system by means of a nonnegative input is impossible [6]. This has to be taken into account while studying the positive stabilization problem. In this chapter, we show in two different ways that a positive linear system is exponentially positively stabilizable by a nonnegative input if and only if the system is already exponentially stable. Then we introduce a classical and relevant example—the pure diffusion (distributed parameter) system—for which the input nonnegativity issue is considered in two different ways, depending on whether the boundary control appears in the boundary conditions or in the dynamics [9]. The system is discretized and all positively stabilizing feedbacks are parameterized [7] by use of classical positive control theory [11, 15]. Finally, the discretized system is positively stabilized with a suitable feedback, and convergence issues are discussed.

14.2 Preliminaries

In the following subsections, we provide the reader with the notations, definitions and main concepts used in the chapter.

14.2.1 Terminology

In the sequel, we will use the sets $\mathbb{R}_+ := \{x \in \mathbb{R} \mid x \geq 0\}$, $\mathbb{R}_{0,+} := \{x \in \mathbb{R} \mid x > 0\}$, $\mathbb{R}^n_+ := \{(x_1, \ldots, x_n) \in \mathbb{R}^n \mid x_i \in \mathbb{R}_+, \forall i = 1, \ldots, n\}$ and $\mathbb{R}^n_{0,+} := \{(x_1, \ldots, x_n) \in \mathbb{R}^n \mid x_i \in \mathbb{R}_{0,+}, \forall i = 1, \ldots, n\}$. Similarly, $\mathbb{R}_-$, $\mathbb{R}_{0,-}$, $\mathbb{R}^n_-$ and $\mathbb{R}^n_{0,-}$ denote the sets $\{x \in \mathbb{R} \mid x \leq 0\}$, $\{x \in \mathbb{R} \mid x < 0\}$, $\{(x_1, \ldots, x_n) \in \mathbb{R}^n \mid x_i \in \mathbb{R}_-, \forall i = 1, \ldots, n\}$ and $\{(x_1, \ldots, x_n) \in \mathbb{R}^n \mid x_i \in \mathbb{R}_{0,-}, \forall i = 1, \ldots, n\}$ respectively. For convenience, we use the notations $v \geq 0$ if $v \in \mathbb{R}^n_+$, $v > 0$ if $v \in \mathbb{R}^n_+$ and $v \neq 0$, $v \gg 0$ if $v \in \mathbb{R}^n_{0,+}$. The real part of a complex number $z \in \mathbb{C}$ will be denoted by $\mathscr{R}(z)$. A *nonnegative* vector v has all its components greater or equal to zero (i.e. $v_i \in \mathbb{R}_+$, for all i). The transpose of a matrix A will be denoted by A^T. The ijth entry of a matrix A will be denoted by a_{ij}. The spectrum of a matrix A is the set of its eigenvalues and will be denoted by $\sigma(A)$. A *nonnegative* matrix A (denoted by $A \geq 0$) has all its entries greater or equal to zero (i.e. $a_{ij} \in \mathbb{R}_+$, for all i, j). A *Metzler* matrix A has all its off-diagonal entries greater or equal to zero (i.e. $a_{ij} \in \mathbb{R}_+$, for all $i \neq j$). A *stable* matrix A has all its eigenvalues with negative real parts (i.e. $\mathscr{R}(\lambda) < 0$, $\forall \lambda \in \sigma(A)$). For convenience, lower-case letters when used in an appropriate context will represent scalars or vectors, while upper-case letters will represent matrices.

14.2.2 Main Concepts

Consider a linear time-invariant system

$$\begin{cases} \dot{x} = Ax + Bu \\ y = Cx + Du \end{cases}$$

where $A \in \mathbb{R}^{n\times n}$, $B \in \mathbb{R}^{n\times m}$, $C \in \mathbb{R}^{p\times n}$ and $D \in \mathbb{R}^{p\times m}$. We first recall the concept of positive linear system [10, 11, 13, 15].

Definition 14.1 A linear system $R = [A, B, C, D]$ is positive if for every nonnegative initial state $x_0 \in \mathbb{R}^n_+$ and for every admissible nonnegative input u (i.e. every piecewise continuous function $u : \mathbb{R}_+ \to \mathbb{R}^m_+$) the state trajectory x of the system and the ouput trajectory y are nonnegative (i.e. for all $t \geq 0$, $x(t) \in \mathbb{R}^n_+$ and $y(t) \in \mathbb{R}^p_+$).

It is possible to express the positivity of a system by use of the matrices A, B, C and D only [10, 11].

Theorem 14.1 *A linear system $R = [A, B, C, D]$ is positive if and only if A is a Metzler matrix and B, C and D are nonnegative matrices.*

Now we define the positive stabilizability of positive systems. For convenience, throughout the chapter the notion of stability will refer to asymptotic stability, which is equivalent to exponential stability as we deal with LTI systems.

Definition 14.2 A positive linear system $R = [A, B, C, D]$ is positively (exponentially) stabilizable if there exists a state feedback matrix $K \in \mathbb{R}^{m\times n}$ such that $A + BK$ is a stable Metzler matrix, i.e. such that there exist positive constants M and σ such that for all $t \geq 0$

$$\left\| e^{(A+BK)t} \right\| \leq Me^{-\sigma t}$$

and for all $t \geq 0$, $e^{(A+BK)t} \geq 0$. Such a feedback matrix K is called a positively stabilizing feedback for the system R.

The positive stabilization problem is concerned with existence conditions and the computation of such a matrix K. Finally, we introduce an important result from [4, 11, 15] which provides a necessary and sufficient condition for the stability of a Metzler matrix.

Lemma 14.1 *A Metzler matrix $A \in \mathbb{R}^{n\times n}$ is stable if and only if there exists $v \gg 0$ in $\mathbb{R}^n$ such that $Av \ll 0$.*

Remark 14.1 The sufficiency of the condition can be shown by considering the Lyapunov function $V(x) = v^T x$ which leads to $\dot{V}(x) = v^T Ax < 0$. The necessity follows from the fact that the opposite of the inverse of a stable Metzler matrix is nonnegative: it suffices to define $v = -A^{-1}\tau$ with $\tau \gg 0$. See [11, Lemma 2.2] or [16, Lemma 1.1].

14.3 Positive Stabilization by Nonnegative Input

One obvious way to ensure the nonnegativity of the state trajectory of a positive system is to force the input to remain nonnegative. However, it is impossible to positively stabilize an unstable positive system with such an input. The first subsection provides a classical approach of the problem, while the second one provides an alternative as we work on an extended system.

14.3.1 A Classical Approach

First, let us recall the Perron-Frobenius theorem for Metzler matrices [2, 12]:

Theorem 14.2 *If A is a Metzler matrix, there exist a real number λ and a real vector $v > 0$ such that $Av = \lambda v$ and for every eigenvalue μ of A, $\mathscr{R}(\mu) \leq \lambda$.*

Remark 14.2 The result in [12] is actually shown for nonnegative matrices. However, a Metzler matrix is a nonnegative matrix up to a diagonal shift. It is easy to see that a diagonal shift just shifts the eigenvalues and leaves the eigenvectors unchanged, making the result valid for Metzler matrices.

In [6] it is stated without proof that, in view of [17], if the dominant eigenvalue of A is nonnegative one cannot stabilize the system with a nonnegative input. Then one can conclude that if a positive system is not already stable, it cannot be stabilized by use of a nonnegative input. For the sake of self-containedness, let us briefly formulate and prove that assertion.

Theorem 14.3 *Consider the positive linear system $\dot{x} = Ax + bu$. The system is (exponentially) positively stabilizable by a state feedback $u = Kx$ such that $u \in \mathbb{R}_+$ if and only if it is already (exponentially) stable.*

Proof The sufficiency of the condition is trivial: it suffices to take $K = 0$, hence $u = 0$. Let us prove the necessity. Suppose that the system is unstable, then the dominant eigenvalue λ of A^T is nonnegative (see Theorem 14.2). By [10, 12] there exists an eigenvector $v > 0$ such that $A^T v = \lambda v$. Now let us define $\rho = v^T x$ and focus on the unstable part of the system relative to λ. We then have

$$\dot{\rho} = v^T \dot{x} = v^T Ax + v^T bu = (A^T v)^T x + v^T bu = \lambda\rho + (v^T b)u$$

where $v^T b \geq 0$. If $u = Kx$ was a state feedback such that $u \in \mathbb{R}_+$, then

$$\rho(t) = e^{\lambda t}\rho_0 + \int_0^t e^{\lambda(t-\tau)}(v^T b)u(\tau)d\tau$$

would not tend to zero as $t \to \infty$, since λ, $e^{\lambda t}$, ρ_0, $(v^T b)$ and u are all nonnegative (or positive), thus showing that the system cannot be positively stabilized in this way. □

14.3.2 An Extended System

As we showed the issue of considering a nonnegative input, we try to circumvent the problem by working on an extended system. Consider

$$\begin{cases} \dot{x} = Ax + Bu \\ \dot{u} = v \end{cases}$$

where A is a Metzler matrix, B is nonnegative and v is the new input. This leads to the positive extended system

$$\dot{\begin{bmatrix} x \\ u \end{bmatrix}} = \begin{bmatrix} A & B \\ 0 & 0 \end{bmatrix} \begin{bmatrix} x \\ u \end{bmatrix} + \begin{bmatrix} 0 \\ I \end{bmatrix} v$$

that we will denote by $\dot{\tilde{x}} = \tilde{A}\tilde{x} + \tilde{B}v$ with initial condition

$$\tilde{x}_0 = \begin{bmatrix} x(0) \\ u(0) \end{bmatrix} = \begin{bmatrix} x_0 \\ u_0 \end{bmatrix} \geq 0$$

and with state feedback control

$$v = \tilde{K}\tilde{x} = [K_x \; K_u] \begin{bmatrix} x \\ u \end{bmatrix} = K_x x + K_u u$$

where the new input v has no sign restriction as it represents the variation of u, which allows us to get rid of the input positivity problem. The resulting closed-loop extended system is therefore described by

$$\dot{\begin{bmatrix} x \\ u \end{bmatrix}} = \begin{bmatrix} A & B \\ 0 & 0 \end{bmatrix} \begin{bmatrix} x \\ u \end{bmatrix} + \begin{bmatrix} 0 \\ I \end{bmatrix} [K_x \; K_u] \begin{bmatrix} x \\ u \end{bmatrix} = \begin{bmatrix} A & B \\ K_x & K_u \end{bmatrix} \begin{bmatrix} x \\ u \end{bmatrix}.$$

Note that if one considers a static feedback $v = \tilde{K}\tilde{x}$ for the extended system, it actually corresponds to a dynamic feedback controller $\dot{u} = K_u u + K_x x$ for the initial system. The extended system is positively stabilizable if and only if there exists a state feedback $\tilde{K} = [K_x \; K_u]$ such that

1. the matrix $\begin{bmatrix} A & B \\ K_x & K_u \end{bmatrix}$ is Metzler, i.e. K_u is Metzler and $K_x \geq 0$, and
2. the matrix $\begin{bmatrix} A & B \\ K_x & K_u \end{bmatrix}$ is exponentially stable.

As a consequence of these conditions, the pair $(\tilde{A}, \tilde{B})$ should be exponentially stabilizable. Now, [3, Sect. 10.3] provides necessary and sufficient conditions for the positive stabilizability of a positive system, using LMIs and a Lyapunov equation. We adapt this result to the extended system, leading to the following theorem.

Theorem 14.4 *Consider a linear time-invariant system* $\dot{x} = Ax + Bu$ *and his extended system* $\dot{\tilde{x}} = \tilde{A}\tilde{x} + \tilde{B}v$ *as defined above. The extended system is positively stabilizable if and only if there exist a positive-definite diagonal matrix* $Q = \begin{bmatrix} Q_1 & 0 \\ 0 & Q_2 \end{bmatrix}$ *and a feedback* $\tilde{K}$ *such that, with* $Y = [Y_1\ Y_2] = \tilde{K}Q$, *the matrix*

$$\begin{bmatrix} Q_1 A^T + AQ_1 & Y_1^T + BQ_2 \\ Q_2 B^T + Y_1 & Y_2^T + Y_2 \end{bmatrix}$$

is negative-definite, Y_1 *is nonnegative and* Y_2 *is Metzler.*

Proof By [3] the extended system is positively stabilizable if and only if there exist a positive-definite diagonal matrix Q and a feedback $\tilde{K}$ such that, with $Y = \tilde{K}Q$, $(\tilde{A}Q + \tilde{B}Y)$ is Metzler and $Q\tilde{A}^T + Y^T\tilde{B}^T + \tilde{A}Q + \tilde{B}Y$ is negative-definite. One easily sees that $(\tilde{A}Q + \tilde{B}Y)$ is Metzler if and only if the matrix

$$\begin{bmatrix} A & B \\ K_x & K_u \end{bmatrix}$$

is Metzler, which means (as stated previously) that K_x has to be nonnegative and K_u has to be Metzler. Moreover, as $Y = \tilde{K}Q$,

$$[Y_1\ Y_2] = [K_x\ K_u]\begin{bmatrix} Q_1 & 0 \\ 0 & Q_2 \end{bmatrix}$$

and then

$$\begin{cases} K_x = Y_1 Q_1^{-1} \\ K_u = Y_2 Q_2^{-1} \end{cases}$$

which implies that Y_1 has to be nonnegative and Y_2 has to be Metzler. Now, we can rewrite $Q\tilde{A}^T + Y^T\tilde{B}^T + \tilde{A}Q + \tilde{B}Y$ as

$$\begin{bmatrix} Q_1 & 0 \\ 0 & Q_2 \end{bmatrix}\begin{bmatrix} A^T & 0 \\ B^T & 0 \end{bmatrix} + \begin{bmatrix} Y_1^T \\ Y_2^T \end{bmatrix}[0\ I] + \begin{bmatrix} A & B \\ 0 & 0 \end{bmatrix}\begin{bmatrix} Q_1 & 0 \\ 0 & Q_2 \end{bmatrix} + \begin{bmatrix} 0 \\ I \end{bmatrix}[Y_1\ Y_2]$$

which is equal to

$$\begin{bmatrix} Q_1 A^T + AQ_1 & Y_1^T + BQ_2 \\ Q_2 B^T + Y_1 & Y_2^T + Y_2 \end{bmatrix}.$$

□

Remark 14.3 By the previous theorem, the matrix

$$\begin{bmatrix} Q_1 A^T + AQ_1 & Y_1^T + BQ_2 \\ Q_2 B^T + Y_1 & Y_2^T + Y_2 \end{bmatrix}$$

has to be negative-definite in order to positively stabilize the extended system by means of a feedback $v = \tilde{K}\tilde{x}$. However, it is known that every principal submatrix of a negative-definite matrix is negative-definite. This means that $Q_1 A^T + AQ_1$ is negative-definite and thus the initial system should be stable already. Thus the use of an extended system does not allow to circumvent the obstacle of using a nonnegative input as described in Sect. 14.3.1.

14.4 A Pure Diffusion System

Now we show that one can actually positively stabilize a pure diffusion system—which is a distributed parameter positive system—by use of a nonnegative boundary control, as long as the input appears in the boundary conditions.

14.4.1 Modelization

Consider a standard example of unstable positive distributed parameter system, namely the pure diffusion system described by the partial differential equation (PDE)

$$\frac{\partial x}{\partial t} = D_a \frac{\partial^2 x}{\partial z^2} \tag{14.1}$$

with Neumann boundary conditions

$$\begin{cases} \frac{\partial x}{\partial z}(t, 0) = v(t) \\ \frac{\partial x}{\partial z}(t, L) = 0 \end{cases} \tag{14.2}$$

where v is the input, D_a is the diffusion parameter and L is the domain length. By [9, Example 2.1], this boundary control system is equivalent to the system described by the PDE

$$\frac{\partial x}{\partial t} = D_a \frac{\partial^2 x}{\partial z^2} + \delta_0 u(t) \tag{14.3}$$

with the Dirac delta distribution δ_0 as control operator and with homogeneous Neumann boundary conditions

$$\begin{cases} \frac{\partial x}{\partial z}(t, 0) = 0 \\ \frac{\partial x}{\partial z}(t, L) = 0 \end{cases} \tag{14.4}$$

where the input $u(t) = -v(t)$. This implies that considering a positive input $v(t)$ in the boundary conditions leads to a negative input $u(t)$ in the dynamics, and thus to a potential stabilization of the system.

14.4.2 Discretization

In order to stabilize the system, we discretize it by the finite difference method and we obtain the finite-dimensional system (considering n discretization points z_i, $i = 1, \ldots, n$, with $z_1 = 0$, $z_n = L$ and $\Delta z = L/(n-1)$ the discretization step)

$$\dot{x}^{(n)} = A^{(n)}x^{(n)} + b^{(n)}u \tag{14.5}$$

where

$$A^{(n)} = \begin{bmatrix} -p_2 & p_2 & 0 & \cdots & 0 \\ p_2 & -2p_2 & p_2 & & \vdots \\ 0 & \ddots & \ddots & \ddots & 0 \\ \vdots & & p_2 & -2p_2 & p_2 \\ 0 & \cdots & 0 & p_2 & -p_2 \end{bmatrix} \in \mathbb{R}^{n \times n},$$

$$b^{(n)} = [p_1 \quad 0 \quad \cdots \quad 0]^T \in \mathbb{R}^n$$

and

$$x^{(n)} = [x(z_1) \quad \cdots \quad x(z_n)]^T \in \mathbb{R}^n$$

where

$$p_1 = \frac{1}{\Delta z} \quad \text{and} \quad p_2 = \frac{D_a}{\Delta z^2}.$$

Clearly, this finite-dimensional system is positive (see Theorem 14.1). Moreover, the infinite-dimensional system (14.1)–(14.2) is not exponentially stable [1, 5] as zero is in the spectrum of its generator. Discretizing the system will perturb the spectrum though one easily sees that the finite-dimensional system (14.5) is not exponentially stable as zero is still in the spectrum of $A^{(n)}$. Also note that all eigenvalues are real, $A^{(n)}$ being symmetric.

14.4.3 Positive Stabilization of the System

Now we can provide the reader with a parameterization of all positively stabilizing feedbacks for the distributed pure diffusion system (14.5), using Lemma 14.1 and developing the resulting set of inequalities [7].

Theorem 14.5 *A feedback $k = [k_1 \quad \cdots \quad k_n]$ is positively stabilizing for the discretized pure diffusion system (14.5) if and only if it is such that*

$$k_1 = \frac{D_a v_1 - D_a v_2 - k_2 v_2 \Delta z - \cdots - k_n v_n \Delta z - \Delta z^2 \omega}{v_1 \Delta z},$$

$$k_2 \geq \frac{-D_a}{\Delta z} \quad \textit{and} \quad k_i \geq 0 \;\; i = 3, \ldots, n,$$

with $\omega > 0$ *(free parameter) and such that* $v \gg 0$ *is a positive solution of the strict inequalities set*

$$\begin{aligned} -v_1 + 2v_2 - v_3 &> 0 \\ &\vdots \\ -v_{n-2} + v_{n-1} - v_n &> 0 \\ -v_{n-1} + v_n &> 0. \end{aligned} \tag{14.6}$$

It is actually possible to parameterize all the solutions of the inequalities set (14.6), leading to a full parameterization of all the positively stabilizing feedbacks for the pure diffusion system (see [7]). In order to illustrate the theoretical results, let us design a particular feedback that falls in the class defined in Theorem 14.5. Let us set

$$k_1^{(n)} = -\frac{1}{\Delta z}\kappa \quad \text{and} \quad k_i^{(n)} = 0 \;\; (i = 2, \ldots, n) \tag{14.7}$$

with $\kappa > 0$. Considering $L = 1$, $D_a = 1$, $\kappa = 0.2$ and $n = 11$ and choosing the initial condition $x_0 = 2z^3 - 3z^2 + 1$ (this polynomial respects the boundary conditions and the all-ones eigenvector corresponds to the Frobenius unstable eigenvalue $\lambda = 0$, so the initial condition excites the unstable mode) yields the open-loop state trajectory shown in Fig. 14.1, and the closed-loop state trajectory shown in Fig. 14.2. This illustrates that the closed-loop system is positive and that it is stable unlike the open-loop system. Figure 14.3 shows the nonnegative input trajectory $v(t)$.

Fig. 14.1 Open-loop state trajectory $x^{(n)}(t)$ $(n = 11)$. It converges to a constant non-null value

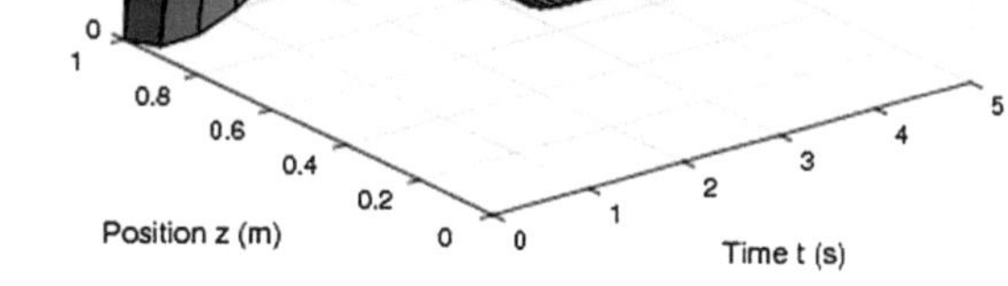

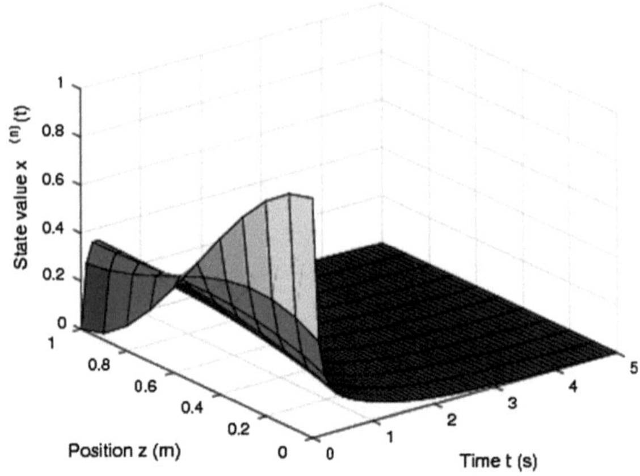

Fig. 14.2 Closed-loop state trajectory $x^{(n)}(t)$ $(n = 11)$. It stabilizes to zero while staying nonnegative at all time

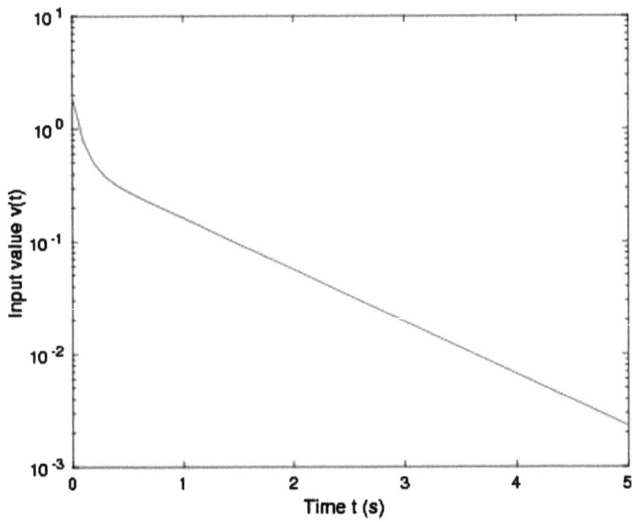

Fig. 14.3 Input trajectory $v(t) = -k^{(n)}x^{(n)}(t)$ $(n = 11)$. The input—as it appears in the boundary conditions—is nonnegative at all time and decreases to zero

14.4.4 Convergence Analysis

Now we focus on convergence issues, whenever the finite difference step tends to zero. Let us introduce the following result (see [7]).

Theorem 14.6 *Applying the feedback $k^{(n)}$ given by (14.7) to the approximate system (14.5) leads to the convergence of the resulting closed-loop system*

$$\dot{x}^{(n)} = (A^{(n)} + b^{(n)}k^{(n)})x^{(n)}, \tag{14.8}$$

as Δz tends to zero, to the system described by the PDE

$$\frac{\partial x}{\partial t} = D_a \frac{\partial^2 x}{\partial z^2} \tag{14.9}$$

with Neumann boundary conditions

$$\begin{cases} \frac{\partial x}{\partial z}(0, t) = \kappa x(0, t) \\ \frac{\partial x}{\partial z}(L, t) = 0. \end{cases} \tag{14.10}$$

Moreover, the approximate closed-loop system (14.8) is positive and (exponentially) stable for n sufficiently large, and the system (14.9)–(14.10) is positive and (exponentially) stable.

One can show the convergence of the system operators by a state space approach, setting the discretized operators in the appropriate spaces and using the related norms [9, Example 2.1]. Positivity of system (14.9)–(14.10) can be proved by standard arguments (positivity of the resolvent operator as in [14] or the maximum principle as in [18]). Also, as it is of Sturm-Liouville type, system (14.9)–(14.10) is a Riesz-spectral system [8]. Its spectrum is thus real and discrete: it is easy to compute all eigenvalues and to show that they are negative, implying the stability of the system. For a complete proof, refer to [7].

14.5 Conclusion

In this chapter, we have studied the issue of considering a nonnegative input while positively stabilizing a positive system, using a classical approach and working on an extended system. Then we have shown via a classical example that the boundary input sign may vary depending on whether it acts in the dynamics or in the boundary conditions, implying that it is technically possible to positively stabilize the system with a nonnegative input. Finally we have provided a convenient way to parameterize all the positively stabilizing feedbacks for a discretized model of the pure diffusion system, we have discussed the convergence of the results and we have produced some numerical simulations. Next steps in this work are—among others—to extend Theorem 14.3 and its proof to infinite-dimensional systems, to find conditions over any discretized feedback so that it converges to a positively stabilizing feedback for the nominal PDE system, to optimize the choice of a positively stabilizing feedback with respect to some given criterion, to design observer based compensators and to extend the results to a specific interesting application in biochemical engineering. These questions are currently under investigation.

Acknowledgements This chapter presents research results of the Belgian network DYSCO (Dynamical Systems, Control and Optimization), funded by the Interuniversity Attraction Poles Programme, initiated by the Belgian state, Science Policy Office (BELSPO). The scientific responsibility rests with its authors.

References

1. Abouzaid, B., Winkin, J.J., Wertz, V.: Positive stabilization of infinite-dimensional linear systems. In: Proceedings of the 49th IEEE Conference on Decision and Control, pp. 845–850 (2010)
2. Arrow, K.J.: A "dynamic" proof of the Frobenius-Perron theorem for Metzler matrices. Techn. Rep. Ser. **542** (1989)
3. Boyd, S., El Ghaoui, L., Feron, E., Balakrishnan, V.: Linear Matrix Inequalities in System and Control Theory. Society for Industrial and Applied Mathematics (1994)
4. Chellaboina, V., Bhat, S.P., Haddad, W.M., Bernstein, D.S.: Modeling and analysis of mass-action kinetics: nonnegativity, realizability, reducibility, and semistability. IEEE Control Syst. Mag. **29**(4), 60–78 (2009)
5. Curtain, R.F., Zwart, H.J.: An Introduction to Infinite-Dimensional Linear Systems Theory. Springer (1995)
6. De Leenheer, P., Aeyels, D.: Stabilization of positive linear systems. Syst. Control Lett. **44**, 259–271 (2001)
7. Dehaye, J.N., Winkin, J.J.: Parameterization of positively stabilizing feedbacks for single-input positive systems. Syst. Control Lett. **98**, 57–64 (2016)
8. Delattre, C., Dochain, D., Winkin, J.J.: Sturm-Liouville systems are Riesz-spectral systems. Int. J. Appl. Math. Comput. Sci. **13**(4), 481–484 (2003)
9. Emirsjlow, Z., Townley, S.: From PDEs with boundary control to the abstract state equation with an unbounded input operator: A tutorial. European Journal of Control **6**, 27–49 (2000)
10. Farina, L., Rinaldi, S.: Positive Linear Systems: Theory and Applications. Wiley (2000)
11. Haddad, W.M., Chellaboina, V., Hui, Q.: Nonnegative and Compartmental Dynamical Systems. Princeton University Press (2010)
12. Horn, R.A., Johnson, C.R.: Matrix Analysis, vol. 1. Cambridge University Press (1990)
13. Kaczorek, T.: Positive 1D and 2D Systems. Springer, London (2002)
14. Laabissi, M., Achhab, M.E., Winkin, J.J., Dochain, D.: Trajectory analysis of nonisothermal tubular reactor nonlinear models. Syst. Control Lett. **42**, 169–184 (2001)
15. Roszak, B., Davison, E.J.: Necessary and sufficient conditions for stabilizability of positive LTI systems. Syst. Control Lett. **58**, 474–481 (2009)
16. Rüffer, B.S.: Monotone inequalities, dynamical systems, and paths in the positive orthant of Euclidean n-space. Positivity **14**(2), 257–283 (2010)
17. Saperstone, S.H.: Global controllability of linear systems with positive controls. SIAM J. Control **11**(3), 417–423 (1973)
18. Smith, H.L.: Monotone Dynamical Systems: An Introduction to the Theory of Competitive and Cooperative Systems. American Mathematical Society (2008)
19. Winkin, J.J., Dochain, D., Ligarius, P.: Dynamical analysis of distributed parameter tubular reactors. Automatica **36**, 349–361 (2000)

Chapter 15
Positive Stabilization of a Class of Infinite-Dimensional Positive Systems

M. Elarbi Achhab and Joseph J. Winkin

Abstract For a class of positive unstable infinite-dimensional linear systems, a method is described for computing a positively stabilizing state feedback, such that the resulting input trajectory remains in an affine cone. This design results in a possibly negative lower bound on the input, which makes the resulting closed-loop system stable and which maintains the nonnegativity of the state trajectory for specific initial states.

Keywords Infinite dimensional systems · Positive linear systems · Positive stabilization · State feedback · Affine cone

15.1 Introduction

Positive linear systems are linear dynamical systems whose state trajectories are nonnegative for every nonnegative initial state and for every admissible nonnegative input function. Equivalently a linear dynamical system is positive whenever the corresponding cone (which defines the considered order) of the state-space is invariant under the state transition map (positive invariance). The positivity (or, more precisely, nonnegativity) property occurs quite frequently in practical applications where the state variables correspond to quantities that do not have real meaning unless they are nonnegative, see e.g. [5, 14, 15, 19, 20], for examples of positive infinite dimensional systems that are described by specific partial differential equations.

The positivity property and the positive stabilization problem have been extensively studied for finite dimensional systems, see e.g. [6, 10, 16, 17] and references

---

M.E. Achhab
Faculté des Sciences El Jadida, Département de Mathématiques,
Université Chouaib Doukkali, BP 20 24000 El Jadida, Morocco
e-mail: elarbi.achhab@gmail.com

J.J. Winkin (✉)
Department of Mathematics and naXys, University of Namur,
Rempart de la Vierge 8, 5000 Namur, Belgium
e-mail: joseph.winkin@unamur.be

F. Cacace et al. (eds.), *Positive Systems*, Lecture Notes in Control and Information Sciences 471, DOI 10.1007/978-3-319-54211-9_15

therein. Concerning the positivity of infinite dimensional linear systems, some new points of view and perspectives are available in the literature. In particular, algebraic conditions of positivity for dynamical systems defined on an ordered Banach space whose positive cone has an empty interior are established in [1].

In this chapter, conditions are derived for the positive stabilization of a class of distributed parameter systems, such that the closed loop system is stable and positive. More specifically, for positive unstable infinite-dimensional linear systems, conditions are established for positive stabilizability and a method is described for computing a positively stabilizing state feedback, which guarantees that the stable closed loop dynamics are nonnegative for specific initial states.

A feedback control is designed such that the unstable finite-dimensional spectrum of the dynamics generator is replaced by the eigenvalues of the stable input dynamics and such that the resulting input trajectory remains in an affine cone, thereby ensuring a possibly negative lower bound on the input, which maintains the nonnegativity of the state trajectory for specific initial states. Moreover a synthesis methodology of a positively stabilizing state feedback is described; see [4]. These results constitute extensions of those obtained in [3], with statements and proofs adapted to the case of an invariant shifted cone for the input values. This extended theory is motivated by the fact that positive stabilization of an unstable system by a nonnegative input (in the dynamics equation) is not possible; see [9].

15.2 Positive Stabilization Problem

Let's consider an infinite dimensional (state-space) system described by

$$\begin{cases} \dot{x}(t) = Ax(t) + Bu(t) \\ x(0) = x_0 \end{cases} \tag{15.1}$$

where the linear operator A is the infinitesimal generator of a positive unstable C_0-semigroup $T(t)$ on an ordered (separable) Hilbert space X with positive cone X_+, [8, 12]. It is known that there exist constants $M \geq 1$ and $\omega \in \mathbb{R}$ such that

$$\text{for all } t \geq 0, \ \|T(t)\| \leq Me^{\omega t}. \tag{15.2}$$

In Eq. (15.1), the operator $B \in \mathscr{L}(\mathbb{R}^m, X)$ is positive, i.e. B is a bounded linear operator from $\mathbb{R}^m$ to X such that

$$B(\mathbb{R}^m_+) \subset X_+ \,, \tag{15.3}$$

where $\mathbb{R}_+$ denotes the set of nonnegative real numbers and the input $u(\cdot)$ is any locally square integrable function. Hence the system (15.1) is *positive*: for every initial state $x_0 \in X_+$ and for every (admissible) nonnegative input u, the corresponding state trajectory $x(\cdot)$ (interpreted as the mild solution of (15.1)) remains in X_+, i.e. for all $t \geq 0$, $x(t) \in X_+$ (see e.g. [1] and references therein).

The following conditions are also assumed to hold:

(C0) the operator A admits a Riesz basis of eigenvectors $(\phi_n)_{n\geq 1}$;

This condition holds e.g. if the operator A is a Riesz spectral operator or if it is similar to a normal operator (see e.g. [11]). It follows from this condition that $(\phi_n)_{n\geq 1}$ is also a Riesz basis of eigenvectors of the operator $K = \lambda R(\lambda, A)$, for all $\lambda > \omega$.

(C1) the system (15.1), i.e. the pair (A, B), is (exponentially) stabilizable;

Condition (C1) implies that the unstable part of the dynamics is a (totally) unstable finite-dimensional system corresponding to all the eigenvalues of the operator A that belong to the closed right-half plane, [8]. In the sequel, we will assume that the number of such eigenvalues (counting multiplicities) is m.

(C2) the C_0-semigroup $T(t)$ has a compact resolvent, i.e. the resolvent operator $R(\lambda, A) := (\lambda I - A)^{-1} \in \mathscr{L}(X)$ is compact for all $\lambda > \omega$.

Definition 15.1 The system (15.1), i.e. the pair (A, B), is said to be **locally positively stabilizable** if there exist a subset $S \subset X$ and a state feedback control law $u = Fx$, where the feedback operator F is in $\mathscr{L}(X, \mathbb{R}^m)$, such that the resulting closed-loop system is stable and positive on S, i.e. the C_0-semigroup $T_F(t)$ generated by $A + BF$ is exponentially stable and $T_F(t)$ is **positive on the subset** S, i.e.

$$T_F(t)(X_+ \cap S) \subset X_+ \, ; \tag{15.4}$$

hence, in particular, (A, B) is stabilizable.

The **positive stabilization problem** consists in finding sufficient and/or necessary conditions for the (local) positive stabilizability of a given system of the form (15.1), leading hopefully to a computational method for designing a positively stabilizing feedback, i.e. a stabilizing feedback operator $F \in \mathscr{L}(X, \mathbb{R}^m)$ such that (15.4) holds for some subset $S \subset X$.

Remark 15.1 Observe that the concept of partial positive stabilizability (which requires that the closed-loop state trajectories starting from an initial state in $X_+ \cap S$ stay in $X_+ \cap S$, see [3]) is stronger than local positive stabilizability.

15.3 Main Result

The approach that is followed here is based on a finite spectrum assignment technique yielding a stable closed-loop system and enforcing conditions on the inputs that guarantee positive corresponding state trajectories starting from well-chosen initial states. This approach, which is analyzed in the next section, goes along the lines of the one introduced in [3]. It leads to the following main result:

Theorem 15.1 *Consider an infinite-dimensional system described by (15.1) under assumption (15.3) and assume that conditions (C0), (C1) and (C2) hold.*

(a) *Consider a feedback operator* $F \in \mathscr{L}(X, I\!R^m)$ *of full rank m. Then the set*

$$\mathscr{P}_q(F) := \{ x \in X \ : \ Fx \geq -q \} \quad , \quad where \quad q \in I\!R^m_+ \ ,$$

is $T_F(t)$*-invariant, i.e.* $T_F(t)\ \mathscr{P}_q(F) \subset \mathscr{P}_q(F)$*, if and only if there exists a Metzler matrix* H*, satisfying* $Hq \leq 0$*, such that*

$$F(A + BF) - HF = 0 \ \ on\ D(A) \tag{15.5}$$

i.e.

$$F(A + BF)x = HFx \quad for\ all\ x \in D(A)\ .$$

(b) *If there exists a stabilizing feedback operator* $F \in \mathscr{L}(X, I\!R^m)$ *of full rank m, such that condition (15.5) holds for some Metzler matrix H satisfying* $Hq \leq 0$ *and if there exists a subset* $X_0 \subset X_+$ *such that, for all* $x_0 \in X_0$ *and for all* $t \geq 0$, $x(t) \in X_+$ *for any input* $u(\cdot) \geq -q$*, then the system (15.1), i.e. the pair* (A, B)*, is locally positively stabilizable and, in particular, for every initial state* $x_0 \in X_0 \cap \mathscr{P}_q(F)$ *and for all* $t \geq 0$*,*

$$x(t) = T_F(t)x_0 \in X_+ \cap \mathscr{P}_q(F)$$

and, for some constants $\mu \geq 1$ *and* $\sigma > 0$*,*

$$\|x(t)\| \ \leq \ \mu\, e^{-\sigma t}, \quad for\ all\ t \geq 0.$$

Remark 15.2 (a) A Metzler matrix is a square matrix whose off-diagonal entries are nonnegative, see e.g. [13].

(b) The fact that the set $\mathscr{P}_q(F)$ is $T_F(t)$-invariant implies that, for any initial state x_0 in $X_0 \cap \mathscr{P}_q(F)$, the corresponding input trajectory $u(\cdot) = Fx(\cdot)$ (generated by the feedback F) satisfies $u(\cdot) \geq -q$. In view of this observation, Theorem 15.1b follows directly from its Part a). In the next section, we will therefore focus on the proof of the first part. In addition, since the analysis is similar to the one developed in [3], we will focus on the new specific arguments resulting from the use of the set $\mathscr{P}_q(F)$ instead of the nonnegative cone, i.e. the set $\mathscr{P}_0(F)$, of the input value set.

15.4 Auxiliary Results and Proofs

Theorem 15.1a is a straightforward consequence of the following result.

Theorem 15.2 *Consider a* C_0*-semigroup* $S(t)$ *of bounded linear operators on a Hilbert space* X*, whose infinitesimal generator is the operator* $\mathscr{A}$*, such that*

$$\text{for all } t \geq 0, \ \|S(t)\| \leq Me^{\omega t}, \tag{15.6}$$

for some constants $M \geq 1$ *and* $\omega \in I\!R$. *Assume that* $\mathscr{A}$ *admits a Riesz basis of eigenvectors* $(\phi_n)_{n\geq 1}$ *and that* $S(t)$ *has a compact resolvent, i.e. the resolvent operator* $R(\lambda, \mathscr{A}) := (\lambda I - \mathscr{A})^{-1} \in \mathscr{L}(X)$ *is compact for all* $\lambda > \omega$. *Consider any given bounded linear operator* $F \in \mathscr{L}(X, I\!R^m)$ *of full rank* m. *Then the set*

$$\mathscr{P}_q(F) := \{\, x \in X \ : \ Fx \geq -q \,\} \quad , \quad \text{where} \quad q \in I\!R^m_+$$

is $S(t)$*-invariant, i.e.* $S(t)\ \mathscr{P}_q(F) \subset \mathscr{P}_q(F)$, *if and only if there exists a Metzler matrix* H *satisfying* $Hq \leq 0$ *such that*

$$F\mathscr{A} - HF = 0 \quad \text{on } D(\mathscr{A}) \ . \tag{15.7}$$

The proof of Theorem 15.2 is detailed in Sect. 15.4.2. It is based on auxiliary results concerning discrete time systems, that are developed in Sect. 15.4.1.

Under the conditions of Theorem 15.2, consider the family $(\Sigma_\lambda)_{\lambda>\omega}$ of discrete-time infinite-dimensional systems

$$(\Sigma_\lambda) \begin{cases} x(k+1) = \lambda R(\lambda, \mathscr{A}) x(k) \\ x(0) = x_0 \ . \end{cases} \tag{15.8}$$

Proposition 15.1 *If the set* $\mathscr{P}_q(F)$ *is* $S(t)$*-invariant, then for all* $\lambda \geq \max\{1, \omega\}$, $\mathscr{P}_q(F)$ *is invariant with respect to the system* (Σ_λ), *i.e.*

$$\lambda R(\lambda, \mathscr{A})\ \mathscr{P}_q(F) \subset \mathscr{P}_q(F) \ .$$

Proof If x belongs to $\mathscr{P}_q(F)$, then for all $t \geq 0$, $S(t)x$ is also in $\mathscr{P}_q(F)$, i.e. $FS(t)x \geq -q$. By using the fact that the resolvent operator $R(\lambda, \mathscr{A})$ is the Laplace transform (interpreted as a Bochner integral, i.e. in the strong sense), of the semigroup $S(t)$, it follows that $FR(\lambda, \mathscr{A})x \geq \dfrac{-q}{\lambda}$ for $\lambda > \omega$.

15.4.1 Invariance of Discrete Time Systems

In order to study the invariance properties of discrete time systems of the form (15.8), let's consider a more general class of discrete-time infinite-dimensional systems:

$$\Sigma^d \begin{cases} x(k+1) = K\ x(k) \\ x(0) = x_0 \in X \ , \end{cases} \tag{15.9}$$

where the bounded linear operator $K \in \mathscr{L}(X)$ is assumed to be compact and to have a Riesz basis of eigenvectors $(\phi_n)_{n\geq 1}$.

For every $N \geq 1$, let X_N denote the K-invariant finite-dimensional linear subspace of X defined by $X_N := \text{span}\{\phi_i \; : \; i = 1, 2, \ldots, N\}$ and let the operator $\Gamma_N := \text{Proj}_{X_N}$ denote the orthogonal projection on X_N. Observe that the operator $K_N := K \; \Gamma_N$ has a finite rank and that the sequence (K_N) converges strongly towards K in $\mathscr{L}(X)$, i.e.

$$\lim_{N \to \infty} \|(K - K_N)x\| = 0 \quad , \quad x \in X.$$

Now we can define the sequence (Σ_N^d) of discrete-time finite-dimensional systems:

$$\Sigma_N^d \begin{cases} x_N(k+1) = K_N \; x_N(k) \\ x_N(0) \in X_N \; . \end{cases} \tag{15.10}$$

Observe that such system is well-defined on X_N. Indeed $K_N \in \mathscr{L}(X)$ is such that $K_N(X_N) \subset X_N$.

Proposition 15.2 *If the set $\mathscr{P}_q(F)$ is invariant with respect to the system Σ^d, i.e. $K \; \mathscr{P}_q(F) \subset \mathscr{P}_q(F)$, then for all $N \geq 1$, the set*

$$\mathscr{P}_q(F) \cap X_N := \{ \, x \in X_N \; : \; Fx + q \in I\!R_+^m \, \}$$

is invariant with respect to the system Σ_N^d.

Proof It suffices to observe that, for any $x \in \mathscr{P}_q(F) \cap X_N$, $K_N x \in X_N$ and $F K_N x = F K x \geq -q$.

Now we are in a position to state and prove the main result of this subsection.

Proposition 15.3 *The set $\mathscr{P}_q(F)$ is invariant with respect to the system Σ^d, i.e. $K \; \mathscr{P}_q(F) \subset \mathscr{P}_q(F)$, if and only if there exists a nonnegative matrix H such that*

$$\begin{cases} FK = HF \quad \text{on } X \\ Hq \leq q \end{cases} \tag{15.11}$$

The proof of the necessity of Conditions (15.11) in Proposition 15.3 is based on the following affine form of the Ferkas lemma; see [18].

Lemma 15.1 *let P be a non-empty polyhedron defined by m inequalities*

$$a_k^T z + b_k \geq 0 \, , \quad k = 1, \ldots, m. \tag{15.12}$$

Then an affine form Ψ is nonnegative everywhere in P if and only if it is a nonnegative linear combination of the faces, i.e. there exist positive reals $\lambda_0, \lambda_1, \ldots, \lambda_m$ such that

$$\Psi(z) = \lambda_0 + \sum_{k=1}^{m} \lambda_k (a_k^T z + b_k). \tag{15.13}$$

Proof of Proposition 15.3: Sufficiency is straightforward.

Necessity: Since the operator F is surjective and the set $\cup\{X_N : N \geq 1\}$ is dense in X, where (X_N) is a monotone increasing sequence of linear subspaces of X, by [2, Lemma 3.1], there exists N_0 such that $F(X_{N_0}) = \mathbb{R}^m$, therefore for all $N \geq N_0$, $F(X_N) = \mathbb{R}^m$. Moreover, by Proposition 15.2, for all $N \geq 1$, the set $\mathscr{P}_q(F) \cap X_N$ is invariant with respect to the system Σ_N^d; thus,

$$z \in \mathscr{P}_q(F) \cap X_N \Longrightarrow K_N z \in \mathscr{P}_q(F) \cap X_N$$

or

$$(F_N)_j z + q_j \geq 0 \Longrightarrow (F_N)_j K_N z + q_j \geq 0 \, , \forall j, 1 \leq j \leq m$$

where $(F_N)_j$ is the jth row-vector of the matrix $F_N := F\Gamma_N$. Putting for $j,\ 1 \leq j \leq m$,

$$\Psi_j(z) = (F_N)_j K_N z + q_j, \tag{15.14}$$

$\Psi_j(.)$ is an affine form which is nonnegative everywhere in the polyhedron $\mathscr{P}_q(F) \cap X_N$. Thus, by Lemma 15.1, for every j, $1 \leq j \leq m$ there exist $m+1$ positive reals $\lambda_{j0}, \lambda_{j1}, \ldots, \lambda_{jm}$ such that

$$\Psi_j(z) = \lambda_{j0} + \sum_{k=1}^{m} \lambda_{jk}((F_N)_k z + q_k), \quad 1 \leq j \leq m$$

This implies that for every $j,\ 1 \leq j \leq m$,

$$\begin{cases} (F_N)_j K_N z = \displaystyle\sum_{k=1}^{m} \lambda_{jk}((F_N)_k z \\ q_j = \lambda_{j0} + \displaystyle\sum_{k=1}^{m} \lambda_{jk} q_k \, . \end{cases} \tag{15.15}$$

Now, define the matrix H_N by $(H_N)_{jk} = \lambda_{jk}$ for $1 \leq j, k \leq m$, one deduces that H_N is a nonnegative matrix and

$$\begin{cases} F_N K_N = H_N F_N \text{ on } X \\ H_N q \leq q \, . \end{cases} \tag{15.16}$$

Consider any vector $y \in \mathbb{R}^m$. For all $N \geq \max\{N_0, m\}$, by the surjectivity of F_N and by the definition of X_N, there exists $x \in X_{\max\{N_0,m\}} \subset X_N$ (hence x is independent of N) such that $y = Fx = F_N x$. It follows by identity (15.15) that,

$$H_N y = H_N F x = H_N F_N x = F_N K_N x = F K_N x \, .$$

Thanks to the convergence of the sequence (K_N) towards the operator K, the sequence $H_N y$ is convergent in $\mathbb{R}^m$. Let's define the matrix operator $H \in \mathbb{R}^{m \times m}$ by

$$Hy := \lim_{N \to \infty} H_N y .$$

Obviously H is nonnegative because H_N is nonnegative for N sufficiently large. In addition, by the convergence of (F_N) towards F, it follows from identity (15.15) that (15.11) holds.

15.4.2 *Proof of Theorem 15.2*

Sufficiency: Using the density of $D(\mathscr{A})$ in X, it follows from (15.7) that, for every $x_0 \in X$, the function $u : \mathbb{R}_+ \to \mathbb{R}^m : t \mapsto u(t) := FS(t)x_0$ is the solution of the finite-dimensional Cauchy problem:

$$\dot{u}(t) = Hu(t), \quad u(0) = Fx_0 ,$$

or equivalently $u(t) = e^{Ht} F x_0$, where H is a Metzler matrix satisfying $Hq \leq 0$. Thus, if in addition x_0 is in $\mathscr{P}_q(F)$, i.e. $Fx_0 \geq -q$, then for all $t \geq 0$, $u(t) \geq -q$, i.e. $S(t)x_0 \in \mathscr{P}_q(F)$; see e.g. [13]. This shows that the set $\mathscr{P}_q(F)$ is $S(t)$-invariant.

Necessity: By Proposition 15.1, for all $\lambda \geq \max\{1, \omega\}$, $\mathscr{P}_q(F)$ is invariant with respect to the system (Σ_λ); it follows by Proposition 15.3 applied to $K = \lambda R(\lambda, \mathscr{A})$, that there exists a nonnegative matrix H_λ such that

$$\begin{cases} F\lambda R(\lambda, \mathscr{A}) = H_\lambda F \quad \text{on } X. \\ H_\lambda q \leq q . \end{cases} \tag{15.17}$$

Now consider the (bounded linear) Yosida approximant of $\mathscr{A}$ (see e.g. [12]):

$$\mathscr{A}_\lambda := \lambda \mathscr{A} R(\lambda, \mathscr{A}) = \lambda^2 R(\lambda, \mathscr{A}) - \lambda I .$$

Observe that Equation (15.17) yields the identity:

$$F\mathscr{A}_\lambda = \lambda(H_\lambda - I)F . \tag{15.18}$$

Moreover, by [12, Lemma II.3.4, p. 65], for all $x \in D(\mathscr{A})$,

$$\lim_{\lambda \to \infty} \mathscr{A}_\lambda x = \mathscr{A} x . \tag{15.19}$$

Besides, since the operator F is onto and $D(\mathscr{A})$ is a dense subspace of X, by using [2, Lemma 3.1], $F(D(\mathscr{A})) = \mathbb{R}^m$. Therefore every $y \in \mathbb{R}^m$ can be written as $y = Fx$ for some $x \in D(\mathscr{A})$. Using this fact, it follows from (15.18) that, for all $y \in \mathbb{R}^m$, the following limit

$$Hy := \lim_{\lambda \to \infty} \lambda(H_\lambda - I)y \tag{15.20}$$

exists (in $\mathbb{R}^m$). Also, by (15.17), the matrix H satisfies the condition $Hq \leq 0$. In addition, identity (15.7) holds. Indeed, for all $x \in D(\mathscr{A})$,

$$HFx = \lim_{\lambda \to \infty} \lambda(H_\lambda - I)Fx = \lim_{\lambda \to \infty} F\mathscr{A}_\lambda x = F\mathscr{A}x \,.$$

It remains to be shown that the matrix H given by (15.20) is a Metzler matrix. Recall that, for $\lambda \geq \max\{1, \omega\}$, the matrix H_λ is nonnegative, hence $\lambda(H_\lambda - I)$ is a Metzler matrix. It follows by (15.20) that so is the matrix H.

15.5 Outlook

The design method which is described in [3] and which is based on the decomposition of the dynamics into a totally unstable finite-dimensional positively stabilizable (positive) subsystem and a stable infinite-dimensional positive subsystem (as in [1]), can be readily extended to the more general framework of this chapter. Its implementation on standard examples is currently under investigation, [4].

It is also worth to mention the recent work (in progress) [7] which investigates the question of designing a positive exponential Luenberger type observer for a class of infinite-dimensional linear positive systems. Such positive observers are very important in applications, since negative estimated values of positive states may not have a physical meaning (think of concentrations, for example). Necessary and sufficient conditions for the existence of such positive observers are established in that paper: under the decomposition spectral assumption, the authors show that the problem is reduced to the design of a positive observer for an unstable finite-dimensional subsystem. The technical mathematical tools are comparable to the ones used in [1]. Finally, the applicability of the proposed estimation approach is illustrated by an example of a parabolic system.

Acknowledgements The authors wish to thank the following persons with whom they have worked jointly on dynamical analysis and control of positive systems for many years: B. Abouzaid (Ecole Nationale des Sciences Appliquées, Université Chouaib Doukkali, El Jadida., Morocco), Ch. Beauthier (Cenaero, Gosselies, Belgium), D. Dochain (Université Catholique de Louvain, Belgium), M. Laabissi (Université Chouaib Doukkali, El Jadida, Morocco) and V. Wertz (Université Catholique de Louvain, Belgium).
This chapter presents research results of the Belgian Network DYSCO (Dynamical Systems, Control, and Optimization), funded by the Interuniversity Attraction Poles Programme initiated by the Belgian Science Policy Office.

References

1. Abouzaid, B., Winkin, J., Wertz, V.: Positive stabilization of infinite-dimensional linear systems. In: Proceedings of the 49th Conference on Decision and Control (CDC), Atlanta, GA, USA, 15–17 Dec 2010, Cd-Rom paper 0859, pp. 845–850
2. Achhab, M.E., Laabissi, M.: Feedback stabilization of a class of distributed parameter systems with control constraints. Syst. Control Lett. **45**, 163–171 (2002)
3. Achhab, M.E., Winkin, J.: Stabilization of infinite dimensional systems by state feedback with positivity constraints. In: Proceedings of the 21st International Symposium on Mathematical Theory of Networks and Systems (MTNS 2014), Groningen, NL, 2014, pp. 379–384
4. Achhab, M.E., Winkin, J.: Work in progress
5. Aksikas, I., Winkin, J., Dochain, D.: Optimal LQ-feedback regulation of a nonisothermal plug flow reactor model by spectral factorization. IEEE Trans. Autom. Control **52**(7), 1179–1193 (2007)
6. Beauthier, Ch., Winkin, J.: LQ-optimal control of positive linear systems. Optim. Control Appl. Methods **31**, 547–566 (2010)
7. Binid, A., Achhab, M.E., Laabissi, M., Abouzaid, B.: Positive observers for infinite dimensional positive linear systems (2016)
8. Curtain, R.F., Zwart, H.J.: An Introduction to Infinite-Dimensional Linear Systems Theory. Springer (1995)
9. Dehaye, J.N., Winkin, J.: Positive stabilization of a diffusion system by nonnegative boundary control. In: Proceedings of POSTA 2016
10. De Leenheer, P., Aeyels, D.: Stabilization of positive linear systems. Syst. Control Lett. **44**, 259271 (2001)
11. Dunford, N., Schwartz, J.T.: Linear Operators. Part II. Wiley Intersciences, New York (1951)
12. Engel, K.J., Nagel, R.: A Short Course on Operator Semigroups. Springer (2006) (especially chapter VI)
13. Haddad, W.M., Chellaboina, V., Hui, Q.: Nonnegative and compartmental dynamical systems. Princeton University Press, Princeton, NJ (2010)
14. Laabissi, M., Achhab, M.E., Winkin, J., Dochain, D.: Trajectory analysis of nonisothermal tubular reactor nonlinear models. Syst. Control Lett. **42**, 169–184 (2001)
15. Laabissi, M., Achhab, M.E., Winkin, J., Dochain, D.: Positivity and invariance properties of nonisothermal tubular reactor nonlinear models. In: Benvenuti, L., De Santis, A., Farina, L. (eds.) Positive Systems (Proceedings of the 1rst Multidisciplinary International Symposium on Positive Systems: Theory and Applications (POSTA 03), Grenoble, France), pp. 159–166. Lecture Notes in Control and Information Sciences. Springer, Berlin (2003)
16. Laabissi, M., Winkin, J., Beauthier, Ch.: On the positive LQ-problem for linear continuous-time systems. In: Commault, Ch., Marchand, N. (eds.) Positive Systems (Proceedings of the 2nd Multidisciplinary International Symposium on Positive Systems: Theory and Applications (POSTA 06), Grenoble, France), pp. 295–302. Lecture Notes in Control and Information Sciences. Springer, Berlin (2006)
17. Roszak, B., Davison, E.J.: Necessary and sufficient conditions for stabilizability of positive LTI systems. Syst. Control Lett. **58**, 474481 (2009)
18. Schrijver, A.: Theory of Linear and Integer Programming. Wiley, New York (2000)
19. Smith, H.L.: Monotone Dynamical Systems : An Introduction to the Theory of Competitive and Cooperative Systems. American Mathematical Society, Providence (1995)
20. Winkin, J., Dochain, D., Ligarius, Ph.: Dynamical analysis of distributed parameter tubular reactors. Automatica **36**(3), 349–361 (2000)

Chapter 16
Positivity Analysis of Continuous 2D Fornasini-Marchesini Fractional Model

Krzysztof Rogowski

Abstract In the chapter continuous Fornasini-Marchesini type model containing partial fractional-order derivatives described by the Caputo definition will be considered. General solution formula to the state-space equations of the model will be given. Using this solution formula the positivity of such system will be analyzed and the conditions under which the system is internally positive will be derived. Considerations will be illustrated by numerical simulations.

Keywords Fractional-order systems · Two-dimensional systems · General solution formula · Positive systems

16.1 Introduction

The most popular models of two-dimensional (2D) linear systems are the one introduced by Roesser [15], Fornasini and Marchesini [3, 4] and Kurek [13]. An overview of 2D linear systems theory is given in [1, 5, 11] and for positive 2D systems in [10].

The notion of fractional-order 2D discrete systems was introduced by Kaczorek in [9]. The continuous 2D fractional-order systems of the Roesser structure have been introduced in [16] and extended for descriptor (nonsingular) case in [12, 17]. In these papers the continuous state-space equations containing partial fractional order derivative described by the Caputo definition have been considered. The Riemann-Liouville definition has been applied for continuous fractional-order 2D Roesser type model in [6].

The fractional-order 2D Fornasini-Marchesini continuous model with Riemann-Liouville definition of fractional-order partial derivative has been considered in [7].

In this chapter the state-space equations of fractional-order 2D linear system of the structure similar to the Fornasini-Marchesini first model will be introduced. The partial fractional-order derivatives of a 2D continuous functions used in the chapter

K. Rogowski (✉)
Faculty of Electrical Engineering, Bialystok University of Technology, Bialystok, Poland
e-mail: k.rogowski@pb.edu.pl

F. Cacace et al. (eds.), *Positive Systems*, Lecture Notes in Control and Information Sciences 471, DOI 10.1007/978-3-319-54211-9_16

are based on the Caputo definition. In Sect. 16.2 definitions of fractional-order partial derivatives and integrals for 2D continuous functions will be given. The state-space equation of introduced system will be presented in Sect. 16.3 and, applying the 2D Laplace transform method, the solution to the system will be derived. A numerical example of the solution to the state-equations will be presented in the same section. Internal positivity of such model will be considered in Sect. 16.4 and necessary conditions for positivity will be derived. Concluding remarks and open problems will be formulated in Sect. 16.5.

16.2 Fractional-Order Partial Derivatives and Integrals of 2D Functions

Let $\mathbb{R}^{n\times m}$ be the set of $n \times m$ real matrices and $\mathbb{R}^n = \mathbb{R}^{n\times 1}$. The set of $n \times m$ matrices with real nonnegative elements will be denoted by $\mathbb{R}_+^{n\times m}$ and $\mathbb{R}_+^n = \mathbb{R}_+^{n\times 1}$. The set of nonnegative integers will be denoted by $\mathbb{Z}_+$ and the $n \times n$ identity matrix will be denoted by $\mathbb{I}_n$. The set of $n \times n$ Metzler matrices (matrices with arbitrary diagonal elements and nonnegative remaining elements) will be denoted by $\mathbb{M}_n$.

The following definitions of partial fractional order derivatives used in the chapter are based on Caputo fractional order derivative definition [12, 14].

Definition 16.1 [16] The fractional α_i-order partial derivative of a 2D continuous function $f(t_1, t_2)$ with respect to variable t_i $(i = 1, 2)$ is given by the formula

$$D_{t_i}^{\alpha_i} f(t_1, t_2) = \frac{\partial^{\alpha_i}}{\partial t_i^{\alpha_i}} f(t_1, t_2) = \frac{1}{\Gamma(N_i - \alpha_i)} \int_0^{t_i} \frac{f_{t_i}^{(N_i)}(t_i)}{(t_i - \tau)^{\alpha_i - N_i + 1}} d\tau, \tag{16.1a}$$

where $\alpha_i \in \mathbb{R}$ is the order of fractional partial derivative, $N_i - 1 < \alpha_i < N_i$, $\Gamma(x)$ is the Euler gamma function and

$$f_{t_i}^{(N_i)}(\tau) = \begin{cases} \dfrac{\partial^{N_1}}{\partial \tau^{N_1}} f(\tau, t_2) \text{ for } i = 1, \\ \dfrac{\partial^{N_2}}{\partial \tau^{N_2}} f(t_1, \tau) \text{ for } i = 2. \end{cases} \tag{16.1b}$$

Using Definition 16.1 the 2D fractional derivative of continuous function $f(t_1, t_2)$ is defined by

$$D_{t_1,t_2}^{\alpha_1,\alpha_2} f(t_1, t_2) = D_{t_1}^{\alpha_1} D_{t_2}^{\alpha_2} f(t_1, t_2) = D_{t_2}^{\alpha_2} D_{t_1}^{\alpha_1} f(t_1, t_2). \tag{16.2}$$

The fractional-order integral with respect to the variable t_1 will be defined by the Riemann-Liouville definition [8, 14]

$$I_{t_1}^{\alpha} f(t_1, t_2) = \frac{1}{\Gamma(\alpha)} \int_0^{t_1} (t_1 - \tau)^{\alpha-1} f(\tau, t_2) \mathrm{d}\tau,$$

where $\alpha > 0$ is the fractional (real) order of the integration. For $\alpha = 0$ we assume that $I_{t_1}^{\alpha} f(t_1, t_2) = f(t_1, t_2)$. In a similar way we may define the fractional-order integral with respect to the second variable.

The 2D fractional order integration with respect to the variables t_1 and t_2 is given by the formula

$$I_{t_1,t_2}^{\alpha,\beta} f(t_1, t_2) = I_{t_1}^{\alpha} \left[I_{t_2}^{\beta} f(t_1, t_2) \right] = I_{t_2}^{\beta} \left[I_{t_1}^{\alpha} f(t_1, t_2) \right],$$

where $\alpha, \beta > 0$.

Let $F(p, t_2)$ $(F(t_1, s))$ be the one-dimensional Laplace transform of a 2D continuous function $f(t_1, t_2)$ with respect to t_1 (t_2) defined by [2, 12]

$$F(p, t_2) = \mathscr{L}_{t_1} [f(t_1, t_2)] = \int_0^{\infty} f(t_1, t_2) e^{-pt_1} \mathrm{d}t_1$$

$$\left(F(t_1, s) = \mathscr{L}_{t_2} [f(t_1, t_2)] = \int_0^{\infty} f(t_1, t_2) e^{-st_2} \mathrm{d}t_2 \right).$$

The 2D Laplace transform of $f(t_1, t_2)$ is defined by

$$F(p, s) = \mathscr{L}_{t_1,t_2} [f(t_1, t_2)] = \mathscr{L}_{t_1} \left\{ \mathscr{L}_{t_2} [f(t_1, t_2)] \right\} = \mathscr{L}_{t_2} \left\{ \mathscr{L}_{t_1} [f(t_1, t_2)] \right\}.$$

It is easy to show that the 2D Laplace transform of partial fractional order derivative of the function $f(t_1, t_2)$ with respect to variable t_1 is given by the formula [12]

$$\mathscr{L}_{t_1,t_2} \left[D_{t_1}^{\alpha_1} f(t_1, t_2) \right] = p^{\alpha_1} F(p, s) - \sum_{k=1}^{N_1} p^{\alpha_1 - k} F_{t_1}^{(k-1)}(0, s), \tag{16.3a}$$

where

$$F_{t_1}^{(k)}(0, s) = \mathscr{L}_{t_2} \left\{ \left[\frac{\partial^k}{\partial t_1^k} f(t_1, t_2) \right]_{t_1=0} \right\}. \tag{16.3b}$$

In a similar way we define the 2D Laplace transform of partial fractional order derivative of the 2D function $f(t_1, t_2)$ with respect to the second variable t_2.

Using (16.2) and (16.3) we obtain

$$\begin{aligned}\mathscr{L}_{t_1,t_2}\left[D_{t_1,t_2}^{\alpha_1,\alpha_2} f(t_1,t_2)\right] =& p^{\alpha_1} s^{\alpha_2} F(p,s) + \sum_{k=1}^{N_1}\sum_{l=1}^{N_2} p^{\alpha_1-k} s^{\alpha_2-l} F^{(k-1,l-1)}(0,0)\\ &- p^{\alpha_1}\sum_{l=1}^{N_2} s^{\alpha_2-l} F_{t_2}^{(l-1)}(p,0) - s^{\alpha_2}\sum_{k=1}^{N_1} p^{\alpha_1-k} F_{t_1}^{(k-1)}(0,s),\end{aligned} \tag{16.4a}$$

where

$$F^{(k,l)}(0,0) = \left[\frac{\partial^k \partial^l}{\partial t_1^k \partial t_2^l} f(t_1,t_2)\right]_{\substack{t_1=0\\ t_2=0}}. \tag{16.4b}$$

16.3 2D Fornasini-Marchesini Fractional-Order Model and Its Solution

Let us consider the continuous 2D Fornasini-Marchesini fractional (α,β)-order model described by the state-space equations

$$D_{t_1,t_2}^{\alpha_1,\alpha_2} x(t_1,t_2) = A_0 x(t_1,t_2) + A_1 D_{t_1}^{\alpha_1} x(t_1,t_2) + A_2 D_{t_2}^{\alpha_2} x(t_1,t_2) + Bu(t_1,t_2), \tag{16.5}$$

where $x(t_1,t_2) \in \mathbb{R}^n$ is the state vector, $u(t_1,t_2) \in \mathbb{R}^m$ is the input vector, matrices $A_k \in \mathbb{R}^{n\times n}$ for $k = 0, 1, 2$; $B \in \mathbb{R}^{n\times m}$ and the fractional derivatives are defined by (16.1) and (16.2).

In this section we will consider the 2D fractional-order Fornasini-Marchesini model (16.5) with $\alpha_1 = \alpha$, $\alpha_2 = \beta$, where $0 < \alpha < 1$, $0 < \beta < 1$. Hence we have $N_1 = 1$, $N_2 = 1$.

The boundary conditions for (16.5) are given in the following form

$$x(t_1,0) \in \mathbb{R}^{n_2} \quad \text{and} \quad x(0,t_2) \in \mathbb{R}^{n_2}.$$

Applying the 2D Laplace transform to the state-space equation (16.5) and taking into account (16.3) and (16.4) we obtain

$$\begin{aligned}& p^{\alpha} s^{\beta} X(p,s) - p^{\alpha} s^{\beta-1} X(p,0) - p^{\alpha-1} s^{\beta} X(0,s) + p^{\alpha-1} s^{\beta-1} x(0,0)\\ =& A_0 X(p,s) + BU(p,s) + A_1\left[p^{\alpha} X(p,s) - p^{\alpha-1} X(0,s)\right]\\ &+ A_2\left[s^{\beta} X(p,s) - s^{\beta-1} X(p,0)\right],\end{aligned} \tag{16.6}$$

where

$$X(p,0) = \mathscr{L}_{t_1}\left[f(t_1,0)\right] \quad \text{and} \quad X(0,s) = \mathscr{L}_{t_2}\left[f(0,t_2)\right]$$

and $U(p,s)$ is the 2D Laplace transform of the input vector $u(t_1,t_2)$.

Premultiplying both sides of equality (16.6) by $p^{-\alpha}s^{-\beta}$ we obtain

$$X(p,s) = G^{-1}(p,s)\left\{s^{-1}\left[\mathbb{I}_n - p^{-\alpha}A_2\right]X(p,0) + p^{-1}\left[\mathbb{I}_n - s^{-\beta}A_1\right]X(0,s) - p^{-1}s^{-1}x(0,0) + p^{-\alpha}s^{-\beta}BU(p,s)\right\},$$

where

$$G(p,s) = \left[\mathbb{I}_n - p^{-\alpha}s^{-\beta}A_0 - s^{-\beta}A_1 - p^{-\alpha}A_2\right]. \tag{16.7}$$

Let

$$G^{-1}(p,s) = \sum_{i=0}^{\infty}\sum_{j=0}^{\infty} T_{ij}p^{-i\alpha}s^{-j\beta}. \tag{16.8}$$

It is well known that

$$G(p,s)G^{-1}(p,s) = G^{-1}(p,s)G(p,s) = \mathbb{I}_n$$

and using (16.7) and (16.8) we may write

$$\sum_{i=0}^{\infty}\sum_{j=0}^{\infty}\left[T_{ij} - A_0T_{i-1,j-1} - A_1T_{i,j-1} - A_2T_{i-1,j}\right]p^{-i\alpha}s^{-j\beta} = \mathbb{I}_n. \tag{16.9}$$

Comparing the coefficients at the same powers of p and s we obtain

$$T_{ij} = \begin{cases} \mathbb{I}_n \quad \text{for} \quad i=0,\ j=0; \\ A_0T_{i-1,j-1} + A_1T_{i,j-1} + A_2T_{i-1,j} = T_{i-1,j-1}A_0 + T_{i,j-1}A_1 + T_{i-1,j}A_2 \\ \quad \text{for} \quad i+j>0; \quad i,j \in \mathbb{Z}_+; \\ 0 \quad \text{for} \quad i<0 \ \text{and/or}\ j<0. \end{cases} \tag{16.10}$$

Using the 2D inverse Laplace transform to the Eq. (16.3) we obtain

$$x(t_1,t_2) = \sum_{i=0}^{\infty}\sum_{j=0}^{\infty}\left\{-\frac{t_1^{i\alpha}}{\Gamma(i\alpha+1)}\frac{t_2^{j\beta}}{\Gamma(j\beta+1)}T_{ij}x(0,0) + T_{i-1,j-1}BI_{t_1,t_2}^{i\alpha,j\beta}u(t_1,t_2) + \frac{t_2^{j\beta}}{\Gamma(j\beta+1)}\left[T_{ij} - T_{i-1,j}A_2\right]I_{t_1}^{i\alpha}x(t_1,0) + \frac{t_1^{i\alpha}}{\Gamma(i\alpha+1)}\left[T_{ij} - T_{i,j-1}A_1\right]I_{t_2}^{j\beta}x(0,t_2)\right\}, \tag{16.11}$$

since [12, 16]

$$\mathscr{L}_{t_1}^{-1}\left[p^{-\alpha}\right] = \frac{t_1^{\alpha-1}}{\Gamma(\alpha)} \quad \text{for} \quad \alpha > 0$$

and

$$\mathscr{L}_{t_1,t_2}^{-1}\left[p^{-\alpha}F(p,s)\right] = I_{t_1}^{\alpha}f(t_1,t_2) \quad \text{for} \quad \alpha > 0.$$

From the above considerations we have the following theorem.

Theorem 16.1 *The solution to the state equation (16.5) with fractional orders* $0 < \alpha < 1, 0 < \beta < 1$ *for arbitrary input* $u(t_1, t_2)$ *and boundary conditions* $x(t_1, 0)$, $x(0, t_2)$ *is given by (16.11) with the transition matrices described by the formula (16.10).*

The following example shows the usefulness of the derived solution to the state-space equations of fractional-order 2D system.

Example 16.1 Consider the fractional-order 2D linear system (16.5) with $\alpha = 0.7$, $\beta = 0.9$ and matrices

$$A_0 = \begin{bmatrix} -0.1 & 0 \\ 0.1 & -0.05 \end{bmatrix}, \; A_1 = \begin{bmatrix} -0.01 & 0.1 \\ 0.1 & -0.05 \end{bmatrix}, \\ A_2 = \begin{bmatrix} -0.05 & 0 \\ 0.1 & -0.01 \end{bmatrix}, \; B = \begin{bmatrix} 0.1 \\ 0.1 \end{bmatrix}. \tag{16.12}$$

Find a step response of the system (16.5) with the matrices (16.12) and zero boundary conditions.

The input of such system has the form

$$u(t_1, t_2) = H(t_1, t_2) = \begin{cases} 0 \text{ for } t_1 < 0 \text{ and/or } t_2 < 0, \\ 1 \text{ for } \quad t_1, t_2 \geq 0. \end{cases} \tag{16.13}$$

The solution to the state equations for zero boundary conditions and input of the form (16.13) is given by

$$x(t_1, t_2) = \sum_{i=0}^{\infty}\sum_{j=0}^{\infty} T_{i-1,j-1} B I_{t_1,t_2}^{i\alpha,j\beta} H(t_1, t_2) = \sum_{i=0}^{\infty}\sum_{j=0}^{\infty} T_{i-1,j-1} B \frac{t_1^{i\alpha} t_2^{j\beta}}{\Gamma(i\alpha+1)\Gamma(j\beta+1)}, \tag{16.14}$$

since it is well-known that [14, 16]

$$I_{t_1,t_2}^{\alpha,\beta} H(t_1, t_2) = \frac{t_1^{\alpha} t_2^{\beta}}{\Gamma(\alpha+1)\Gamma(\beta+1)}.$$

Formula (16.14) describes the step response of the system (16.5) with the matrices (16.12). The gamma function strongly increases for growing i and j, therefore in numerical analysis we may assume that i and j are bounded by some natural numbers L_1 and L_2.

The plots of the state variables, where $L_1 = 25$ and $L_2 = 25$ are shown on Figs. 16.1 and 16.2.

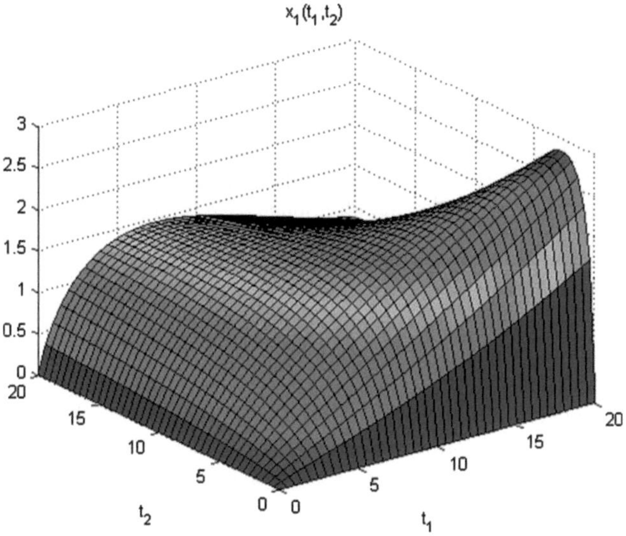

Fig. 16.1 State variable $x_1(t_1, t_2)$

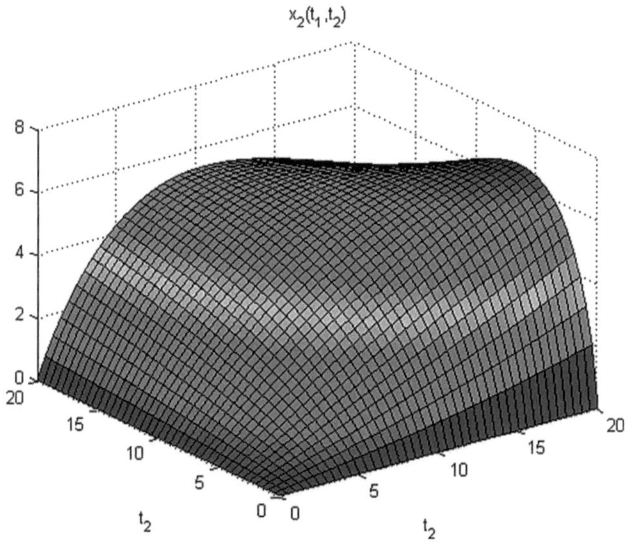

Fig. 16.2 State variable $x_2(t_1, t_2)$

16.4 Positivity Analysis

In this section we will consider the internal positivity of the introduced model.

Definition 16.2 The fractional 2D system (16.5) is called internally positive if the state vector $x(t_1, t_2) \in \mathbb{R}_+^n$ $(t_1, t_2 > 0)$ for all nonnegative boundary conditions $x(t_1, 0) \in \mathbb{R}_+^n$ and $x(0, t_2) \in \mathbb{R}_+^n$ and all nonnegative inputs $u(t_1, t_2) \in \mathbb{R}_+^m$.

Theorem 16.2 *The fractional-order 2D continuous Fornasini-Marchesini system with $0 < \alpha < 1$ and $0 < \beta < 1$ is internally positive if*

$$A_0 \in \mathbb{R}_+^{n\times n}; \quad A_1, A_2 \in \mathbb{M}_n \quad \textit{and} \quad B \in \mathbb{R}_+^{n\times m}.$$

Proof Let us consider the fractional-order 2D system (16.5) with zero input vector $u(t_1, t_2) = 0$ for $t_1, t_2 \geq 0$ and zero boundary conditions $x(0, t_2) = 0$ for $t_2 \geq 0$. We assume that only $x(t_1, 0)$ is nonnegative and nonzero for $t_1 > 0$. Using the solution formula (16.11) we have

$$x(t_1, t_2) = \sum_{i=0}^{\infty}\sum_{j=0}^{\infty} \frac{t_2^{j\beta}}{\Gamma(j\beta + 1)} \left[T_{ij} - T_{i-1,j}A_2\right] I_{t_1}^{i\alpha} x(t_1, 0). \tag{16.15}$$

For a small value of $t_1 > 0$ and taking into account (16.10) and that the fractional order integral of arbitrary function in a very small interval is close to zero we obtain

$$x(t_1, t_2) \approx \sum_{j=0}^{\infty} \frac{t_2^{j\beta}}{\Gamma(j\beta + 1)} A_1^j x(t_1, 0) = E_\beta(A_1 t_2^\beta) x(t_1, 0), \tag{16.16}$$

where $E_\beta(A_1 t_2^\beta)$ is the one-parameter Mittag-Leffler function [8, 12]. There always exists such small t_1 for which the approximation (16.16) occurs.

It is well known [8, 12] that the one-parameter Mittag-Leffler function $E_\beta(A_1 t_2^\beta)$ is nonzero matrix function for all $t_2 > 0$ and $0 < \beta < 1$ if and only if the matrix A_1 is the Metzler matrix, i.e. $A_1 \in \mathbb{M}_n$.

In similar way, assuming that only $x(0, t_2)$ is nonzero and nonnegative for $t_2 > 0$, we may show the necessity of $A_2 \in \mathbb{M}_n$.

Now we will consider the solution of the fractional order 2D system (16.5) with zero boundary conditions $x(t_1, 0) = 0$ for $t_1 \geq 0$, $x(0, t_2) = 0$ for $t_2 \geq 0$ and nonzero inputs $u(t_1, t_2) > 0$ for $t_1, t_2 \geq 0$. In such case we have

$$x(t_1, t_2) = \sum_{i=0}^{\infty}\sum_{j=0}^{\infty} T_{i-1,j-1} B I_{t_1,t_2}^{i\alpha, j\beta} u(t_1, t_2). \tag{16.17}$$

For a very small values of the variables t_1 and t_2 we obtain

$$x(t_1, t_2) \approx B I_{t_1,t_2}^{\alpha,\beta} u(t_1, t_2). \tag{16.18}$$

From the fact that the fractional-order integral of nonnegative inputs $u(t_1, t_2)$ is always nonnegative for $t_1, t_2 \geq 0$ follows the necessity of nonnegativity of the matrix B, i.e. $B \in \mathbb{R}_+^{n\times m}$.

Finally we will consider the solution of the fractional order 2D system (16.5) with nonzero boundary condition with respect to zeros variables $t_1, t_2 = 0$ and zero boundary conditions for $t_1, t_2 > 0$ and zero inputs $u(t_1, t_2) = 0$ for $t_1, t_2 \geq 0$. In such case we have

$$
\begin{aligned}
x(t_1, t_2) = \sum_{i=0}^{\infty}\sum_{j=0}^{\infty}\Bigg\{ & -\frac{t_1^{i\alpha}}{\Gamma(i\alpha+1)}\frac{t_2^{j\beta}}{\Gamma(j\beta+1)}T_{ij}x(0,0) \\
& + \frac{t_2^{j\beta}}{\Gamma(j\beta+1)}\left[T_{ij} - T_{i-1,j}A_2\right] I_{t_1}^{i\alpha}x(t_1, 0) \\
& + \frac{t_1^{i\alpha}}{\Gamma(i\alpha+1)}\left[T_{ij} - T_{i,j-1}A_1\right] I_{t_2}^{j\beta}x(0, t_2)\Bigg\}.
\end{aligned}
\tag{16.19}
$$

Substituting (16.10) into (16.19) we obtain

$$
\begin{aligned}
x(t_1, t_2) &= x(t_1, 0) + x(0, t_2) \\
&+ \sum_{\substack{i=0\\ i+j\neq 0}}^{\infty}\sum_{j=0}^{\infty}\frac{t_2^{j\beta}}{\Gamma(j\beta+1)}\left(T_{i-1,j-1}A_0 + T_{i,j-1}A_1\right) I_{t_1}^{i\alpha}x(t_1, 0) \\
&+ \sum_{\substack{i=0\\ i+j\neq 0}}^{\infty}\sum_{j=0}^{\infty}\frac{t_1^{i\alpha}}{\Gamma(i\alpha+1)}\left(T_{i-1,j-1}A_0 + T_{i-1,j}A_2\right) I_{t_2}^{j\beta}x(0, t_2) \\
&- \sum_{i=0}^{\infty}\sum_{j=0}^{\infty}\frac{t_1^{i\alpha}}{\Gamma(i\alpha+1)}\frac{t_2^{j\beta}}{\Gamma(j\beta+1)}T_{ij}x(0,0).
\end{aligned}
\tag{16.20}
$$

For a small value of t_1 and t_2 and taking into account that $A_1, A_2 \in \mathbb{M}_n$ and (16.16) we obtain the necessity of nonnegativity of matrix A_0, i.e. $A_0 \in \mathbb{R}_+^{n\times n}$, since there always exists such small t_1 and t_2 that the state vector described by (16.20) requires nonnegative matrix A_0.

Note that the fractional-order 2D system from Example 16.1 is not positive, since the matrix A_0 has negative elements. On a Figs. 16.1 and 16.2 we can see that for greater values of the variables t_1 and t_2 the components of the state vector will take negative values.

Example 16.2 Consider the fractional order 2D system from Example 16.1 with

$$
A_0 = \begin{bmatrix} 0.1 & 0 \\ 0.1 & 0.05 \end{bmatrix}. \tag{16.21}
$$

Note that for such case all conditions of Theorem 16.2 are met and the fractional order 2D system is internally positive.

For the same boundary conditions and inputs we obtain the plots of the state variables shown on Figs. 16.3 and 16.4.

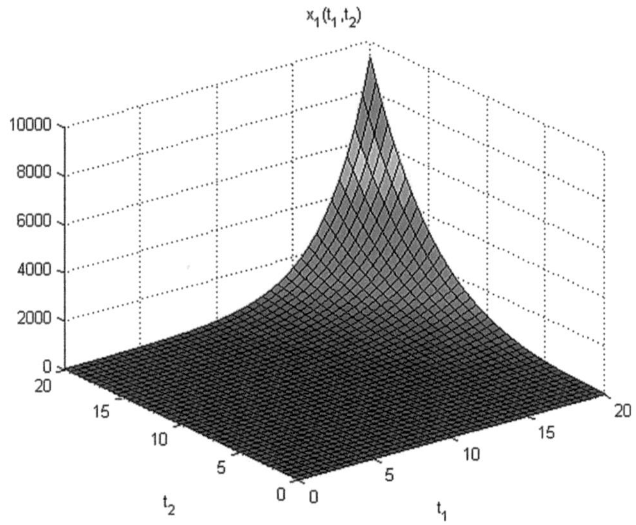

Fig. 16.3 State variable $x_1(t_1, t_2)$

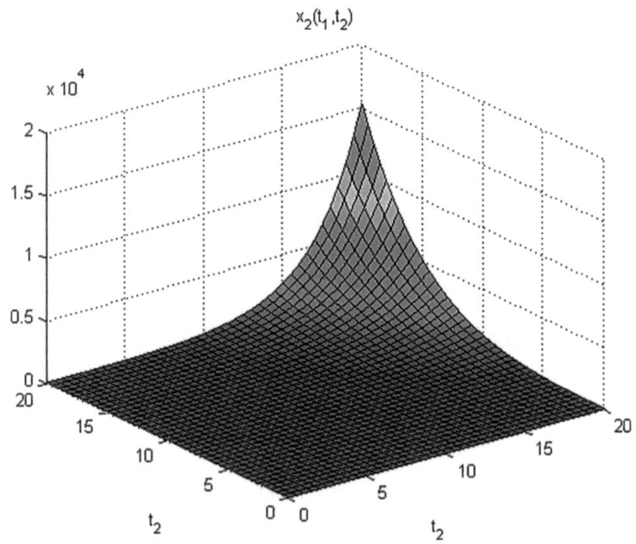

Fig. 16.4 State variable $x_2(t_1, t_2)$

16.5 Concluding Remarks

In the chapter continuous Fornasini-Marchesini type model containing partial fractional-order derivatives described by the Caputo definition has been considered. The general solution formula has been derived using the inverse 2D Laplace transform method. Using this solution the internal positivity of such system has been analyzed. The necessary conditions under which the system is internally positive has been given. The usefulness of the solution formula has been illustrated by numerical examples of positive and not positive systems.

An open problem is to give necessary and sufficient conditions for the internal positivity of considered in this chapter system as well as generalization for descriptor (nonsingular) case.

Acknowledgements This work was supported by National Science Centre in Poland under work No. 2014/13/B/ST7/03467.

References

1. Bose, N.K.: Multidimensional Systems Theory and Applications. Springer (1995)
2. Debnath, J., Dahiya, R.S.: Theorems on multidimensional Laplace transform for solution of boundary value problems. Comput. Math. Appl. **18**(12), 1033–1056 (1989)
3. Fornasini, E., Marchesini, G.: Double indexed dynamical systems. Math. Syst. Theory **12**(1), 59–72 (1978)
4. Fornasini, E., Marchesini, G.: State-space realization theory of two-dimensional filters. IEEE Trans. Autom. Control **21**(4), 484–491 (1976)
5. Galkowski, K.: State-Space Realizations of Linear 2-D Systems with Extensions to the General nD ($n>2$) case. Springer, London (2001)
6. Idczak, D., Kamocki, R., Majewski, M.: Fractional continuous Roesser model with Riemann-Liouville derivative. In: Proceedings of 8th International Workshop on Multidimensional Systems (nDS'13), Sept 9–11, Erlangen, Germany, pp. 33–38 (2013)
7. Idczak, D., Kamocki, R., Majewski, M.: On a fractional continuous counterpart of Fornasini-Marchesini model. In: Proceedings of 8th International Workshop on Multidimensional Systems (nDS'13), Sept 9–11, Erlangen, Germany, pp. 45–49 (2013)
8. Kaczorek, T.: Selected Problems in Fractional Systems Theory. Springer, London (2011)
9. Kaczorek, T.: Fractional 2D linear systems. J. Autom. Mob. Robotics Intell. Syst. **2**(2), 5–9 (2008)
10. Kaczorek, T.: Positive 1D and 2D Systems. Springer, London (2001)
11. Kaczorek, T.: Two-Dimensional Linear Systems. Springer, London (1985)
12. Kaczorek, T., Rogowski, K.: Fractional linear systems and electrical circuits. Stud. Syst. Decis. Control **13** (2015) (Springer)
13. Kurek, J.E.: The general state-space model for a two-dimensional linear digital system. IEEE Trans. Autom. Control **30**(2), 600–602 (1985)
14. Podlubny, I.: Fractional Differential Equations. Academic Press, London (1999)
15. Roesser, R.P.: A discrete state-space model for linear image processing. IEEE Trans. Autom. Control **20**(1), 1–10 (1975)
16. Rogowski, K.: General response formula for fractional 2D continuous-time linear systems described by the Roesser model. Acta Mech. et Autom. **5**(2), 112–116 (2011)
17. Rogowski, K.: Selected problems of theory of 2D noninteger order systems described by the Roesser model. Ph.D. thesis, Bialystok University of Technology (in Polish), Bialystok (2011)

Part V
Theory and Applications of Positive Systems

Chapter 17
Access Time Eccentricity and Diameter

Gabriele Oliva, Antonio Scala, Roberto Setola and Luigi Glielmo

Abstract In this chapter we study the access time on random walks, i.e., the expected time for a random walk starting at a node v_i to reach a node v_j, an index that can be easily calculated resorting to the powerful tools of positive systems. In particular, we argue that such an index can be the base for developing novel topological descriptors, namely access time eccentricity and diameter. While regular eccentricities and diameter are defined considering minimum paths, the indices defined in this chapter are related to random movements across the network, which may follow inefficient paths, and are thus a complementary measure to identify central and peripheral nodes and to set adequate time-to-live for the packets in a network of distributed agents, where few or no routing information is available. A simulation campaign aimed at showing the characteristics of the proposed indices concludes the chapter.

Keywords Random walk · Access time · Diameter · Eccentricity

17.1 Introduction

Random walk over a graph [1, 2] is a powerful tool that finds application in several contexts, ranging from computer science to biology and from economics to psychology (see [3–9] for recent applications in these fields).

G. Oliva (✉) · R. Setola
Università Campus Bio-Medico di Roma, via Álvaro del Portillo 21,
00128 Rome, Italy
e-mail: g.oliva@unicampus.it

R. Setola
e-mail: r.setola@unicampus.it

A. Scala
ISC-CNR UoS "Sapienza", Piazzale Moro 5, 00185 Roma, Italy
e-mail: antonio.scala.phys@gmail.com

L. Glielmo
Universitá Degli Studi del Sannio, 21-82100 Benevento, Italy
e-mail: glielmo@unisannio.it

F. Cacace et al. (eds.), *Positive Systems*, Lecture Notes in Control
and Information Sciences 471, DOI 10.1007/978-3-319-54211-9_17

Within a random walk, we assume a walker visits the nodes in a graph at random starting from a given node, and that at each turn the walker moves from a node to one of its neighbors with uniform probability. In this view, a random walk can conveniently be represented as a Markov chain. Although the resulting matrix of transition probabilities is in general not symmetric, it is possible to operate a transformation into a symmetric matrix, so that the powerful tools of spectral analysis and positivity can be used to prove convergence results and to characterize several noteworthy indicators.

Among others, in this chapter we focus on the *access time* for a pair of nodes in the graph, that is, the expected time in which a random walk starting at a given node i reaches a node j for the first time. We argue that this parameter can be related to topological properties such as the *eccentricity* (i.e., the maximum distance among a node i and any other node by using minimum paths) and the *diameter* (i.e., the maximum among the eccentricities).

Eccentricities and diameter play a pivotal role in several contexts, such as distributed consensus algorithms, where having insights on such indicators can help reduce computational effort [10–13], or can be used to set time-to-live parameters in routing protocols [14].

However, such indices are related to communications along the minimum paths, which can be a rough assumption in a completely distributed context, where the nodes have little or no information about other nodes. For instance, suppose a node i in a network of distributed agents has to transmit a message to a node j, and that the message is encrypted with a public-private key scheme, so that only node j is able to decrypt the message. If no routing information is available, a possible strategy is to send the message at random to a neighbor and so on, thus obtaining a random walk. In this scenario, therefore, it is interesting to find adequate time-to-live for the message, to avoid unnecessary retransmissions and to prevent the message from remaining indefinitely in the system.

The idea of using random walks to derive topological indicators is not new in the literature; for instance in [15] the authors introduce the *random walk closeness centrality* as a measure of centrality of the nodes, in terms of the average of the expected times to reach a node from all the other nodes; in [16], similarly, a measure of betweenness centrality over a graph is defined by counting how many times a random walk passes through a specific node.

The outline of the chapter is as follows: In Sect. 17.2 we provide some preliminary definitions, while in Sect. 17.3 we briefly review random walks and access time; in Sect. 17.4 we discuss the adoption of alternative measures of diameter and eccentricity based on access time and random walks, while in Sect. 17.5 we provide some simulation results. We conclude the chapter with some conclusive remarks in Sect. 17.6.

17.2 Preliminaries

Let $G = \{V, E\}$ denote a *graph* with a finite number n of nodes $v_i \in V$ and m edges $(v_i, v_j) \in E \subset V \times V$ from node v_i to node v_j. A graph is said to be *undirected* if $(v_i, v_j) \in E$ whenever $(v_j, v_i) \in E$, and it is said to be *directed* otherwise. In the following we will consider undirected graphs.

Let the *neighborhood* $\mathcal{N}_i$ of a node i over a graph $G = \{V, E\}$ be the set of nodes $\{v_j | (v_j, v_i) \in E\}$. Let the *degree* d_i of a node v_i be the number of its incident edges, i.e., $d_i = |\mathcal{N}_i|$.

A graph $G = \{V, E\}$ is *bipartite* if the set of nodes can be partitioned in two disjoint sets V_a, V_b such that for all $(v_i, v_j) \in E$ $v_i \in V_a$ and $v_j \in V_b$ (or vice versa).

A *path* over a graph $G = \{V, E\}$, starting from a node $v_i \in V$ and ending in a node $v_j \in V$, is a subset of links in E that connects v_i and v_j. The *length* of the path is the cardinality of such set. A graph is *connected* if for each pair of nodes v_i, v_j there is a path over G that connects them.

A *minimum path* that connects v_i and v_j is the path from v_i to v_j of minimum length, which we call the *distance* of the two nodes. The *eccentricity* ε_i of a node $v_i \in V$ is the maximum distance from node v_i to any other node. The *diameter* δ of a graph G is the maximum distance between each possible pair of distinct nodes $v_i, v_j \in V$. In other terms

$$\delta = \max_{i=1,\ldots,n} \{\varepsilon_i\}.$$

The *radius* r of a graph G is defined as

$$r = \min_{i=1,\ldots,n} \{\varepsilon_i\}.$$

Let the *adjacency matrix* A of a graph G be the $n \times n$ matrix such that

$$A_{ij} = \begin{cases} 1, & \text{if } (v_i, v_j) \in E \\ 0, & \text{otherwise;} \end{cases}$$

and such that $A_{ii} = 0$ for all i, i.e., no self loops are allowed. Moreover, let the *inverse degree matrix* be the $n \times n$ diagonal matrix D whose diagonal entries are

$$D_{ii} = \frac{1}{d_i}.$$

17.3 Random Walks

In this section we briefly review random walks over undirected graphs (see [2] and references therein), with particular reference to a parameter, namely *access time*, that is fundamental for the developments of this chapter. The results reviewed in this section, in fact, are the basis for the definition of the metrics introduced in Sect. 17.4. Notice that we assume the graph is connected and undirected.

A random walk w: $\{0, 1, 2, \ldots\} \rightarrow V$ is a path such that at each step t the next node is randomly chosen among the neighbors of $w(t)$ with equal probabilities $1/d_{w(t)}$.

The random walk can be described in a convenient way as a Markov chain [17, 18], by considering a matrix M of transition probabilities[1] such that

$$M_{ij} = \begin{cases} \frac{1}{d_i}, & \text{if } (v_i, v_j) \in E \\ 0, & \text{otherwise.} \end{cases}$$

In other words, we assume that, while being in a node v_i, we have no preference for the next move, thus visiting any of the neighbors of v_i is equally probable.

It can be easily shown that $M = DA$. The Markov chain representing the random walk can be expressed as

$$p_{t+1}^T = p_t^T M,$$

where $p_t \in \mathbb{R}^n$ is the probability distribution at time t, i.e., a vector whose components $p_{t,i}$ represent the probability that the random walker is in node v_i at time t. A probability distribution $\tilde{p}$ is *stationary* if $\tilde{p}^T = \tilde{p}^T M$; in this case, we refer to the random walk as a *stationary walk*.

It is a well known result that for every graph G the distribution $\pi = [\pi_1, \ldots \pi_n]^T$, where

$$\pi_i = \frac{d_i}{2m}, \forall i = 1, \ldots n$$

is stationary [2]. Moreover, if G is not bipartite, any distribution tends to the stationary distribution as $t \rightarrow \infty$ [2]. For bipartite graphs, the distribution may oscillate between the partitions.

17.3.1 Spectral Decomposition

Notice that matrix M is, in general, not symmetric as the probability of moving from node v_i to v_j is $1/d_i$ while the converse is $1/d_j$ and $d_i \neq d_j$ unless for *d-regular graphs*.[2] It is, however, simple to bring M to a symmetric form; such a symmetric form can be used in order to ease the spectral analysis, and most of the results in the literature are given with respect to the symmetric representation described below.

Let $N = D^{-1/2} M D^{1/2}$, it follows that

$$N = D^{-1/2} DAD^{1/2} = D^{1/2} AD^{1/2};$$

[1]Each entry M_{ij} represents the probability to move from node i to node j at a given time instant.

[2]A graph is d-regular if the degree of each node is d.

hence, N is symmetric and it has non-negative entries. As a consequence, N can be written in a spectral form as

$$N = \sum_{k=1}^{n} \lambda_k q_k q_k^T,$$

where $\lambda_1 \geq \lambda_2 \geq \ldots \geq \lambda_n$ are the eigenvalues of N and $q_i, \cdots, q_n$ are the corresponding eigenvectors of unit length.

Notice that

$$q_1 = \frac{1}{\sqrt{2m}} \left[\sqrt{d_1}, \ldots, \sqrt{d_n}\right]^T$$

is an eigenvector of N with eigenvalue 1, and that all components of q_1 are positive. Notice that the graph underlying the random walk is connected and undirected, hence it can be shown that N is primitive. Therefore, as a consequence of the Frobenius-Perron Theorem, it holds

$$\lambda_1 = 1 > \lambda_2 \geq \cdots \geq \lambda_n \geq -1.$$

If G is not bipartite, moreover, it can be shown that $\lambda_n > -1$.

17.3.2 Access Time

The *access time* H_{ij} for a random walk over a graph G is the expected number of steps required for a random walk starting at node v_i to reach node v_j (see, for instance, [19]). The calculation of H_{ij} is greatly simplified resorting to the above spectral decomposition of N, as it holds [2]

$$H_{ij} = 2m \sum_{k=2}^{n} \frac{1}{1-\lambda_k} \left(\frac{q_{kj}^2}{d_j} - \frac{q_{ki} q_{kj}}{\sqrt{d_i d_j}} \right).$$

17.4 Access Time Eccentricities

In this section, we define an alternative and complementary notion of eccentricity, based on access times over a random walk.

Let us define the *access time eccentricity* ε_i^H of node v_i in G as

$$\varepsilon_i^H = \max_{j=1,\ldots,n} \{H_{ij}\}.$$

Notice that standard eccentricity ε_i models the maximum distance among a node v_i and any other node, using minimum paths, and thus it is well suited to upper-bound the time required for node v_i to send a message to any other node in an efficient way, e.g., when the message routing is known.

Here, conversely, we are representing a case where no routing information is available and messages are forwarded on a purely random basis. In this case, therefore, ε_i^H is an upper bound on the expected time required to reach any node from v_i when no routing information is available. This index, therefore, is of particular interest in fields where the nodes represent entities or agents attempting to communicate with little or no information about other agents.

Similarly to standard diameter, we can define an *access time diameter* δ^H as

$$\delta^H = \max_{i=1,\ldots,n} \{\varepsilon_i^H\} = \max_{i,j=1,\ldots,n} \{H_{ij}\},$$

which represents the maximum expected time required for sending a message between a pair of nodes in the graph.

Let us describe next a few classical results that support the idea of adopting δ^H as an alternate measure of the diameter of G. It should be noted, in fact, that δ^H is closely related to the *cover time* c_i, i.e., the expected number of steps required to visit all nodes starting from a node v_i. Although c_i is complex to calculate exactly, in fact, there is a nice result that links cover time and access time; in [20], Matthews proved that

$$\min_{i,j} H_{ij} \sum_{k=1}^{n} \frac{1}{k} \leq c_i \leq \max_{i,j} H_{ij} \sum_{k=1}^{n} \frac{1}{k}; \tag{17.1}$$

hence, the access time diameter can be used to derive an upper bound the cover time. Another interesting result from Lovász [2] is that the expected number of steps b before a random walk visits half of the nodes is such that

$$b \leq 2\delta_H.$$

17.5 Experimental Results

Figure 17.1 shows a comparison between eccentricities and access time eccentricities on a particular graph with $n = 100$ nodes. Specifically, we consider a *random geometric graph*, i.e., a graph whose nodes are taken in the unit square $[0, 1]^2$ in a uniformly random way, and such that two nodes v_i, v_j are connected by an edge if their Euclidean distance d_{ij} is less than a *communication radius* ρ. For the graph in Fig. 17.1 we take $\rho = 0.25$. In the left plot we color the nodes in the graph according to a heat-map, so that nodes with large eccentricities are red while nodes with comparatively small eccentricities are blue, while we do the same for the access time eccentricities in the right plot. According to the figure, it can be noted that eccentricities and access time eccentricities are distributed in quite a different way; we

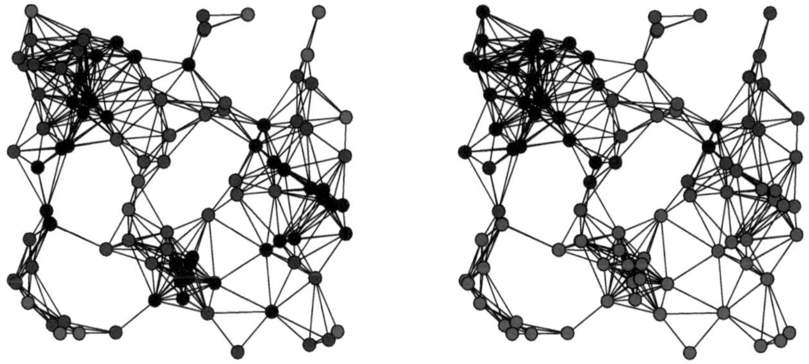

Fig. 17.1 Heat-map comparison (i.e., nodes with higher values are plotted in *red*, while nodes with lower values are in *blue*) between the eccentricity of the nodes (*left plot*) and the access time eccentricity (*right plot*) over a random geometric graph with $n = 100$ nodes and $\rho = 0.35$ (please refer to the online version for colors)

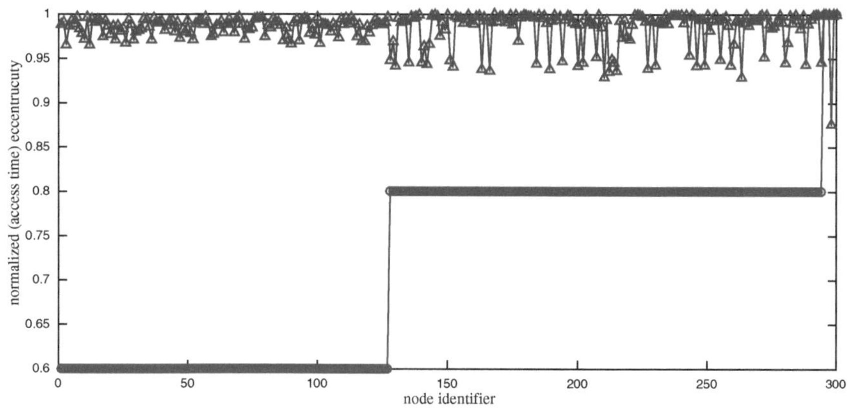

Fig. 17.2 Example showing that the access time eccentricities do not vary monotonically with the eccentricities. In this example we consider a random geometric graph with $n = 300$ nodes and $\rho = 0.15$. We plot the eccentricities (normalized by the maximum one, shown in *red* as the lowermost curve) for each node and the corresponding access time eccentricity (normalized by the maximum one, shown in blue as the uppermost curve)

notice, in fact, that while some of the nodes around the center of the graph have the smallest eccentricities, in the case of access time eccentricity the blue zone is focused in the central upper part. This suggests that there is no trivial dependency relation among the two indicators. In fact, as highlighted in Fig. 17.2, where we order the eccentricities (normalized by the maximum one) for a random geometric graph with $n = 300$ nodes and $\rho = 0.15$ and we shown the corresponding access time eccentricities (normalized by the maximum one), there is no ordering preservation while moving from eccentricities to access time eccentricities.

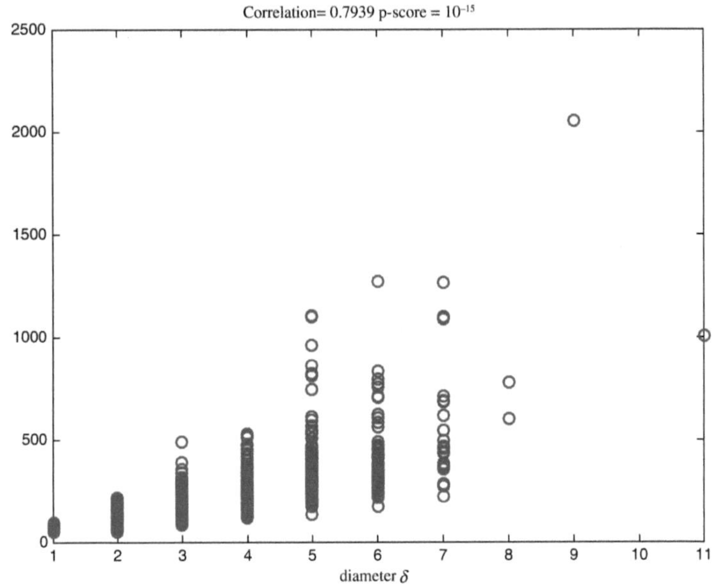

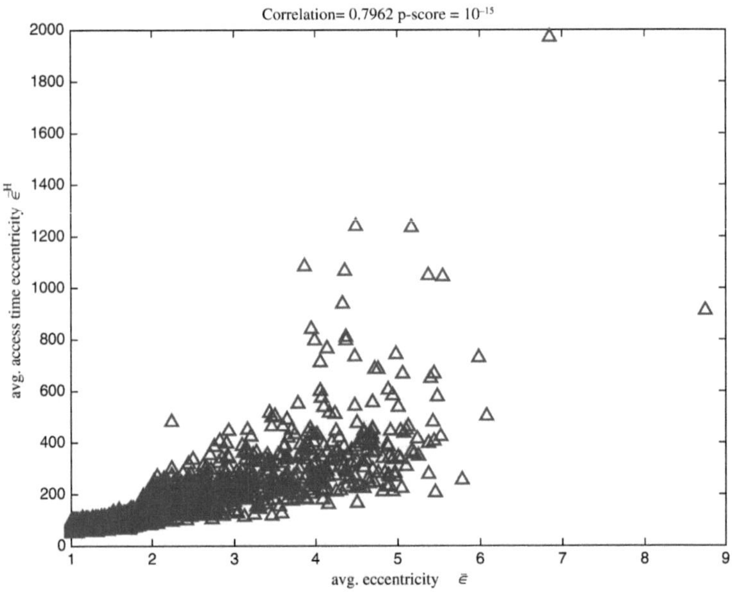

Fig. 17.3 *Upper plot* correlation between the diameter δ and the access time diameter δ^H. *Lower plot* correlation between the average eccentricity $\bar{\varepsilon}$ and the average access time eccentricity $\bar{\varepsilon}^H$. The results are obtained for $R = 2000$ random geometric graphs whose nodes are sampled in the unit square in a uniformly random way; each graph has a random choice of the network size $n \in [50, 100]$ and of the communication radius $\rho \in [0.25, \sqrt{2}]$

Although having relevant differences, we notice that these indicators are indeed closely related. In Fig. 17.3 we show in the upper plot the correlation between the diameter δ and the access time diameter δ^H and in the lower plot the correlation between the average eccentricity

$$\bar{\varepsilon} = \frac{1}{n} \sum_{i=1}^{n} \varepsilon_i$$

and the average access time eccentricity

$$\bar{\varepsilon}^H = \frac{1}{n} \sum_{i=1}^{n} \varepsilon_i^H.$$

Specifically, we consider $R = 2000$ random geometric graphs, each with a uniformly random network size $n \in [50, 100]$ and a uniformly random communication radius $\rho \in [0.25, \sqrt{2}]$. According to both plots, the indices based on access time have high correlation with their counterpart based on the minimum paths, i.e., in both cases we have a correlation around 0.79, with almost negligible p-score, which suggest that the correlation in place is indeed significant.

In Fig. 17.4 we compare the access times H_{ij} (recall that H_{ij} is the expected time for a random walk starting at v_i to reach v_j for the first time) with the associated standard deviation $\sigma_{H_{ij}}$. Specifically, we consider a particular random geometric graph with $n = 100$ nodes and $\rho = 0.35$ (left plot in Fig. 17.4), and we sample $R = 1000$ random walks starting from each node v_i (thus, a total of Rn random walks), and each random walk is run for a sufficient time so that all nodes are reached once. For each random

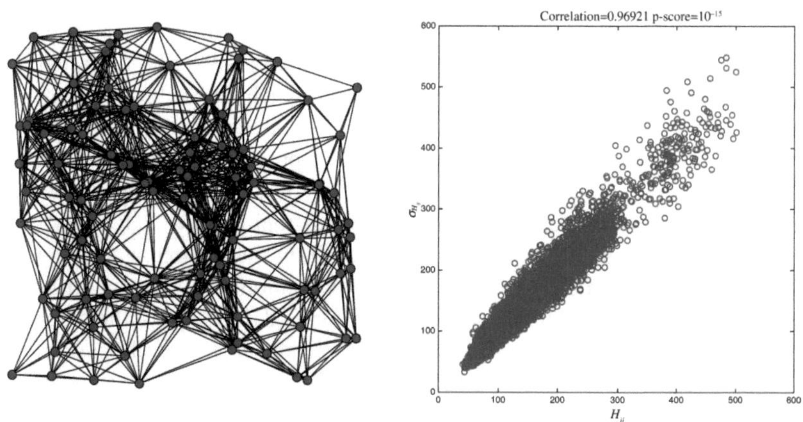

Fig. 17.4 Comparison between the access times and the standard deviation for each pair of nodes (*right plot*) for a particular instance of random geometric graph with $n = 100$ nodes and $\rho = 0.35$ (*left plot*)

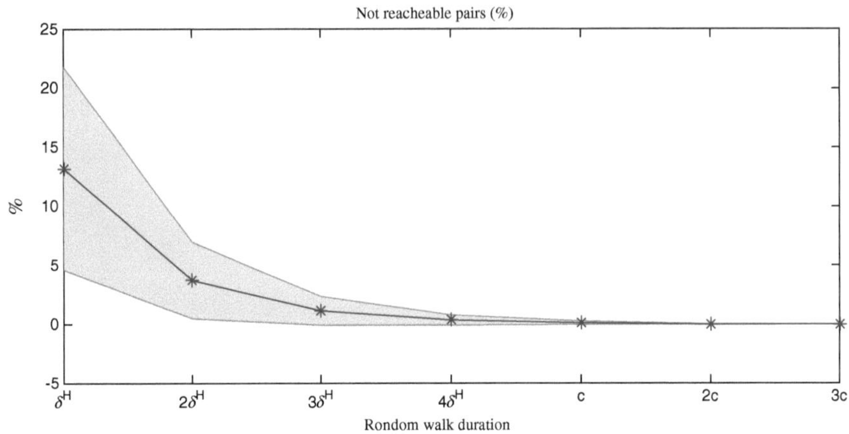

Fig. 17.5 Percentage of pairs of nodes that are not in reach in a random walk for different durations of the random walk, with respect to a random geometric graph with $n = 100$ nodes and $\rho = 0.35$. Results are the average and standard deviation for $R = 1000$ trials for each duration

walk starting at v_i, we get the time t_{ij} in which each node v_j is reached by the random walk for the first time. We then calculate an approximation of the standard deviation $\sigma_{H_{ij}}$ in terms of the R samples of t_{ij} obtained. Notice that we do this due to the difficulty of finding an exact closed form for $\sigma_{H_{ij}}$.

According to the left plot in Fig. 17.4, there is an almost linear relation between H_{ij} and $\sigma_{H_{ij}}$ (the correlation is about 0.97 with a p-score of about 10^{-15}). This suggests that the standard deviation of H_{ij} tends to have the same magnitude as H_{ij}. A consequence of that is that, while using the exact access-time eccentricities/diameter to set the time-to-live of packets routed at random may fail (e.g., due to the high standard deviation), it is reasonable to have time-to-live which is $O(\delta^H)$ (e.g., by choosing $\delta^H + 3\sigma_{\delta^H} = 4\delta^H$ we reach all nodes in about the 99.73% of cases). This evidence calls for broader analyses and simulations, which are a valuable future work direction.

In Fig. 17.5, finally, we evaluate if δ_H or the upper bound on the cover time are sufficient to let a random walk starting at any node v_i reach any other node v_j. Specifically, we consider a random geometric graph with $n = 100$ nodes and $\rho = 0.35$, and we chose different duration times T for the random walks. For each choice of the duration time T we execute $R = 1000$ random walks starting from each node, and we plot in the figure the average and standard deviation of the percentage of pairs v_i, v_j such that a random walk starting in v_i is not able to reach v_j. According to the figure when $T = \delta^H$ there is about 13.2% of pairs which are not in reach. The percentage drops quickly as T grows, and for $T = 4\delta^H$ we have the 0.37% of pairs not in reach. If $T = c$, where c is the maximum among the upper bounds for the cover time c_i discussed in Inequality (17.1), we have 0.12% while for $T = 2c$ we get 0.0009%. It should be noted that only for $T = 3c$ we have that all the nodes are in reach in all trials.

17.6 Conclusions

In this chapter we focus on the access time for a random walk in a graph, which can be conveniently calculated using the tools of spectral analysis and positive systems. Specifically, we develop some indicators, namely access time eccentricity and diameter, which play a role in random walks which is similar to standard eccentricity and diameter.

Instead of giving bounds based on minimum paths, in fact, we consider random movements across the graph and we characterize the maximum expected time for one node to reach all other nodes (access time centrality) and the maximum expected time to reach any node from any other node (access time diameter). We then inspect the relations among the indices and the possibility to use them to tune the time of live of packets transmitted at random in the network.

Future work will be mainly devoted to further characterize the indices with respect to different topologies and to provide applications in the field of wireless sensor networks; for instance, we will inspect the applications of access time eccentricity in distributed network localization problems (e.g., [21, 22]).

References

1. Pearson, K.: The problem of the random walk. Nature **72**(1867), 342 (1905)
2. Lovász, L.: Random walks on graphs: a survey. Combinatorics, special volume "Paul Erdos is Eighty" no. 2 pp. 353–398 (1996)
3. Erdene-Ochir, O., Abdallah, M., Qaraqe, K., Minier, M., Valois, F.: A theoretical framework of resilience: biased random walk routing against insider attacks. In: Wireless Communications and Networking Conference (WCNC), 2015 IEEE, pp. 1602–1607. IEEE, March (2015)
4. Verma, R.K., Das, A.X., Jaiswal, A.K.: Effective performance of location aided routing protocol on random walk (RW) mobility model using constant bit rate (CBR). Int. J. Comput. Appl. **122**(14) (2015)
5. Varadarajan, A., Oswald, F., Bollen, Y.J., Peterman, E.J.: Membrane-protein diffusion in E. coli: a random walk in a heterogeneous landscape. Biophys. J. **108**(2), 323a (2015)
6. Jones, P.J.M., Sim, A., Taylor, H.B., Bugeon, L., Dallman, M.J., Pereira, B., Stumpf, M.P.H., Liepe, J.: Inference of random walk models to describe leukocyte migration. Phys. Biol. **12**(6), 066001 (2015)
7. Sadorsky, P.: Forecasting Canadian mortgage rates. Appl. Econ. Lett. 1–4 (2015)
8. Ballester, C., Vorsatz, M.: Random walk-based segregation measures. Rev. Econ. Stat. **96**(3), 383–401 (2014)
9. Nosofsky, R.M., Palmeri, T.J., Nosofsky, R.: An exemplar-based random-walk model of categorization and recognition. The Oxford Handbook of Computational and Mathematical Psychology, vol. 142 (2015)
10. Han-Lim, C., Brunet, L., How, J.P.: Consensus-based decentralized auctions for robust task allocation. IEEE Trans. Robot. **25**(4), 912–926 (2009)
11. Olfati-Saber, R., Murray, R.M.: Consensus problems in networks of agents with switching topology and time-delays. IEEE Trans. Autom. Control **49**(9), 1520–1533 (2004)
12. Oliva, G., Setola, R., Hadjicostis, C.: Distributed finite-time calculation of node eccentricities, graph radius and graph diameter. Syst. Control Lett. **92**, 20–27 (2016)
13. Oliva, G., Setola, R., Hadjicostis, C.: Distributed finite-time average-consensus with limited computational and storage capability. IEEE Trans. Control Netw. Syst. (early access article)

14. Lee, S., Belding-Royer, E.M., Perkins, C.E.: Scalability study of the ad hoc on-demand distance vector routing protocol. Int. J. Netw. Manage. **13**(2), 97–114 (2003)
15. Noh, J.D., Rieger, H.: Random walks on complex networks. Phys. Rev. Lett. **92**(11), 118701–118704 (2004)
16. Newman, M.E.: A measure of betweenness centrality based on random walks. Soc. Netw. **27**(1), 39–54 (2005)
17. Doyle, P.G., Snell, J.L.: Random Walks and Electric Networks. MAA (1984)
18. Diaconis, P.: Group Representations in Probability and Statistics. Institute of Mathematical Statistics, Hayward, California (1988)
19. Tetali, P.: Random walks and the effective resistance of networks. J. Theor. Probab. **4**(1), 101–109 (1991)
20. Matthews, P.: Covering problems for Brownian motion on spheres. The Annals of Probability, pp. 189–199 (1988)
21. Oliva, G., Panzieri, S., Pascucci, F., Setola, R.: Simultaneous localization and routing in sensor networks using shadow edges. In: 2013 IFAC Intelligent Autonomous Vehicles Symposium (IAV2013). Gold Coast, Australia, 26–28, pp. 199–204 (2013)
22. Oliva, G., Pascucci, F., Panzieri, S., Setola, R.: Sensor network localization: extending trilateration via shadow edges. IEEE Trans. Autom. Control **60**(10), 2752–2755 (2015)

Chapter 18
Nonlinear Left and Right Eigenvectors for Max-Preserving Maps

Björn S. Rüffer

Abstract It is shown that max-preserving maps (or join-morphisms) on the positive orthant in Euclidean n-space endowed with the component-wise partial order give rise to a semiring. This semiring admits a closure operation for maps that generate stable dynamical systems. For these monotone maps, the closure is used to define suitable notions of left and right eigenvectors that are characterized by inequalities. Some explicit examples are given and applications in the construction of Lyapunov functions are described.

Keywords Monotone systems · Join-morphisms · Perron-Frobenius theory · Positive eigenvectors · Small-gain condition · Lyapunov functions

18.1 Introduction

Classical Perron-Frobenius theory asserts the existence of nonnegative left and right eigenvectors corresponding to the dominant eigenvalue of a nonnegative matrix [3–5, 9, 10]. For (nonlinear) monotone mappings from a positive cone into itself, various extensions to this theory have been developed, see [8] and the references therein. While most of the nonlinear extensions consider some form of right eigenvalue problem for monotone cone mappings, the question of left eigenvectors has not found a lot of attention. One reason that left eigenvectors do not have obvious counterparts in the world of nonlinear mappings may be that they are naturally elements of the (linear) dual of the underlying vector space in the classical spectral theory of linear operators. Linear duals are not very natural places to look for nonlinear eigenvectors.

In this chapter we consider a class of monotone mappings defined on the positive cone in $\mathbb{R}^n$ equipped with the component-wise partial order. It admits a suitable notion of left eigenvectors. This class consists of *max-preserving* mappings from $\mathbb{R}^n_+$ into itself, i.e., continuous, monotone maps $A\colon \mathbb{R}^n_+ \to \mathbb{R}^n_+$ with $A0 = 0$ for which

B.S. Rüffer (✉)
School of Mathematical and Physical Sciences, The University of Newcastle (UON), Callaghan, NSW 2308, Australia
e-mail: bjorn.ruffer@newcastle.edu.au

F. Cacace et al. (eds.), *Positive Systems*, Lecture Notes in Control and Information Sciences 471, DOI 10.1007/978-3-319-54211-9_18

$\max\{Ax, Ay\} = A \max\{x, y\}$. Instead of a numerical maximal eigenvalue, we consider the case when a nonlinear extension of the spectral radius is less than one, which can be characterised by the requirement that $A^k x \to 0$ as $k \to \infty$ for any $x \in \mathbb{R}^n_+$, or alternatively by the inequality

$$Ax \not\geq x \text{ for all } x \in \mathbb{R}^n_+, x \neq 0.$$

Given this starting point, it is not surprising that our nonlinear left and right "eigenvectors" are characterised by inequalities rather than equations. The terms "sub-eigenvectors" and spectral inequalities have been suggested as alternative terms for the objects introduced here. Both are (nonlinear) functions $l\colon \mathbb{R}^n_+ \to \mathbb{R}_+$ and $r\colon \mathbb{R}_+ \to \mathbb{R}^n_+$ that are continuous, zero at zero, monotone and unbounded in every component. They satisfy

$$l(Ax) < l(x)$$

for all $x \in \mathbb{R}^n_+$, $x > 0$ as well as

$$A\big(r(t)\big) < r(t)$$

for all $t > 0$.

Both, l and r are defined via the closure A^* of A in the semiring of max-preserving maps on $\mathbb{R}^n_+$.

This chapter is organised as follows. The next section provides a little more background on our interest in left eigenvectors. In Sect. 18.3 we recall some necessary notation and preliminary results. Section 18.4 contains our main results with formulas for left and right eigenvectors in Theorems 18.2 and 18.3, respectively. Two explicit examples are given in Sect. 18.5. In Sect. 18.6 we explain how these eigenvectors can be used to construct Lyapunov functions. Section 18.7 concludes this chapter.

18.2 Motivation

Our interest in left eigenvectors is rooted in the stability analysis of interconnected systems, where the construction of Lyapunov functions for monotone comparison systems is of special interest [2].

For a dynamical system $x(k+1) = Ax(k)$, evolving on $\mathbb{R}^n_+$, a Lyapunov function $V\colon \mathbb{R}^n_+ \to \mathbb{R}_+$ is an energy function that decreases along trajectories. Lyapunov functions are used to prove that trajectories converge to zero, to prove stability, or to compute regions of attraction. Finding Lyapunov functions, however, is notoriously hard. Basic properties they need to satisfy are continuity, positive definiteness, radial unboundedness (i.e., $\|x\| \to \infty$ implies $V(x) \to \infty$) and descent along trajectories, i.e., $V(Ax) < V(x)$ whenever $x \neq 0$.

If $A \in \mathbb{R}^{n\times n}_+$ has spectral radius less than one, one can find a positive vector r (even in the case that A is merely nonnegative [11, Lemma 1.1]) so that $Ar \ll r$, i.e., the image under A of r is less than the vector r in every component. Such a vector determines a Lyapunov function via $V(x) = \max_i x_i/r_i$, and this Lyapunov function is called max-separable.

Max-separable Lyapunov functions exist for various monotone but nonlinear systems as well, but not for all [2]. In some of these nonlinear cases one can instead find sum-separable Lyapunov functions, which are of the form $V(x) = \sum_i v_i(x_i)$. If again $A \in \mathbb{R}_+^{n\times n}$ has spectral radius less than one, i.e., in the linear case, there exists a positive vector $l \in \mathbb{R}_+^n$, so that $l^T A \ll l^T$. This vector, too, determines a Lyapunov function, $V(x) = l^T x$, and this one is sum-separable. For general monotone systems however, these sum-separable Lyapunov functions are not well understood yet, although progress has been made in some special cases [2, 6].

As left Perron eigenvectors do determine (sum-) separable Lyapunov functions in the linear case, there is hope that a suitable notion of left eigenvectors will also provide Lyapunov functions in more general scenarios. It turns out, however, that while the present definition of left-eigenvectors does yield Lyapunov functions given by explicit formulas, these Lyapunov functions are not separable in the above sense.

18.3 Preliminaries

In this work we consider $\mathbb{R}^n$ equipped with the component-wise partial order, which generates the positive cone $\mathbb{R}_+^n = [0, \infty)^n$. We use the following notation.

$$\begin{aligned}
&x \le y \text{ if } y - x \in \mathbb{R}_+^n,\\
&x < y \text{ if } x \le y \text{ and } x \ne y,\\
&x \ll y \text{ if } y - x \text{ is in the interior of } \mathbb{R}_+^n.
\end{aligned}$$

Note that $\max\{x, y\}$ is the component-wise maximum of the two vectors $x, y \in \mathbb{R}^n$. For notational convenience we use the binary symbol $x \oplus y$ to denote the same thing. We also write $\bigoplus\{x_k\}$ to denote the component-wise supremum of a possibly infinite set $\{x_k\}$ of vectors $x_k \in \mathbb{R}^n$.

By $\|x\| = \max_i |x_i|$ we denote the maximum-norm of $x \in \mathbb{R}^n$. We note that for $x, y \in \mathbb{R}^n$ we have $\|x \oplus y\| \le \max\{\|x\|, \|y\|\}$ and equality holds if $x, y \in \mathbb{R}_+^n$.

The vector $(1, \ldots, 1)^T \in \mathbb{R}^n$ will be denoted by $\mathbf{1}$. The standard unit vectors in $\mathbb{R}^n$ are denoted by $e_1, \ldots, e_n$.

In this work we will restrict our attention to continuous and monotone mappings. A mapping A is monotone if it preserves the partial order, i.e., $Ax \le Ay$ whenever $x \le y$. The set of *max-preserving* mappings from $\mathbb{R}_+^n$ into itself is given by

$$\mathrm{MP} = \mathrm{MP}(\mathbb{R}_+^n) = \Big\{A\colon \mathbb{R}_+^n \to \mathbb{R}_+^n \text{ such that } A(x \oplus y) = (Ax) \oplus (Ay) \text{ for all } x, y \in \mathbb{R}_+^n\Big\}.$$

The term max-preserving map has been coined in [7] in the context of stability analysis of interconnected control systems. It coincides with the notion of join-morphisms in lattice theory [1]. It is immediate that max-preserving mappings are also monotone.

For $A \in \mathrm{MP}$ we define non-decreasing functions $a_{ij} : \mathbb{R}_+ \to \mathbb{R}_+, i, j = 1, \ldots, n$, by $a_{ij}(t) = \big(A(te_j)\big)_i$ for $t \in \mathbb{R}_+$. It is immediate that A can be represented as

$$Ax = \begin{pmatrix} a_{11}(x_1) \oplus \ldots \oplus a_{1n}(x_n) \\ \vdots \\ a_{n1}(x_1) \oplus \ldots \oplus a_{nn}(x_n) \end{pmatrix},$$

so it is natural to think of A as the matrix (a_{ij}).

We state the following observation, where $\circ$ refers to composition.

Lemma 18.1 *The set MP is a $(\circ, \oplus)$-semiring with identity element $\mathrm{id}_{\mathbb{R}^n_+}$ and neutral element $0_{\mathbb{R}^n_+}$.*

Proof If $A, B \in \mathrm{MP}$ then we verify

$$(A \circ B)(x \oplus y) = A(Bx \oplus By) = (A \circ B)x \oplus (A \circ B)y,$$

so MP is closed under composition and

$$\begin{aligned}(A \oplus B)(x \oplus y) = A(x \oplus y) \oplus B(x \oplus y) = \\ (Ax \oplus Ay) \oplus (Bx \oplus By) = (Ax \oplus Bx) \oplus (Ay \oplus By) = \\ (A \oplus B)x \oplus (A \oplus B)y,\end{aligned}$$

so MP is closed under the maximum operation as well.

Clearly the identity $\mathrm{id}_{\mathbb{R}^n_+}$ is a member of MP and it is the identity element for composition. The function $0 = 0_{\mathbb{R}^n_+}$, which sends all of $\mathbb{R}^n_+$ to $0 \in \mathbb{R}^n_+$, is in MP, and it serves as neutral element for the maximum operation.

For convenience we will write compositions simply as products, i.e.,

$$A^k = A \circ A \circ \ldots \circ A.$$

We make the convention that $A^0 = \mathrm{id}$.

We now further restrict our attention to continuous mappings $A \in \mathrm{MP}(\mathbb{R}^n_+)$ that satisfy $A0 = 0$. We have the following characterisation.

Theorem 18.1 ([11]) *Let $A \in MP(\mathbb{R}^n_+)$ be continuous and satisfy $A0 = 0$. Then the following are equivalent.*

1. *For every $x \in \mathbb{R}^n_+$,*
$$A^k x \longrightarrow 0 \text{ as } k \to \infty. \tag{18.1}$$

2. *For every* $x \in \mathbb{R}_+^n$, $x \neq 0$,

$$Ax \ngeq x.$$

3. *Every cycle in the matrix* A *is a contraction, i.e.,*

$$\left(a_{i_1 i_2} \circ a_{i_2 i_3} \circ \ldots \circ a_{i_k i_1}\right)(t) < t$$

for every $t > 0$ *and all finite sequences* $(i_1, \ldots, i_k) \in \{1, \ldots, n\}^k$.

4. *All minimal cycles in* A *are contractions, i.e., those that do not contain shorter cycles.*
5. *For every* $b \in \mathbb{R}_+^n$ *there is a unique maximal solution* $x \in \mathbb{R}_+^n$ *to the inequality*

$$x \leq Ax \oplus b.$$

Along with an alternative construction of a right eigenvector, a slightly weaker version of this result has been proven in [11, Theorem 6.4], where the functions a_{ij} were assumed to be either strictly increasing or zero. However, the proof is essentially the same in the current framework and thus omitted.

18.4 Main Results

Our main technical ingredient for the construction of left and right eigenvectors is the closure of max-preserving maps in the semiring MP.

Lemma 18.2 *Let* $A \in MP(\mathbb{R}_+^n)$ *be continuous and satisfy* $A0 = 0$. *Let any of the conditions 1–5 of Theorem 18.1 hold. Then the* closure of A, *given by*

$$A^* x = \bigoplus_{k=0}^{\infty} A^k x \tag{18.2}$$

is a continuous and max-preserving map $A^* \colon \mathbb{R}_+^n \to \mathbb{R}_+^n$ *with* $A^* 0 = 0$ *that satisfies*

$$A^* = ID \oplus AA^* = ID \oplus A^* A. \tag{18.3}$$

Proof The identities (18.3) follow immediately from writing out (18.2). That A^* is well-defined is mostly a consequence of (18.1), once we note that (18.1) implies that the supremum in (18.2) is a maximum that is attained after a finite number of iterates of A.

The (i, j)th entry of the matrix A^* consists of the supremum over all possible paths from node j to node i in the weighted graph with n vertices and directed edges weighted with the functions a_{ij}. Because any path longer than n edges will contain a cycle, which in turn is a contraction, the infinite supremum in the definition of A^*, cf. (18.2), is in fact a maximum over at most n powers of A.

Thus A^* is max-preserving. In particular, only a finite number of terms $\|A^k x\|$ can be larger than $\|x\|$ and they depend continuously on $\|x\|$.

Remark 18.1 From the proof we see that in fact

$$A^*x = \bigoplus_{k=0}^{n-1} A^k x,$$

a finite maximum of only n vectors instead of a supremum. This will be demonstrated in Sect. 18.5.

Lemma 18.3 *Let $A \in MP(\mathbb{R}^n_+)$ be continuous and satisfy $A0 = 0$. Let any of the conditions 1–5 of Theorem 18.1 hold. Then the* closure of *A satisfies*

$$A\big(A^*(x)\big) = A^*\big(A(x)\big) < A^*(x) \tag{18.4}$$

for all $x > 0$.

Proof First we note that from the definition (18.2) it follows that $A^*A = AA^*$.

We have $A^*A \leq A^*A \oplus \mathrm{id} = A^*$ from (18.3), so we only need to show that equality does not hold. To this end assume there is an $x \in \mathbb{R}^n_+$, $x > 0$, with $A^*Ax = A^*x$. Denoting $z = A^*x$, we have

$$Az = AA^*x = A^*Ax = A^*x = z,$$

which contradicts property 2 of Theorem 18.1, as $z \geq x > 0$. Hence no such x can exist, proving that indeed $AA^*x = A^*Ax < A^*x$ for all $x > 0$.

Our main result is the following.

Theorem 18.2 (left eigenvectors for max-preserving maps) *Let $A \in MP(\mathbb{R}^n_+)$ be continuous and satisfy $A0 = 0$. Let any of the conditions 1–5 of Theorem 18.1 hold.*

Then $l \colon \mathbb{R}^n_+ \to \mathbb{R}_+$ given by

$$x \mapsto \mathbf{1}^T A^*(x) \tag{18.5}$$

is continuous, monotone, satisfies $l(0) = 0$, as well as

1. *$l(x) \to \infty$ whenever $\|x\| \to \infty$,*
2. *the left eigenvector inequality*

$$lAx \leq lx$$

for all $x \in \mathbb{R}^n_+$, and, moreover, $lAx < lx$ whenever $x \neq 0$.

Proof That the map l is well defined, continuous, monotone, and satisfies $l(0) = 0$ is an immediate consequence of Lemma 18.2. Assertion 1 follows from the fact that $A^* \geq \mathrm{id}$. Assertion 2 is a direct consequence of Lemma 18.3.

Remark 18.2 Instead of a summation of the components of $A^*(x)$ in (18.5) we could have taken their maximum instead, at the expense of loosing the strict inequality in in Assertion 2 of the theorem. In the context of Sect. 18.6, this would in general give rise to a weak Lyapunov function, i.e., one that is merely non-increasing along trajectories.

Our notion of left eigenvectors is complemented by right eigenvectors that are given by a similar construction, which, to the best of our knowledge, was first demonstrated in [7]. A different construction is given in [11].

Theorem 18.3 (right eigenvectors for max-preserving maps [7]) *Let $A \in MP(\mathbb{R}^n_+)$ be continuous and satisfy $A0 = 0$. Let any of the conditions 1–5 of Theorem 18.1 hold.*

Then $r\colon \mathbb{R}_+ \to \mathbb{R}^n_+$ given by

$$t \mapsto A^*(t\mathbf{1}) \tag{18.6}$$

is continuous, monotone, satisfies $r(0) = 0$ as well as

1. *$r_i(t) \to \infty$ when $\|t\| \to \infty$ for every $i = 1, \ldots, n$,*
2. *the right eigenvector inequality*

$$A\big(r(t)\big) \leq r(t) \tag{18.7}$$

for all $t \geq 0$, and, moreover, $A\big(r(t)\big) < r(t)$ when $t > 0$.

Proof That r is well defined, continuous, monotone and satisfies $r(0) = 0$ follows again from Lemma 18.2. Assertion 1 is a consequence of the fact that $A^* \geq \mathrm{id}$, see (18.3), so $r(t) \geq t\mathbf{1}$. Assertion 2 follows from Lemma 18.3 applied to $x = t\mathbf{1}$.

Remark 18.3 In both, Theorems 18.2 and 18.3, instead of the vector $\mathbf{1}$ in the definition of l, respectively, r, any strictly positive vector could have been taken instead.

18.5 Examples

We demonstrate with two examples that the left and right eigenvectors obtained in the previous section are given by finite expressions, cf. Remark 18.1, not as limits as the definition in (18.2) might suggest. The examples are borrowed from [12]. To this end we define

$$\mathcal{K}_\infty = \big\{a\colon \mathbb{R}_+ \to \mathbb{R}_+ \,\big|\, a \text{ is continuous, unbounded,} \\ \text{strictly increasing and satisfies } a(0) = 0\big\},$$

which is the set of homeomorphisms from $\mathbb{R}_+$ into itself.

First we consider the case $n = 2$. In this case A takes the form

$$A = \begin{pmatrix} a_{11} & a_{12} \\ a_{21} & a_{22} \end{pmatrix}$$

with $a_{ij} \in (\mathcal{K}_\infty \cup \{0\})$. The associated max-preserving mapping $A\colon \mathbb{R}^2_+ \to \mathbb{R}^2_+$ is given by

$$\begin{pmatrix} x_1 \\ x_2 \end{pmatrix} \mapsto \begin{pmatrix} a_{11}(x_1) \oplus a_{12}(x_2) \\ a_{21}(x_1) \oplus a_{22}(x_2) \end{pmatrix}.$$

The conditions of Theorem 18.1 are satisfied if and only if

$$a_{11} < \mathrm{id}$$
$$a_{22} < \mathrm{id}$$

and

$$a_{12} \circ a_{21} < \mathrm{id}. \tag{18.8}$$

Note that (18.8) holds if and only if

$$a_{21} \circ a_{12} < \mathrm{id}$$

holds. This can be seen by observing that every $\mathcal{K}_\infty$ function has an inverse which is again a $\mathcal{K}_\infty$ function.

Writing $x = (x_1, x_2)^T$ and under the above assumptions we compute

$$\begin{aligned} A^*(x) &= \bigoplus_{k=0}^{\infty} A^k(x) \\ &= x \oplus Ax = \left(\mathrm{id}_{\mathbb{R}^2_+} \oplus A\right)(x) \\ &= \begin{pmatrix} \mathrm{id} & a_{12} \\ a_{21} & \mathrm{id} \end{pmatrix}(x) = \begin{pmatrix} a^*_{11} & a^*_{12} \\ a^*_{21} & a^*_{22} \end{pmatrix}(x) \end{aligned} \tag{18.9}$$

as already

$$A^2 = \begin{pmatrix} a_{11}^2 \oplus a_{12} \circ a_{21} & a_{12} \circ a_{22} \oplus a_{11} \circ a_{12} \\ a_{21} \circ a_{11} \oplus a_{22} \circ a_{21} & a_{22}^2 \oplus a_{21} \circ a_{12} \end{pmatrix}$$

is component-wise less than the matrix $(\mathrm{id}_{\mathbb{R}^2_+} \oplus A)$ computed above.

From (18.9) we obtain

$$l(x) = x_1 \oplus a_{12}(x_2) + x_2 \oplus a_{21}(x_1)$$

Notably, this function is in general not smooth and neither sum- nor max-separable.

For the case $n = 3$ things are essentially the same.

Starting from

$$A = \begin{pmatrix} a_{11} & a_{12} & a_{13} \\ a_{21} & a_{22} & a_{23} \\ a_{31} & a_{32} & a_{33} \end{pmatrix}$$

and under the assumption that all cycles in A are contractions, we can compute A^* simply by

$$\begin{aligned} A^* &= \mathrm{id}_{\mathbb{R}^3_+} \oplus A \oplus A^2 \\ &= \begin{pmatrix} \mathrm{id} & a_{12} \oplus a_{13} \circ a_{32} & a_{13} \oplus a_{12} \circ a_{23} \\ a_{21} \oplus a_{23} \circ a_{31} & \mathrm{id} & a_{23} \oplus a_{21} \circ a_{13} \\ a_{31} \oplus a_{32} \circ a_{21} & a_{32} \oplus a_{31} \circ a_{12} & \mathrm{id} \end{pmatrix}, \end{aligned} \tag{18.10}$$

where we note that the simplifications used to obtain (18.10) are possible because all cycles are contractions.

From (18.10) we obtain

$$\begin{aligned} l(x) = x_1 &\oplus (a_{12} \oplus a_{13} \circ a_{32})(x_2) \oplus (a_{13} \oplus a_{12} \circ a_{23})(x_3) \\ &+(a_{21} \oplus a_{23} \circ a_{31})(x_1) \oplus x_2 \oplus (a_{23} \oplus a_{21} \circ a_{13})(x_3) \\ &+(a_{31} \oplus a_{32} \circ a_{21})(x_1) \oplus (a_{32} \oplus a_{31} \circ a_{12})(x_2) \oplus x_3. \end{aligned}$$

18.6 Application

Let $A \in \mathrm{MP}(\mathbb{R}^n_+)$ be continuous and satisfy $A0 = 0$. If $A^k x \to 0$ for $k \to \infty$, two types of Lyapunov functions can be defined based on the eigenvectors introduced in the previous section. Let $l\colon \mathbb{R}^n_+ \to \mathbb{R}_+$ and $r\colon \mathbb{R}_+ \to \mathbb{R}^n_+$ denote the left and right eigenvectors of A, respectively.

Under some additional regularity assumptions, or rather, regularisation of r, a max-separable Lyapunov function $V\colon \mathbb{R}^n_+ \to \mathbb{R}_+$ is given by

$$V(x) = \max_i r_i^{-1}(x_i),$$

where r_i denotes the ith component function of r. We refer the interested reader to [7] or to [2] and the references therein for further details.

The left eigenvector l also yields a Lyapunov function $V\colon \mathbb{R}^n_+ \to \mathbb{R}_+$ simply by

$$V(x) = l(x).$$

Theorem 18.2 establishes that this is indeed a Lyapunov function for the system $x(k+1) = A(x(k))$.

We note that this Lyapunov function is in general neither sum- nor max-separable. However, it has the advantage that no additional regularity has to be assumed to make the components of the eigenvector invertible and that it can be computed directly from the problem data.

Example 18.1 Consider the matrix

$$A = \begin{pmatrix} \frac{1}{2} & 2 & 0 \\ \frac{1}{3} & \frac{1}{2} & 3 \\ \frac{1}{7} & 0 & \frac{1}{2} \end{pmatrix}$$

where we take the entries as linear functions $t \mapsto a_{ij}t$ and compute Ax in max algebra, making the associated map $A\colon \mathbb{R}^n_+ \to \mathbb{R}^n_+$ max-preserving.

There are five cycles in this matrix. Three of them are "self-loops" of weight $1/2$. The other two are from node 2 to 1 with weight 2 and back to node 2 with weight $1/3$, as well as from 2 to 1 with weight 2, from there to 3 with weight $1/7$ and back to 2 with weight 3. All of the loop-weights (products) are less than one, so this matrix satisfies the equivalent conditions of Theorem 18.1.

A simple computation yields

$$A^* = \begin{pmatrix} 1 & 2 & 6 \\ \frac{3}{7} & 1 & 3 \\ \frac{1}{7} & \frac{2}{7} & 1 \end{pmatrix}.$$

From here we obtain

$$l(x) = \max\{x_1, 2x_2, 6x_3\} + \max\left\{\frac{3}{7}x_1, x_2, 3x_3\right\} + \max\left\{\frac{1}{7}x_1, \frac{2}{7}x_2, x_3\right\}$$

and we verify that for $x > 0$ the expression

$$l(Ax) = \max\left\{\frac{6}{7}x_1, 2x_2, 6x_3\right\} + \max\left\{\frac{3}{7}x_1, \frac{6}{7}x_2, 3x_3\right\} + \max\left\{\frac{1}{7}x_1, \frac{2}{7}x_2, \frac{6}{7}x_3\right\}$$

is indeed smaller.

18.7 Conclusion

For max-preserving maps A on $\mathbb{R}^n_+$ we have shown that left and right eigenvectors can be defined in a natural sense based on the closure of the map A, extending the classical Perron-Frobenius theory appropriately to nonlinear dominant eigenvalues. In this work the dominant eigenvalue was assumed to be less than the identity, but via suitable scaling this could be extended to more general scenarios.

Our results have been presented on $\mathbb{R}^n_+$, however, an extension to join-morphisms acting on Banach lattices is a natural next step.

The construction of left-eigenvectors and corresponding Lyapunov functions for general monotone systems that are not generated by elements of a semiring remains a challenge.

References

1. Birkoff, G.: Lattice Theory, 3rd edn. American Mathematical Society (1973)
2. Dirr, G., Ito, H., Rantzer, A., Rüffer, B.S.: Separable Lyapunov functions: constructions and limitations. Discrete Contin. Dyn. Syst. Ser. B **20**(8), 2497–2526 (2015)
3. Frobenius, G.: Über Matrizen aus positiven Elementen. Sitzungsberichte der Königlich Preussischen Akademie der Wissenschaften, pp. 471–476 (1908)
4. Frobenius, G.: Über Matrizen aus positiven Elementen. ii. Sitzungsberichte der Königlich Preussischen Akademie der Wissenschaften, vol. 1909, pp. 514–518 (1909)
5. Frobenius, G.: Über Matrizen aus nicht negativen Elementen. Sitzungsberichte der Königlich Preussischen Akademie der Wissenschaften, pp. 456–477 (1912)
6. Ito, H., Jiang, Z.P., Dashkovskiy, S., Rüffer, B.S.: Robust stability of networks of iISS systems: construction of sum-type Lyapunov functions. IEEE Trans. Autom. Control **58**(5), 1192–1207 (2013)
7. Karafyllis, I., Jiang, Z.P.: A vector small-gain theorem for general non-linear control systems. IMA J. Math. Control Inf. **28**(3), 309–344 (2011)
8. Lemmens, B., Nussbaum, R.: Nonlinear Perron-Frobenius Theory. Cambridge Tracts in Mathematics, vol. 189. Cambridge University Press, Cambridge (2012)
9. Perron, O.: Grundlagen für eine Theorie des Jacobischen Kettenbruchalgorithmus. Math. Ann. **64**(1), 1–76 (1907)
10. Perron, O.: Zur Theorie der Matrices. Math. Ann. **64**(2), 248–263 (1907)
11. Rüffer, B.S.: Monotone inequalities, dynamical systems, and paths in the positive orthant of Euclidean n-space. Positivity **14**(2), 257–283 (2010)
12. Rüffer, B.S., Ito, H.: Sum-separable Lyapunov functions for networks of ISS systems: a gain function approach. In: Proceedings of the 54th IEEE Conference on Decision Control, pp. 1823–1828 (2015)

Chapter 19
Positive Consensus Problem: The Case of Complete Communication

Maria Elena Valcher and Irene Zorzan

Abstract In this chapter the positive consensus problem for homogeneous multi-agent systems is investigated, by assuming that agents are described by positive single-input and continuous-time systems, and that each agent communicates with all the other agents. Under certain conditions on the Laplacian of the communication graph, that arise only when the graph is complete, some of the main necessary conditions for the problem solvability derived in [17–19] do not hold, and this makes the problem solution more complex. In this chapter we investigate this specific problem, by providing either necessary or sufficient conditions for its solvability and by analysing some special cases.

Keywords Multi agent system · Continuous time positive system · Consensus · Complete communication graph

19.1 Introduction

Research on multi-agent systems and consensus problems has been flourishing in the last decades [2, 7, 9, 11, 13, 14, 16], strongly stimulated by the large number of different applications areas where practical problems that can be formalized as consensus problems among autonomous agents/units arise. Just to mention the most popular ones, flocking and swarming in animal groups, dynamics of opinion forming, coordination in sensor networks, clock synchronization, distributed tasks among mobile robots/vehicles. These apparently different set-ups share some common features: in each of them there is a group of individuals/units (the agents), whose behavior can be regarded as homogeneous. Each agent performs tasks and updates

M.E. Valcher (✉) · I. Zorzan
Dip. di Ingegneria dell'Informazione, Univ. di Padova, via Gradenigo 6/B, 35131 Padova, Italy
e-mail: meme@dei.unipd.it

I. Zorzan
e-mail: irene.zorzan@studenti.unipd.it

F. Cacace et al. (eds.), *Positive Systems*, Lecture Notes in Control and Information Sciences 471, DOI 10.1007/978-3-319-54211-9_19

a vector of describing parameters (its state) based on the information received from neighbouring agents, with the final goal of agreeing on a common value for such a vector.

In a number of contexts, the information vector that the agents update (based on communication exchange with their neighbours), aiming to achieve consensus, is the value of variables that are intrinsically nonnegative. For instance, wireless sensor networks in greenhouses [1] exchange information regarding physical parameters as temperature, humidity, and CO_2 concentration, and the sensors must converge to some common values for these parameters, based on which ventilation/heating systems will be activated, shading or artificial lights will be controlled, CO_2 will be injected, and so on.

Another interesting problem, that is formalized as a consensus problem with positivity constraint, is the emissions control for a fleet of Plugin Hybrid Vehicles [8]. Each vehicle has a parallel power-train configuration that allows for any arbitrary combination of the power generated by the combustion engine and the electric motor. Moreover, the vehicles can communicate. Under these assumptions, an algorithm is proposed to regulate in a cooperative way the CO_2 emissions, so that no vehicle has a higher emission level than the others.

In a series of recent papers [17–19] we have investigated the consensus problem for homogeneous multi-agent systems, whose agents are modelled as continuous-time, single-input, positive state-space models. We assumed that interactions among agents are cooperative and the communication graph regulating the agents' mutual interactions is weighted, undirected and connected but not complete, namely not every agent directly exchanges information with all the other agents. As the agents' states are intrinsically nonnegative, a natural requirement to introduce, in addition to consensus, is the positivity of the overall controlled multi-agent system and hence that the state feedback law adopted to achieve consensus constrains all the state trajectories to remain in the positive orthant. A rather complete characterization of the problem solvability has been provided, and special cases, arising under special conditions either on the agents' description or on the communication graph, have been discussed.

The simple assumption that the communication graph is connected but not complete allowed to exclude the rather peculiar situation when the maximum weighted degree of an agent, namely the largest of the diagonal entries of the Laplacian associated with the communication graph, is smaller than all the positive eigenvalues of the Laplacian. By ruling out this case, we were able to derive some powerful necessary conditions for the solvability of the positive consensus problem that provided the backbone of the analysis carried on in [17–19]. This chapter addresses the critical case, namely the situation when the communication among the agents is described by a complete graph and all the positive eigenvalues of its Laplacian are greater than its diagonal entries. As we will see, the necessary conditions derived in this context are weaker, and conditions that in the previous investigation turned out to be necessary and sufficient for the problem solvability under the current assumptions are only sufficient.

In detail, Sect. 19.2 provides some background material. In Sect. 19.3 the positive consensus problem is formalized. A set of necessary or sufficient conditions for the problem solvability is provided in Sect. 19.4. The case when the input to state matrix involved in the agents' description is monomial is investigated in Sect. 19.5. Finally, in Sect. 19.6, we address the case of two-dimensional agents.

19.2 Background Material

Given a positive integer N, we let $[1, N]$ denote the set $\{1, 2, \dots, N\}$. $\mathbf{e}_i$ is the ith *canonical vector* (whose size is always clear from the context). The (i, j)th entry of a matrix A will be denoted either by a_{ij} or by $[A]_{ij}$, the ith entry of a vector $\mathbf{v}$ by v_i or $[\mathbf{v}]_i$.A vector $\mathbf{v} = v_i \mathbf{e}_i$, $v_i > 0$, is called ith *monomial vector*. $\mathbf{1}_N$ is the N-dimensional vector whose entries are all unitary. The *Kronecker product* of two matrices $A \in \mathbb{R}^{m\times n}$ and $B \in \mathbb{R}^{p\times q}$ is the matrix

$$C = [A \otimes B] := \begin{bmatrix} a_{11}B & a_{12}B & \dots & a_{1n}B \\ a_{21}B & a_{22}B & \dots & a_{2n}B \\ \vdots & \vdots & \ddots & \vdots \\ a_{m1}B & a_{m2}B & \dots & a_{mn}B \end{bmatrix} \in \mathbb{R}^{pm\times qn}.$$

Given a matrix $A \in \mathbb{R}^{n\times n}$, we denote by $\sigma(A)$ its *spectrum*. A is *Hurwitz* if all its eigenvalues lie in the open left complex halfplane, i.e. $\lambda \in \sigma(A)$ implies $\Re(\lambda) < 0$. $\mathbb{R}_+$ is the set of nonnegative real numbers. A matrix (in particular, a vector) A_+ with entries in $\mathbb{R}_+$ is a *nonnegative matrix* ($A_+ \geq 0$); if $A_+ \geq 0$ and at least one entry is positive, A_+ is a *positive matrix* ($A_+ > 0$), while if all its entries are positive it is a *strictly positive matrix* ($A_+ \gg 0$). A matrix $A \in \mathbb{R}^{n\times n}$ is a *Metzler matrix* if its off-diagonal entries are nonnegative.

Given $A \in \mathbb{R}^{n\times n}$, we define the *spectral abscissa* of A as

$$\lambda_{\max}(A) := \max\{\Re(\lambda), \lambda \in \sigma(A)\}. \tag{19.1}$$

For a Metzler matrix, the spectral abscissa is always an eigenvalue (namely the eigenvalue with maximal real part is always real) and it is called *Frobenius eigenvalue*. Also, Metzler matrices exhibit a monotonicity property [15]: if A and $\bar{A} \in \mathbb{R}^{n\times n}$ are Metzler matrices and $A \leq \bar{A}$, then $\lambda_{\max}(A) \leq \lambda_{\max}(\bar{A})$.

An *undirected, weighted graph* is a triple [10, 12] $\mathscr{G} = (\mathscr{V}, \mathscr{E}, \mathscr{A})$, where $\mathscr{V} = \{1, \dots, N\}$ is the set of vertices, $\mathscr{E} \subseteq \mathscr{V} \times \mathscr{V}$ is the set of arcs. $(i, j) \in \mathscr{E}$ if and only if $(j, i) \in \mathscr{E}$. Finally, $\mathscr{A} \in \mathbb{R}_+^{N\times N}$ is the (positive and symmetric) *adjacency matrix* of the weighted graph $\mathscr{G}$. We assume that the graph $\mathscr{G}$ has no self-loops, i.e. $[\mathscr{A}]_{ii} = 0$ for every index $i \in [1, N]$. If $\mathscr{A}$ is irreducible, the graph is *connected*. If $[\mathscr{A}]_{ij} > 0$ for every $i, j \in \mathscr{V}, i \neq j$, the graph $\mathscr{G}$ is *complete*. If $[\mathscr{A}]_{ij} > 0$ implies $[\mathscr{A}]_{ij} = 1$ the graph is called *unweighted*. We define the *Laplacian matrix*

$\mathscr{L} \in \mathbb{R}^{N\times N}$ associated with the graph $\mathscr{G}$ as $\mathscr{L} := \mathscr{C} - \mathscr{A}$, where $\mathscr{C} \in \mathbb{R}_+^{N\times N}$ is a diagonal matrix with $[\mathscr{C}]_{ii} := \sum_{j=1}^N [\mathscr{A}]_{ij}$, $\forall\, i \in [1, N]$. Accordingly, the Laplacian matrix $\mathscr{L} = \mathscr{L}^\top$ takes the following form:

$$\mathscr{L} = \begin{bmatrix} \ell_{11} & \ell_{12} & \dots & \ell_{1N} \\ \ell_{12} & \ell_{22} & \dots & \ell_{2N} \\ \vdots & \vdots & \ddots & \vdots \\ \ell_{1N} & \ell_{2N} & \dots & \ell_{NN} \end{bmatrix} = \begin{bmatrix} \sum_{j=1}^N [\mathscr{A}]_{1j} & -[\mathscr{A}]_{12} & \dots & -[\mathscr{A}]_{1N} \\ -[\mathscr{A}]_{12} & \sum_{j=1}^N [\mathscr{A}]_{2j} & \dots & -[\mathscr{A}]_{2N} \\ \vdots & \vdots & \ddots & \vdots \\ -[\mathscr{A}]_{1N} & -[\mathscr{A}]_{2N} & \dots & \sum_{j=1}^N [\mathscr{A}]_{Nj} \end{bmatrix}.$$

If $\mathscr{G}$ is connected then $\ell_{ii} > 0$ for every $i \in [1, N]$, and hence $\ell^* := \max_{i\in[1,N]} \ell_{ii} > 0$. Notice that all rows of $\mathscr{L}$ sum up to 0, and hence $\mathbf{1}_N$ is always a right eigenvector of $\mathscr{L}$ corresponding to the null eigenvalue [3].

Lemma 19.1 *[3, 13, 20] If the undirected, weighted graph $\mathscr{G}$ is connected, then $\mathscr{L}$ is a symmetric positive semidefinite matrix, and 0 is a simple eigenvalue of $\mathscr{L}$.*

Therefore, if we denote by $\{\lambda_1, \lambda_2, \dots, \lambda_N\}$ the spectrum $\sigma(\mathscr{L})$, then $\lambda_i \in \mathbb{R}_+$ for every $i \in [1, N]$, and we can always assume that the λ_i's are sorted in non-decreasing order, namely

$$0 = \lambda_1 \le \lambda_2 \le \dots \le \lambda_N. \tag{19.2}$$

It is well-known that if the eigenvalues of $\mathscr{L}$ are sorted as in (19.2), then [4, 5] $\ell^* \le \lambda_N$. In addition, if $\mathscr{L}$ is irreducible, then $\ell^* < \lambda_N$ (see Theorem 3 in [4]).

Lemma 19.2 *(1) Let $\mathscr{G}$ be an undirected, weighted graph with N vertices. If $\ell^* < \lambda_2$, then [12] $\mathscr{G}$ is complete.*

(2) If $\mathscr{G}$ is the undirected, unweighted graph with N vertices, then [3, 5, 10] $\ell^* < \lambda_2$ if and only if $\mathscr{G}$ is complete. Moreover, in this case $\ell^* = N - 1$ and $\lambda_2 = \dots = \lambda_N = N$.

Notice that, differently from the unweighted case, completeness of a weighted graph $\mathscr{G}$ does not imply $\ell^* < \lambda_2$. Consider, e.g., the weighted Laplacian matrix

$$\mathscr{L} = \begin{bmatrix} 3 & -1 & -2 \\ -1 & 2 & -1 \\ -2 & -1 & 3 \end{bmatrix},$$

and notice that $\lambda_2 = 3$ and hence $\lambda_2 = \ell^* = 3$ even if $\mathscr{G}$ is complete. In the following the complete, undirected and unweighted graph will be denoted by $\mathscr{G}_N$. Clearly, its Laplacian can be expressed as $\mathscr{L} = N I_N - \mathbf{1}_N \mathbf{1}_N^\top$ and its eigenvalues are $\lambda_2 = \dots = \lambda_N = N$, while $\ell^* = N - 1$.

19.3 Problem Statement

We consider a homogeneous multi-agent system consisting of N identical agents whose dynamics is described by the continuous-time positive single-input system:

$$\dot{\mathbf{x}}_i(t) = A\mathbf{x}_i(t) + Bu_i(t), \qquad t \in \mathbb{R}_+,$$

where $\mathbf{x}_i \in \mathbb{R}^n$ and $u_i \in \mathbb{R}$ are the state vector and the (scalar) input, respectively, of the ith agent. $A = [a_{ij}] \in \mathbb{R}^{n\times n}$ is a non-Hurwitz Metzler matrix, and $B = [b_i] \in \mathbb{R}^n_+$ is a positive vector. The mutual interactions among agents are described by a (connected, undirected, weighted) *communication graph* $\mathcal{G} = (\mathcal{V}, \mathcal{E}, \mathcal{A})$, where $\mathcal{V} = \{1, \dots, N\}$ and $\mathcal{A} = \mathcal{A}^\top \in \mathbb{R}_+^{N\times N}$. Note that we assume that the mutual interactions are cooperative and hence $\mathcal{A}$ is a nonnegative matrix. Differently from what we did in [17–19], we assume that the graph $\mathcal{G}$ is complete, namely each agent communicates with all the other agents, and that $\ell^* < \lambda_2$. As we will see, this apparently more restrictive situation makes the problem solution more difficult. In this scenario, $\mathcal{A} \in \mathbb{R}_+^{N\times N}$ is irreducible (in fact, primitive if $N > 2$), and if we sort the eigenvalues of $\mathcal{L}$ as in (19.2), then

$$0 = \lambda_1 < \ell^* < \lambda_2 \leq \dots \leq \lambda_N.$$

Let $K \in \mathbb{R}^{1\times n}$ be a state-feedback matrix (to be designed) and assume that each ith agent adopts the following DeGroot type control law [20]:

$$u_i(t) = K \sum_{j=1}^{N} [\mathcal{A}]_{ij} [\mathbf{x}_j(t) - \mathbf{x}_i(t)].$$

Define $\mathbf{x}(t) \in \mathbb{R}^{Nn}$ and $\mathbf{u}(t) \in \mathbb{R}^N$ as

$$\mathbf{x}(t) := \left[\mathbf{x}_1^\top(t) \dots \mathbf{x}_N^\top(t)\right]^\top \qquad \mathbf{u}(t) := \left[u_1(t) \dots u_N(t)\right]^\top$$

respectively. The state-space description of the overall system becomes:

$$\begin{aligned} \dot{\mathbf{x}}(t) &= (I_N \otimes A)\mathbf{x}(t) + (I_N \otimes B)\mathbf{u}(t) \\ \mathbf{u}(t) &= -(\mathcal{L} \otimes K)\mathbf{x}(t) \end{aligned}$$

and the resulting autonomous closed-loop system is described by

$$\dot{\mathbf{x}}(t) = [(I_N \otimes A) - (I_N \otimes B)(\mathcal{L} \otimes K)]\mathbf{x}(t). \tag{19.3}$$

The *positive consensus problem* is naturally posed as follows: determine, if possible, a state-feedback matrix $K = [k_i] \in \mathbb{R}^{1\times n}$ such that the (closed-loop multi-agent) system (19.3) satisfies the following conditions:

(I) *positivity*: $\mathbb{A} := (I_N \otimes A) - (I_N \otimes B)(\mathscr{L} \otimes K)$ is a Metzler matrix;
(II) *consensus*: meaning that

$$\lim_{t\to+\infty} \mathbf{x}_i(t) - \mathbf{x}_j(t) = 0, \quad \forall i, j \in [1, N].$$

As well-known in the literature [2, 20], a necessary and sufficient condition for the homogeneous agents to reach consensus is that all matrices $A - \lambda_i BK, i \in [2, N]$, are Hurwitz. A necessary condition for this to happen is that the pair (A, B) is stabilizable, a steady assumption from now onward.

As far as condition (I) is concerned, once we explicitly write the expression of the overall state matrix $\mathbb{A}$:

$$\mathbb{A} = \begin{bmatrix} A - \ell_{11}BK & -\ell_{12}BK & \dots & -\ell_{1N}BK \\ -\ell_{12}BK & A - \ell_{22}BK & \dots & -\ell_{2N}BK \\ \vdots & \vdots & \ddots & \vdots \\ -\ell_{1N}BK & -\ell_{2N}BK & \dots & A - \ell_{NN}BK \end{bmatrix}$$

it is easy to see [17–19] that $\mathbb{A}$ is Metzler if and only if (a) the off-diagonal blocks $-\ell_{ij}BK$, $i, j \in [1, N], i \neq j$, are non-negative; and (b) the diagonal blocks $A - \ell_{ii}BK$, $i \in [1, N]$, are Metzler. So, keeping in mind the assumptions on A and B, once we define the vector $K^* = [k_i^*] \in \mathbb{R}_+^{1\times n}$ as:

$$k_i^* := \begin{cases} \min_{\substack{j=1,\dots,n \\ j\neq i}} \dfrac{a_{ji}}{b_j}\dfrac{1}{\ell^*}, & \text{if } \exists\, j \neq i \text{ s.t. } b_j \neq 0; \\ +\infty, & \text{otherwise,} \end{cases}$$

it is immediate to prove that condition (I) holds if and only if $0 \leq K \leq K^*$. Note that in the special case when B is a monomial vector, say $B = b_i\mathbf{e}_i$, for some $i \in [1, n]$ and $b_i > 0$, the ith entry of K^* is infinite. In all the other cases (namely if B has at least two non-zero entries) K^* is always finite.

To summarize, the positive consensus problem can be equivalently posed as follows:

Positive consensus problem: determine, if possible, $K \in \mathbb{R}_+^{1\times n}, 0 \leq K \leq K^*$, such that all matrices $A - \lambda_i BK, i \in [2, N]$, are Hurwitz.

19.4 Necessary and/or Sufficient Conditions

A major consequence of the apparently more restrictive assumption that all the agents communicate with each other and $\ell^* < \lambda_i, i \in [2, N]$, is that one of the main necessary conditions for the positive consensus problem solvability we exploited in the previous analysis, namely the fact that the matrix $A - \lambda_2 BK^*$ is Metzler and Hurwitz, does not hold anymore. As ℓ^* is smaller than λ_2, by the way K^* is defined the matrix $A - \lambda_2 BK^*$ (and hence all matrices $A - \lambda_i BK^*, i \in [2, N]$) is not Metzler, and the case may occur that $A - \lambda_2 BK$ is Hurwitz even if $A - \lambda_2 BK^*$ is not.

Some necessary conditions for the problem solvability, however, can be determined, as they are independent of the relationship between ℓ^* and λ_2.

Proposition 19.1 *Assume that A is an $n \times n$ Metzler non-Hurwitz matrix, $B \in \mathbb{R}^n_+$ is a positive vector and $0 < \ell^* < \lambda_i, i \in [2, N]$. If the positive consensus problem is solvable, then*

(i) $\lambda_{\max}(A)$ *is a simple nonnegative eigenvalue;*
(ii) $K^* B > \text{tr}(A)/\lambda_2$.

Proof (i) The fact that $\lambda_{\max}(A)$ is a simple eigenvalue follows from Proposition 1 and Remark 1 in [18], since those results are independent of the relationship between ℓ^* and λ_2. The fact that it is real and nonnegative follows from the assumption that A is a Metzler non-Hurwitz matrix.
(ii) As the trace of a matrix equals the sum of its eigenvalues, a necessary condition for the matrices $A - \lambda_i BK, i \in [2, N]$, to be Hurwitz is that their traces are negative, i.e., $\text{tr}(A - \lambda_i BK) = \text{tr}(A) - \lambda_i KB < 0, \forall i \in [2, N]$. However, since both B and K are positive vectors, if there exists a matrix K such that $0 \le K \le K^*$ and $A - \lambda_i BK$ is Hurwitz, then $K^* B \ge KB > \frac{\text{tr}(A)}{\lambda_i}, \forall\, i \in [2, N]$. Finally, note that if $\text{tr}(A) < 0$ the previous condition is trivial. If $\text{tr}(A) \ge 0$ then

$$\frac{\text{tr}(A)}{\lambda_2} \ge \frac{\text{tr}(A)}{\lambda_i}$$

for every $i \in [2, N]$. So, in both cases, condition $K^* B > \frac{\text{tr}(A)}{\lambda_i}$ holds for every $i \in [2, N]$ if and only if $K^* B > \frac{\text{tr}(A)}{\lambda_2}$.

Conditions (i) and (ii) of the above proposition are not sufficient, not even when dealing with $N = 2$ agents described by a two-dimensional ($n = 2$) model, as the following elementary example shows.

Example 19.1 Consider the positive single-input agent

$$\dot{\mathbf{x}}_i(t) = A\mathbf{x}_i(t) + Bu_i(t) = \begin{bmatrix} 3 & 1 \\ 1 & -1 \end{bmatrix} \mathbf{x}_i(t) + \begin{bmatrix} 1 \\ 1 \end{bmatrix} u_i(t)$$

A is a Metzler and non-Hurwitz matrix and the pair (A, B) is stabilizable. The matrix A has a simple positive eigenvalue and a negative one. Assume that there are $N = 2$ agents and assume that the interconnection topology is described by the complete, undirected and unweighted graph $\mathscr{G}_2$, namely

$$\mathscr{L} = \begin{bmatrix} 1 & -1 \\ -1 & 1 \end{bmatrix}.$$

Then (see Lemma 19.2) $0 = \lambda_1 < \ell^* = 1 < \lambda_2 = 2$. The matrix K^* is easily proved to be $K^* = \begin{bmatrix} 1 & 1 \end{bmatrix}$, and hence condition $2 = K^*B > \text{tr}(A)/\lambda_2 = 1$ holds. Yet, for every $K = \begin{bmatrix} k_1 & k_2 \end{bmatrix}$, with $0 \leq k_i \leq 1, i \in [1, 2]$, $A - \lambda_2 BK$ is not Hurwitz. So, the positive consensus problem is not solvable. ♣

Example 19.2 Consider the positive single-input agent

$$\dot{\mathbf{x}}_i(t) = A\mathbf{x}_i(t) + Bu_i(t) = \begin{bmatrix} -1 & 1 & 1 \\ 1 & -1 & 0 \\ 1 & 1 & 6 \end{bmatrix} \mathbf{x}_i(t) + \begin{bmatrix} 1 \\ 1 \\ 0 \end{bmatrix} u_i(t)$$

Notice that A is a Metzler and non-Hurwitz matrix and that the pair (A, B) is stabilizable. Consider $N = 3$ agents and assume that the interconnection topology is described by the complete, undirected and unweighted graph $\mathscr{G}_3$. In this case (see Lemma 19.2) $\ell^* = 2$ and the eigenvalues of $\mathscr{L}$ are $\lambda_1 = 0$ and $\lambda_2 = \lambda_3 = 3$. The matrix K^* is easily proved to be $K^* = \begin{bmatrix} \frac{1}{2} & \frac{1}{2} & 0 \end{bmatrix}$. As $K^*B = 1 < \frac{4}{3} = \frac{\text{tr}(A)}{\lambda_2}$, we conclude that the positive consensus problem is not solvable. ♣

In order to investigate the problem solvability, let us define the set of solutions of the positive consensus problem as

$$\mathscr{K}^H := \{K : 0 \leq K \leq K^*,\ A - \lambda_i BK \text{ Hurwitz},\ i \in [2, N]\}.$$

A sufficient condition for the solvability of the positive consensus problem is represented by the case when there is a matrix K, satisfying the given bounds, that makes all matrices $A - \lambda_i BK, i \in [2, N]$, Metzler and Hurwitz. To investigate this situation, we define

$$\mathscr{K}^{MH} := \{K \in \mathscr{K}^H : A - \lambda_i BK \text{ Metzler},\ i \in [2, N]\}.$$

The following result provides, in the case when $\ell^* < \lambda_2$, an analysis that parallels the one carried on in Sect. 6 of [18]. In the case we are currently investigating the matrix $A - \lambda_2 BK^*$ is no longer Metzler and Hurwitz. However, $A - \ell^* BK^*$ is necessarily Metzler and hence we can ensure that all matrices taking the form $K = \alpha K^*$, with $\alpha \in [0, 1]$, make $A - \ell^* BK$ Metzler. So, we focus on this class of state feedback matrices to determine whether some of them belong to $\mathscr{K}^{MH}$.

Proposition 19.2 *Assume that A is an $n \times n$ Metzler non-Hurwitz matrix, $B \in \mathbb{R}^n_+$ is a positive vector and $0 < \ell^* < \lambda_i$, $i \in [2, N]$. The following conditions are equivalent:*

(i) $\mathscr{K}^{MH} \neq \emptyset$;
(ii) $\frac{\ell^*}{\lambda_N} K^* \in \mathscr{K}^{MH}$;
(iii) *the set* $\{\alpha \in (0, 1] : A - \alpha \ell^* BK^* \textit{ is Hurwitz}\}$ *is not empty and*

$$\tilde{\alpha} := \inf\{\alpha \in (0, 1] : A - \alpha \ell^* BK^* \textit{ is Hurwitz}\} \tag{19.4}$$

satisfies $\tilde{\alpha} < \frac{\lambda_2}{\lambda_N}$.

Proof (i) ⇒ (ii) Suppose that $\mathcal{K}^{MH} \neq \emptyset$ and let $K \in \mathcal{K}^{MH}$. As $K \in \mathcal{K}^{MH}$ then $A - \lambda_N BK$ is Metzler (and Hurwitz) and this implies that $\lambda_N K \leq \ell^* K^*$, namely $K \leq \frac{\ell^*}{\lambda_N} K^*$. On the other hand, the Metzler matrix $A - \ell^* BK^* \leq A - \lambda_N BK$, being upper bounded by a Metzler and Hurwitz matrix, is Hurwitz in turn. Therefore, for every $k \in [2, N]$, $A - \lambda_k BK \geq A - \lambda_k \frac{\ell^*}{\lambda_N} BK^* \geq A - \ell^* BK^*$. Since $A - \ell^* BK^*$ is Metzler, then $A - \lambda_k \frac{\ell^*}{\lambda_N} BK^*$ is Metzler, too, and being upper-bounded by a Metzler Hurwitz matrix, it is Hurwitz, in turn. This proves that $A - \lambda_k \frac{\ell^*}{\lambda_N} BK^*$ is Metzler and Hurwitz for every $k \in [2, N]$, namely $\frac{\ell^*}{\lambda_N} K^* \in \mathcal{K}^{MH}$.
(ii) ⇒ (iii) If $\frac{\ell^*}{\lambda_N} K^* \in \mathcal{K}^{MH}$, then $A - \lambda_2 \frac{\ell^*}{\lambda_N} BK^*$ is Metzler and Hurwitz, and hence $\frac{\lambda_2}{\lambda_N} \in \{\alpha \in (0, 1] : A - \alpha \ell^* BK^* \text{ is Hurwitz}\}$. This also implies that $\tilde{\alpha} < \frac{\lambda_2}{\lambda_N}$.
(iii) ⇒ (i) Observe, first, that if $\{\alpha \in (0, 1] : A - \alpha \ell^* BK^* \text{ is Hurwitz}\}$ is not empty and $\tilde{\alpha}$ is the infimum value of the set, then for every $\alpha \in (\tilde{\alpha}, 1]$ the matrix $A - \alpha \ell^* BK^*$ satisfies $A - \ell^* BK^* \leq A - \alpha \ell^* BK^* < A - \tilde{\alpha} \ell^* BK^*$ and hence it is Metzler Hurwitz. Set $K = \frac{\ell^*}{\lambda_N} K^*$. By assumption, $\tilde{\alpha} < \frac{\lambda_2}{\lambda_N}$, and hence $A - \lambda_2 BK$ is Hurwitz. On the other hand, $A - \lambda_N BK = A - \ell^* BK^*$ is Metzler. This implies that $A - \lambda_2 BK \geq A - \lambda_3 BK \geq \cdots \geq A - \lambda_N BK$ are all Metzler matrices, and since the largest one is Hurwitz, by the monotonicity property of the spectral abscissa we can claim that they are all Hurwitz. So, $K \in \mathcal{K}^{MH}$.

Remark 19.1 It is easy to see that since $A - \ell^* BK^*$ is Metzler, then the set $\{\alpha \in (0, 1] : A - \alpha \ell^* BK^* \text{ is Hurwitz}\}$ coincides with the set $\{\alpha \in (0, 1] : A - \alpha \ell^* BK^* \text{ is Metzler and Hurwitz}\}$. Moreover, if the set is not empty then the Metzler matrix $A - \ell^* BK^*$ satisfies $A - \ell^* BK^* \leq A - \tilde{\alpha} \ell^* BK^*$ and hence it is necessarily Hurwitz. So, Proposition 19.2 above, essentially states that the set $\mathcal{K}^{MH}$ is not empty, namely there exists a state feedback matrix K, satisfying the usual bounding conditions, that makes all matrices $A - \lambda_i BK, i \in [2, N]$, Metzler and Hurwitz, if and only if such a solution can be found in the set of matrices $\{\alpha K^* : \alpha \in (0, 1]\}$. Note that not only the set $\{\alpha \in (0, 1] : A - \alpha \ell^* BK^* \text{ is Hurwitz}\}$ must be not empty, and hence the parameter $\tilde{\alpha}$ well defined, but the interval $(\tilde{\alpha}, 1]$ must be sufficiently "large" to include the interval $\left[\frac{\lambda_2}{\lambda_N}, 1\right]$. Only in this way we can determine a matrix of the form $K = \alpha K^*$ that makes $A - \lambda_i BK$ Metzler and Hurwitz for every $\lambda \in [\lambda_2, \lambda_N]$.

19.5 *B* Is a Monomial Vector

We consider now the case when B is a monomial vector. Without loss of generality we assume that $B = \mathbf{e}_1$, since we can always reduce ourselves to this case by resorting to a permutation and a rescaling that do not influence the problem solvability, only the value of the specific solution.

Proposition 19.3 *Assume that* $B = \mathbf{e}_1$ *and denote by* A_{22} *the principal submatrix obtained from A by deleting its first row and column.*

(i) If the positive consensus problem is solvable then every eigenvalue of A_{22} with nonnegative real part has geometric multiplicity equal to 1*;*
(ii) If A_{22} is Hurwitz, then the positive consensus problem is solvable.

Proof (i) Assume that the positive consensus problem is solvable and suppose by contradiction that there exists $\mu \in \sigma(A_{22})$ with $\Re\{\mu\} \geq 0$ and geometric multiplicity $d > 1$. Partition the matrix A as:

$$A = \begin{bmatrix} a_{11} & \mathbf{r}^\top \\ \mathbf{c} & A_{22} \end{bmatrix},$$

where $a_{11} \in \mathbb{R}$, $\mathbf{r}, \mathbf{c} \in \mathbb{R}_+^{n-1}$ are nonnegative vectors, and $A_{22} \in \mathbb{R}^{(n-1)\times(n-1)}$ is a Metzler matrix. Partition the feedback matrix $K \in \mathbb{R}_+^{1\times n}$, $0 \leq K \leq K^*$, in a consistent way, namely as $K = \begin{bmatrix} k_1 & \mathbf{k}_2 \end{bmatrix}$, where $\mathbf{k}_2 \in \mathbb{R}_+^{1\times(n-1)}$. Now, notice that for every $i \in [2, N]$ the characteristic polynomial of $A - \lambda_i BK$ can be written as

$$\begin{aligned} \det(sI_n - A + \lambda_i BK) &= \det(sI_n - A) + \lambda_i K \operatorname{adj}(sI_n - A)B \\ &= \det(sI_{n-1} - A_{22}) \left[s - a_{11} - \mathbf{r}^\top (sI_{n-1} - A_{22})^{-1} \mathbf{c} \right] \\ &+ \lambda_i \begin{bmatrix} k_1 & \mathbf{k}_2 \end{bmatrix} \begin{bmatrix} \det(sI_{n-1} - A_{22}) \\ \operatorname{adj}(sI_{n-1} - A_{22})\mathbf{c} \end{bmatrix} \\ &= (s - a_{11} + \lambda_i k_1) \det(sI_{n-1} - A_{22}) \\ &+ \left(\lambda_i \mathbf{k}_2 - \mathbf{r}^\top\right) \operatorname{adj}(sI_{n-1} - A_{22})\mathbf{c}. \end{aligned}$$

If $\mu \in \sigma(A_{22})$, then $\det(\mu I_{n-1} - A_{22}) = 0$ and, since the geometric multiplicity of μ as an eigenvalue of A_{22} is $d > 1$, it also holds that $\operatorname{adj}(\mu I_{n-1} - A_{22}) = 0$, and hence $\det(\mu I_n - A + \lambda_i BK) = 0$ for every $K \in \mathbb{R}_+^{1\times n}$, which contradicts the assumption of the solvability of the positive consensus problem.
(ii) It is the same as the proof of the sufficiency part of Proposition 7 in [18].

Differently from the case $\lambda_2 \leq \ell^*$, the Hurwitz condition on the submatrix A_{22} is sufficient for the problem solvability, but it is not necessary, as shown in Example 19.3 below.

Example 19.3 Consider the positive single-input agent

$$\dot{\mathbf{x}}_i(t) = A\mathbf{x}_i(t) + Bu_i(t) = \begin{bmatrix} -1 & 1 & 0 \\ 0 & 0 & 1 \\ 0 & 2 & -1 \end{bmatrix} \mathbf{x}_i(t) + \begin{bmatrix} 0 \\ 0 \\ 1 \end{bmatrix} u_i(t)$$

Notice that A is a Metzler and non-Hurwitz matrix and that the pair (A, B) is stabilizable. Consider $N = 3$ agents and the same adjacency matrix as in Example 19.2, so that $\ell^* = 2$ and $\lambda_2 = \lambda_3 = 3$. $B = \mathbf{e}_3$ and the matrix A_{11}, obtained from A by deleting the third row and the third column, is non-Hurwitz, however this does not preclude the problem solvability. If we consider

$$A - \ell^* BK = \begin{bmatrix} -1 & 1 & 0 \\ 0 & 0 & 1 \\ -2k_1 & 2-2k_2 & -1-2k_3 \end{bmatrix}$$

we notice that $K^* = \begin{bmatrix} 0 & 1 & +\infty \end{bmatrix}$. It is easy to verify that the positive consensus problem is solvable since for $K = \begin{bmatrix} 0 & 1 & 0 \end{bmatrix} \in \mathbb{R}^{1\times 3}$, with $0 \leq K \leq K^*$, we get

$$A - \lambda_2 BK = A - \lambda_3 BK = \begin{bmatrix} -1 & 1 & 0 \\ 0 & 0 & 1 \\ 0 & -1 & -1 \end{bmatrix}$$

which is Hurwitz. ♣

19.6 Second-Order Agents

We investigate now the case when the agents are modelled by a second-order (positive) linear system, i.e.

$$\dot{\mathbf{x}}_i(t) = A\mathbf{x}_i(t) + Bu_i(t) = \begin{bmatrix} a_{11} & a_{12} \\ a_{21} & a_{22} \end{bmatrix} \mathbf{x}_i(t) + \begin{bmatrix} b_1 \\ b_2 \end{bmatrix} u_i(t), \qquad (19.5)$$

with a_{12}, a_{21}, b_1 and b_2 nonnegative real numbers. Recalling that any matrix $M \in \mathbb{R}^{2\times 2}$ is Hurwitz if and only if $\mathrm{tr(M)} < 0$ and $\det(M) > 0$, after elementary manipulations it can be seen that for every $A \in \mathbb{R}^{2\times 2}$, $B \in \mathbb{R}^2$ and $K \in \mathbb{R}^{1\times 2}$, the matrix $M := A - \lambda BK$ is Hurwitz if and only if

$$\begin{cases} \lambda KB > \mathrm{tr(A)}; \\ \lambda K \mathrm{adj(A)B} < \det(A). \end{cases} \qquad (19.6)$$

This simple observation leads to the following Lemma.

Lemma 19.3 *[18] Given $A \in \mathbb{R}^{2\times 2}$ and $B \in \mathbb{R}^2$ and $K \in \mathbb{R}^{1\times 2}$, for every choice of the $N-1$ positive real numbers $0 < \lambda_2 \leq \lambda_3 \leq \cdots \leq \lambda_N$, the following facts are equivalent:*

(i) $A - \lambda BK$ is Hurwitz for every $\lambda \in [\lambda_2, \lambda_N]$;
(ii) $A - \lambda_i BK$ is Hurwitz for every $i \in [2, N]$;
(iii) $A - \lambda_i BK$ is Hurwitz for $i = 2, N$.

As a straightforward consequence of Lemma 19.3 and of the fact that $KB \geq 0$ (and hence $\lambda_N KB \geq \lambda_2 KB$), it follows that for two-dimensional agents the set of feedback matrices that solve the positive consensus problem is the set of matrices $K \in \mathbb{R}^{1\times 2}$ that satisfy the following LMIs:

$$K^* \geq K \quad \geq 0;$$

$$\begin{aligned}&\lambda_2 KB > \text{tr(A)};\\ &\det(A) > \lambda_i K \text{adj(A)B}, \qquad \text{i} = 2, \text{N}.\end{aligned}$$

This ensures that the set of solutions is necessarily convex.

When the agents are described by second-order state-space models the case of B monomial can be completely solved. To this aim recall that from Proposition 19.3 part (ii) it follows that condition $a_{22} < 0$ ensures the solvability of the positive consensus problem, but as we have shown this is not a necessary condition. So, in the following we assume $a_{22} \geq 0$, $\ell^* < \lambda_2$, and investigate under which additional conditions on the matrix A and on the interconnection topology the positive consensus problem is solvable.

Proposition 19.4 *Assume that $B = \mathbf{e}_1$, $A_{22} = a_{22} \geq 0$ and $\ell^* < \lambda_2$. Then, the positive consensus problem for second-order agents is solvable if and only if $a_{21} > 0$ and the following condition holds:*

$$\max\left\{0, \frac{\text{tr}(A)a_{22}}{\lambda_2}\right\} < \frac{a_{12}a_{21}}{\ell^*} + \min\left\{\frac{\det(A)}{\lambda_2}, \frac{\det(A)}{\lambda_N}\right\}. \tag{19.7}$$

When so, there is always a solution of the form $K = \left[\max\left\{0, \frac{\text{tr}(A)a_{22}}{\lambda_2}\right\} + \epsilon \ \frac{a_{12}}{\ell^*}\right]$, with $\epsilon > 0$ and arbitrarily small.

Proof Note first that as $B = \mathbf{e}_1$ and $a_{22} \geq 0$, if the positive consensus problem is solvable, then a_{21} must be positive, otherwise a_{22} would be an eigenvalue of every matrix $A - \lambda_i BK$, $i \in [2, N]$. Conversely, it is easy to see that condition (19.7) implies $a_{21} > 0$. So, in the following we will assume $a_{21} > 0$. Set $K = \left[k_1 \ k_2\right]$. Then $KB = k_1$, $K^* = \left[+\infty \ \frac{a_{12}}{\ell^*}\right]$, and the previous LMIs become

$$k_1 \geq 0, \quad k_1 > \frac{\text{tr}(A)}{\lambda_2}, \quad \frac{a_{12}}{\ell^*} \geq k_2 \geq 0, \tag{19.8}$$

$$\left[k_1 \ k_2\right]\begin{bmatrix} a_{22} \\ -a_{21}\end{bmatrix} < \min\left\{\frac{\det(A)}{\lambda_2}, \frac{\det(A)}{\lambda_N}\right\}. \tag{19.9}$$

It is clear that, as $a_{21} > 0$, inequality (19.9) holds if and only if it holds for $k_2 = k_2^* = \frac{a_{12}}{\ell^*}$. So, inequality (19.9) becomes

$$k_1 a_{22} < \frac{a_{12}a_{21}}{\ell^*} + \min\left\{\frac{\det(A)}{\lambda_2}, \frac{\det(A)}{\lambda_N}\right\}. \tag{19.10}$$

If $\text{tr}(A) < 0$ then the only constraint on k_1 is the nonnegativity and condition (19.10) holds if and only if it holds for $k_1 = 0$. And if this is the case it also holds for $k_1 = \epsilon$, with $\epsilon > 0$ and arbitrarily small. On the other hand, if $\text{tr}(A) \geq 0$, then the problem is solvable if and only if it is solvable by assuming $k_1 = \frac{\text{tr}(A)}{\lambda_2} + \epsilon$, with $\epsilon > 0$ arbitrarily small, and this happens if and only if

$$\frac{\text{tr}(A)}{\lambda_2}a_{22} < \frac{a_{12}a_{21}}{\ell^*} + \min\left\{\frac{\det(A)}{\lambda_2}, \frac{\det(A)}{\lambda_N}\right\}.$$

When the N agents are described by a second-order state-space model, $B = \mathbf{e}_1$, $A_{22} = a_{22} > 0$, and the communication among them is described by $\mathcal{G}_N$, Proposition 19.4 allows us to draw the following conclusion concerning the number of agents.

Corollary 19.1 *Assume that $B = \mathbf{e}_1$, $A_{22} = a_{22} > 0$ and the communication graph is described by the complete undirected and unweighted graph $\mathcal{G}_N$ (and hence $\ell^* < \lambda_2$). Then, there exists $\bar{N}$ such that for every $N \geq \bar{N}$ positive consensus cannot be reached.*

Proof The Laplacian of $\mathcal{G}_N$ has $\ell^* = N - 1$ and eigenvalues $\lambda_2 = \cdots = \lambda_N = N$. So, condition (19.7) becomes

$$\max\left\{0, \frac{\text{tr}(A)a_{22}}{N}\right\} < \frac{a_{12}a_{21}}{N-1} + \frac{\det(A)}{N},$$

and it implies $a_{22}^2 < \frac{1}{N-1}a_{12}a_{21}$. Clearly, the term on the right goes to 0 as N tends to $+\infty$, while $a_{22}^2 > 0$. So, there exists $\bar{N}$ such for every $N \geq \bar{N}$ the previous inequality and hence condition (19.7) do not hold, i.e. positive consensus cannot be reached.

Example 19.4 Consider the positive single-input agent

$$\dot{\mathbf{x}}_i(t) = A\mathbf{x}_i(t) + Bu_i(t) = \begin{bmatrix} -1 & 1 \\ 3 & 1 \end{bmatrix}\mathbf{x}_i(t) + \begin{bmatrix} 1 \\ 0 \end{bmatrix}u_i(t)$$

Assume that the communication among the agents is described by $\mathcal{G}_N$: it follows from (19.7) that for every $N \geq \bar{N} = 4$ the positive consensus problem is not solvable.

References

1. Chaudhary, D.D., Nayse, S.P., Waghmare, L.M.: Application of wireless sensor networks for greenhouse parameter control in precision agriculture. Int. J. Wirel. Mobile Netw. (IJWMN) **3**(1) (2011)
2. Fax, J.A., Murray, R.M.: Information flow and cooperative control of vehicle formations. IEEE Trans. Autom. Control **49**(9), 1465–1476 (2004)
3. Fiedler, M.: Algebraic connectivity of graphs. Czechoslovak Math. J. **23**, 298–305 (1973)
4. Fu, E., Markham, T.: On the eigenvalues and diagonal entries of a hermitian matrix. Linear Algebra Appl. **179**, 7–10 (1993)
5. Goldberg, F.: Bounding the gap between extremal Laplacian eigenvalues of graphs. Linear Algebra Appl. **416**, 68–74 (2006)
6. Hinrichsen, D., Plischke, E.: Robust stability and transient behaviour of positive linear systems. Vietnam J. Math. **35**, 429–462 (2007)
7. Jadbabaie, A., Lin, J., Morse, A.S.: Coordination of groups of mobile autonomous agents using nearest neighbor rules. IEEE Trans. Autom. Control **48**(6), 988–1001 (2003)

8. Knorn, F., Corless, M.J., Shorten, R.N.: A result on implicit consensus with application to emissions control. In: Proceedings of the 2011 CDC-ECC, pp. 1299–1304, Orlando, FL (2011)
9. Lin, J., Morse, A.S., Anderson, B.D.O.: The multi-agent rendezvous problem. In: Proceedings of the 42nd IEEE Conference on Decision and Control, pp. 1508–1513, Maui, Hawaii (2003)
10. Mohar, B.: The Laplacian spectrum of graphs. Graph Theory Comb. Appl. **2**, 871–898 (1991)
11. Olfati-Saber, R., Fax, J.A., Murray, R.M.: Consensus and cooperation in networked multi-agent systems. Proc. IEEE **95**(1), 215–233 (2007)
12. Pejic, S.: Algebraic graph theory in the analysis of frequency assignment problems. PhD thesis, London School of Economics and Political Science (2008)
13. Ren, W., Beard, R.W.: Consensus seeking in multiagent systems under dynamically changing interaction topologies. IEEE Trans. Autom. Control **50**(5), 655–661 (2005)
14. Smith, H.L.: Monotone Dynamical Systems: An Introduction to the Theory of Competitive and Cooperative Systems, vol. 41. AMS, Mathematical Surveys and Monographs, Providence, RI (1995)
15. Son, N.K., Hinrichsen, D.: Robust stability of positive continuous time systems. Numer. Funct. Anal. Optimiz. **17**(5–6), 649–659 (1996)
16. Tsitsiklis, J.N.: Problems in Decentralized Decision Making and Computation. PhD thesis, Department of EECS, MIT (1984)
17. Valcher, M.E., Zorzan, I.: New results on the solution of the positive consensus problem. In: Proceedings of the 55th IEEE Conf. on Decision and Control, pp. 5251–5256, Las Vegas, Nevada, December 12–14 (2016)
18. Valcher, M.E., Zorzan, I.: On the consensus of homogeneous multi-agent systems with positivity constraints (2017, under review)
19. Valcher, M.E., Zorzan, I.: On the consensus problem with positivity constraints. In: Proceedings of the 2016 American Control Conference, pp. 2846–2851, Boston, MA (2016)
20. Wieland, P., Kim, J.-S., Allgover, F.: On topology and dynamics of consensus among linear high-order agents. Int. J. Syst. Sci. **42**(10), 1831–1842 (2011)

MIX
Papier aus verantwortungsvollen Quellen
Paper from responsible sources
FSC® C105338

If you have any concerns about our products,
you can contact us on
ProductSafety@springernature.com

In case Publisher is established outside the EU,
the EU authorized representative is:
Springer Nature Customer Service Center GmbH
Europaplatz 3, 69115 Heidelberg, Germany

Printed by Libri Plureos GmbH
in Hamburg, Germany